# Microorganisms in the Deterioration and Preservation of Cultural Heritage

Edith Joseph

Editor

# Microorganisms in the Deterioration and Preservation of Cultural Heritage

 Springer

*Editor*
Edith Joseph
Institute of Chemistry
University of Neuchâtel
Neuchâtel, Switzerland

Haute Ecole Arc Conservation Restauration
University of Applied Sciences and Arts HES-SO
Neuchâtel, Switzerland

ISBN 978-3-030-69413-5          ISBN 978-3-030-69411-1    (eBook)
https://doi.org/10.1007/978-3-030-69411-1

This Springer imprint is published by the registered company Springer Nature Switzerland AG.
The registered company address is: Gewerbestrasse 11, 6330 Cham, Switzerland

# Preface

Awareness regarding our environmental impact is greater than before. In recent years, the development of environmentally friendly methods has become a significant alternative towards more sustainable practices. This has been encouraged by environmental and socio-economic policies for land development and tourism. In the heritage field as well, there is an increased research interest towards green approaches. Different initiatives related to this matter developed such as the international conferences in Green Conservation of Cultural Heritage (Roma 2015; Palermo 2017; Porto 2019), *Sustainability in Conservation* founded in 2016 or *Ki Culture,* a non-profit organization founded in 2019 and that provides sustainable solutions for cultural heritage. Confronted with the transmission of heritage to the next generations, stakeholders in the field have also a major role to assume towards a global societal change. In direct link to this topic, this book gives a comprehensive overview of sustainable conservation and in particular on biotechnology applied to the preservation of cultural heritage. Using microorganisms offers both opportunities and challenges and potential of microorganism's pro- and against- deterioration of cultural materials (e.g. stones, metals, graphic documents, textiles, paintings) is presented. The chapters are organized into three main sections: (1) examples of microorganisms involved in biodeterioration, (2) green control methods in the heritage field and (3) microorganisms involved in the preservation and protection of heritage using green materials.

Microorganisms are often considered harmful for the preservation of cultural heritage. Indeed, microorganisms are a major cause of deterioration on cultural artefacts, both in the case of outdoor monuments and archaeological finds. Microbial processes, such as bioweathering (rocks and minerals), biodeterioration (organic substrates) or biocorrosion (metals), thus contribute to irreversible changes and loss of valuable heritage. In the first section of this book "Occurrence of Microorganisms in Heritage Materials", emphasize was given to stone (Chaps. 1–3) as one of the most representative inorganic substrates but also includes inputs to wall paintings, subterranean environments, stained glass and metals (Chapter 1). As well graphic documents (paper and parchment) were chosen as examples of organic materials

(Chap. 4). Nonetheless, Chap. 9 addresses biodeterioration of textiles with its section "Microbial Growth and Metabolism as Degradative and Deterioration Agents". Chapter 6 briefly refers to the biodeterioration of wood, textiles and easel paintings. It is worth mentioning that biodeterioration affects a large variety of substrates and that due to space constraints not all the tremendous work done on wood, ceramics (e.g. deterioration of Chinese terracotta statues), underground cultural heritage ( e.g. tombs, rock art) could be considered for a contribution to this book.

To reduce the impact of microbial activity on heritage, conservation strategies are mainly devoted to control microbial development. Hence, preventive or remedial methods, such as controlled environmental conditions, mechanical removal, application of biocides, fumigation or ultraviolet radiation, are commonly adopted. The use of biocides is controverted nowadays as this can be detrimental to non-target populations as well as lead to resistance development and green alternatives to biocides are presented in the section entitled "Green Methods Again Biodeterioration", in particular Chap. 5. In addition, the design of mitigation strategies with the help of bacterial secondary metabolites is developed in Chap. 6. Chapter 7 illustrates how an appropriate use of public lighting can control biological colonization.

Over the last decades, a completely opposite perspective has emerged: microbes can be used to safeguard heritage. This creates new opportunities for the development of methods and materials for the conservation of cultural artefacts, with real progress in terms of sustainability, effectiveness and toxicity. In particular, microbial mechanisms are exploited aiming to consolidate, clean, stabilize or even protect surfaces of cultural items. For instance, biological methods using different soil bacteria and resulting in carbonate mineralization or sulphate reduction have been used as alternative treatments for stone conservation. Another example is the development of biological cleaning agents that use microorganisms and enzymes, conferring significant advantages in terms of efficiency, impact on the surface texture, environmental safety and risk for operators. These cleaning agents have been applied for the removal of undesirable organic substances or inorganic deposits on stone, paintings, ceramic, paper and even concrete substrates. Some additional methods include the formation of passivating biogenic layers that can be applied for preserving copper- and iron-based heritage, in particular sculptures but also archaeological objects, as well as the development of bacterial extraction methods of iron species from waterlogged wood. Such examples are showed in the section entitled "Biocleaning and Bio-based Conservation Methods". Interestingly, Chap. 11 provides a comprehensive proof of concept for the biocleaning of organic substances and inorganic compounds using microorganisms and plant extracts of renewable origin. In addition, biocleaning is demonstrated on different substrates with the bioremoval of graffiti (Chap. 8), residual organic matter on ancient textiles (Chap. 9) and on wall paintings (Chap. 10), efflorescence salts on stone (Chap. 12) as well as iron staining from wood and textiles (Chap. 14). Two examples of protection on stone and metals and achieved by bacterial carbonatogenesis and bio-based corrosion inhibitors, respectively, are discussed in Chaps. 13 and 15.

At the same time, a large-scale transfer into real practice faces different challenges, such as the negative perception of heritage stakeholders towards microorganisms, the eventual additional cost and prolonged time required in the case of biological treatments, the safety and potential risks of undesired microbial growth and regulatory barriers. There are however encouraging signs of alternative approaches to address these issues. For instance, alternative modes of application are currently been explored, which include the identification of active extracellular metabolites (i.e. enzymes) to be applied directly on the substrate, the use of dead cells or cellular fractions or the enhancement of the activity of indigenous microorganisms. By dealing with the challenges cited above, significant steps are being made towards unsealing the unexploited potential of microorganisms as miniature chemical factories. The contributions collected here attend to illustrate but also to inspire green and sustainable strategies in heritage conservation.

Neuchâtel, Switzerland                                                                Edith Joseph

# Acknowledgement

This book was published as an open-access resource, thanks to the financial support provided by The Swiss National Science Foundation, grant number 10BP12_200221/1.

# Contents

**Part I    Occurrence of Microorganisms in Heritage Materials**

**1    Microbial Growth and its Effects on Inorganic Heritage Materials** . . . . . . . . . . . . . . . . . . . . . . . . . . . . . . . . . . . . .    3
Daniela Pinna

**2    Microbiota and Biochemical Processes Involved in Biodeterioration of Cultural Heritage and Protection** . . . . . . . . . . .    37
Ji-Dong Gu and Yoko Katayama

**3    Molecular-Based Techniques for the Study of Microbial Communities in Artworks** . . . . . . . . . . . . . . . . . . . . . . . . . . . . .    59
Katja Sterflinger and Guadalupe Piñar

**4    Extreme Colonizers and Rapid Profiteers: The Challenging World of Microorganisms That Attack Paper and Parchment** . . . . . . . . . .    79
Flavia Pinzari and Beata Gutarowska

**Part II    Green Methods Again Biodeterioration**

**5    Novel Antibiofilm Non-Biocidal Strategies** . . . . . . . . . . . . . . . . . . .    117
Francesca Cappitelli and Federica Villa

**6    Green Mitigation Strategy for Cultural Heritage Using Bacterial Biocides** . . . . . . . . . . . . . . . . . . . . . . . . . . . . . . . . . . . . .    137
Ana Teresa Caldeira

**7    New Perspectives Against Biodeterioration Through Public Lighting** . . . . . . . . . . . . . . . . . . . . . . . . . . . . . . . . . . . . . . . . . .    155
Patricia Sanmartín

**Part III   Biocleaning and Bio-Based Conservation Methods**

8    **Bioremoval of Graffiti in the Context of Current Biocleaning
     Research** . . . . . . . . . . . . . . . . . . . . . . . . . . . . . . . . . . . . . . . . . . 175
     Pilar Bosch-Roig and Patricia Sanmartín

9    **Ancient Textile Deterioration and Restoration: Bio-Cleaning
     of an Egyptian Shroud Held in the Torino Museum** . . . . . . . . . . . 199
     Roberto Mazzoli and Enrica Pessione

10   **Advanced Biocleaning System for Historical Wall Paintings** . . . . . . 217
     Giancarlo Ranalli and Elisabetta Zanardini

11   **Sustainable Restoration Through Biotechnological Processes:
     A Proof of Concept** . . . . . . . . . . . . . . . . . . . . . . . . . . . . . . . . . . 235
     Anna Rosa Sprocati, Chiara Alisi, Giada Migliore, Paola Marconi,
     and Flavia Tasso

12   **The Role of Microorganisms in the Removal of Nitrates
     and Sulfates on Artistic Stoneworks** . . . . . . . . . . . . . . . . . . . . . . . 263
     Giancarlo Ranalli and Elisabetta Zanardini

13   **Protection and Consolidation of Stone Heritage by Bacterial
     Carbonatogenesis** . . . . . . . . . . . . . . . . . . . . . . . . . . . . . . . . . . . . 281
     Fadwa Jroundi, Maria Teresa Gonzalez-Muñoz,
     and Carlos Rodriguez-Navarro

14   **Siderophores and their Applications in Wood, Textile,
     and Paper Conservation** . . . . . . . . . . . . . . . . . . . . . . . . . . . . . . . . 301
     Stavroula Rapti, Stamatis C. Boyatzis, Shayne Rivers,
     and Anastasia Pournou

15   **Organic Green Corrosion Inhibitors Derived from Natural
     and/or Biological Sources for Conservation of Metals
     Cultural Heritage** . . . . . . . . . . . . . . . . . . . . . . . . . . . . . . . . . . . . . 341
     Vasilike Argyropoulos, Stamatis C. Boyatzis, Maria Giannoulaki,
     Elodie Guilminot, and Aggeliki Zacharopoulou

# Part I
# Occurrence of Microorganisms in Heritage Materials

# Chapter 1
# Microbial Growth and its Effects on Inorganic Heritage Materials

**Daniela Pinna**

**Abstract** Cultural heritage objects composed of inorganic materials, such as metals and stones, support microbial life. Many factors affect the growth of microorganisms: moisture, pH, light, temperature, nutrients. Their colonization relates closely to the nature of the substrata as well as to the characteristic of the surrounding environment. This chapter contains an overview of the complex relationships among microbial growth, materials, and the environment. It emphasizes issues on bioreceptivity of stones and the factors influencing biological colonization, focusing on the biological alteration of inorganic heritage objects and on the agents of biodeterioration. It outlines the effect of biofilms and lichens in terms of degradation of substrata and includes a discussion on an important topic, the bioprotection of stones by biofilms and lichens. In summary, this chapter aims to discuss these issues and review the recent literature on (i) biofilms and lichens colonizing inorganic materials, (ii) the limiting factors of this colonization, (iii) the deteriorative aspects, and (iv) the protective effects of the colonization.

**Keywords** Biofilms · Lichens · Stone bioreceptivity · Substratum pH · Eutrophication · Environmental factors · Biodeteriorative processes of stones, metals and stained-glass windows · Bioprotection of stones by biofilms and lichens

## 1 Introduction

Cultural heritage objects that are composed of inorganic materials, such as metals and stones, support microbial life. Bacteria, algae, fungi, lichens colonize the surfaces of historic buildings, archaeological sites, stone and metal sculptures, rock art sites, caves, catacombs. Microbial growth has been detected on wall paintings, ceramics, mosaics, glass, mortars, concrete. Many factors affect the

D. Pinna (✉)
Department of Chemistry, University of Bologna, Bologna, Italy
e-mail: daniela.pinna@unibo.it

© The Author(s) 2021
E. Joseph (ed.), *Microorganisms in the Deterioration and Preservation of Cultural Heritage*, https://doi.org/10.1007/978-3-030-69411-1_1

colonization of microorganisms: moisture, pH, light, temperature, nutrients. Variable ecological spatial patterns can occur on monuments because of changes in these environmental factors. When a microbial colonization is evident, its relevance to degradation and weathering of inorganic materials should be carefully evaluated as biotic and abiotic agents interact in quantitatively variable relations (Siegesmund and Snethlage 2014). The detection of organisms on cultural objects does not indicate necessarily that they are modifying the chemical composition or physical properties of the materials (Pinna 2017). Only in particular conditions and in combination with other factors they can initiate, facilitate, or accelerate deterioration processes. Moreover, the growth of some organisms is very slow, and the damage becomes visible only after years or even decades.

At present, the importance of biodeterioration processes on historical objects of art has reached growing attention of people in charge of the conservation of cultural heritage. A large set of relevant studies have documented and discussed the interaction between biological colonization and cultural heritage objects. Despite considerable research efforts, there are still general issues that need to be addressed. Regarding stones, many aspects of the interaction between microbial communities, lichens and these materials are still unknown (Di Martino 2016). Not surprisingly, in recent times many papers report that the biological colonization of outdoor stones may act as a protective layer shielding the materials from other factors that cause decay, such as wind and rainwater. In addition, the species and their amount within biofilm-forming microbial communities can change over time. Progress in microbial ecology and genomics, in parallel with developments in biological imaging and analytical surface techniques, can promote a comprehensive insight into the dynamics of the structured microbial community within biofilms. This evidence will contribute to foresee their responses after short- and long-term disturbances and in future changes of environmental conditions.

Regarding metals, the importance of microbial ecology in microbiologically influenced corrosion (i.e. the differences between actively corroding and non-corroding microbes present on metals exposed in the same environment) and the effect of the biofilm matrix on the electrochemical behavior of metals are issues that need to be further examined. Elucidating them will also facilitate the development of more efficient prevention and protection measures.

This chapter aims to discuss these issues and review the recent literature on (i) biofilms and lichens colonizing inorganic materials, (ii) the limiting factors of this colonization, (iii) the deteriorative aspects, and (iv) the protective effects of the colonization.

## 2 Biofilms and Lichens

Scientific observations of a wide variety of natural habitats have established that the majority of microbes predominantly live in complex sessile communities known as biofilms (Bridier et al. 2017). **Biofilms** are highly structured assemblages of microbial cells attached to a surface and entwined in a matrix of self-produced extracellular polymeric substances (EPS) (López et al. 2010). The structured biofilm ecosystem enables microbial cells to resist stress and increase their tolerance to stressors, both at the level of individual microorganisms and collectively as a community (Flemming et al. 2016). The microorganisms within a biofilm can survive and thrive in harsh environments characterized by desiccation, low-nutrient concentrations, large temperature variations, and high exposure to wind, UV radiation, and physical damage (Jacob et al. 2018). Biofilms are present in natural, industrial, medical, household environments and, from the human point of view, they can be either beneficial or detrimental.

The characteristics of microbial cells forming biofilms are distinct from those of their planktonic counterparts; their higher resistance to antimicrobial agents and ultraviolet radiation (UV), the development of physical and social interactions, an enhanced rate of gene exchange, and the selection for phenotypic variants are traits related to the structural characteristics of the community (Hentzer et al. 2003; Flemming et al. 2016; Di Martino 2016; Bridier et al. 2017; Mittelmann 2018). Indeed, both the microbial development and the matrix lead to the growth of a cooperative consortium that offers a protective structure able to hinder diffusion and action of antimicrobials (Bridier et al. 2017).

The process of biofilm development is coordinated by molecular pathways while the spatial structure of the biofilms is dependent on the species as well as on the environmental conditions (Tolker-Nielsen 2015). Cell-to-cell biochemical signals influence each step in the process of biofilm formation and enhance the persistence of both individual species and the biofilm (Katharios-Lanwermeyer et al. 2014; Vega et al. 2014). Multispecies biofilm is a result of cell–cell and cell–environment interactions such as cooperation, competition or exploitation (Liu et al. 2016). The spatial organization of biofilms is driven by the specific interactions between species.

The first step in the development of biofilms is the formation of a conditioning layer on noncolonized surfaces (Fig. 1.1). The layer is formed by organic substances like polysaccharides and proteins which improve the attachment of initial colonizers. In the second step, microbial cells attach to these substances due to their size and net negative charge by nonspecific interactions, such as electrostatic, hydrophobic, and van der Waals forces (Mittelmann 2018). Hydrophobicity and electric charge of the conditioning substances are of importance in bacterial adhesion (Diao et al. 2014). The electrostatic forces are weak and reversible at this stage. At a later stage, cells firmly adhere to the surface and to each other, producing EPS that entrap cells and debris within a glue-like matrix (Fig. 1.1). EPS can increase the rate of coaggregation of cells (Mittelmann 2018). Polysaccharides are characteristic components of the EPS, but proteins, nucleic acids, lipids, and humic substances have also been identified, sometimes in substantial amount. As mentioned above, the matrix—a three-dimensional, gel-like, highly hydrated and locally charged (often anionic) material—is composed of EPS the proportion of which, in general, can vary between roughly 50% and 90% of the total organic matter. EPS play significant roles in the attachment of microorganisms to the mineral surfaces and in the protection of the microbial community from toxic compounds (Diao et al. 2014). At this stage, the

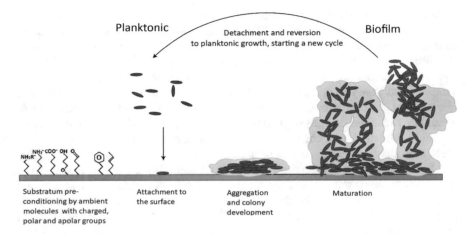

Planktonic                    Detachment and reversion                    Biofilm
                           to planktonic growth, starting a new cycle

Substratum pre-          Attachment to        Aggregation        Maturation
conditioning by ambient  the surface          and colony
molecules with charged,                       development
polar and apolar groups

**Fig. 1.1** Stages leading to microbial formation on a surface. Once cells attach to surfaces irreversibly, they replicate and grow forming microcolonies. The microbial cells secrete extracellular polymeric substances (EPS) and become encapsulated in a hydrogel layer, which forms a physical barrier between the community and the extracellular environment. The community forms a three-dimensional structure and matures into a biofilm. Cells in an established biofilm are "glued" together by the EPS, which resist mechanical stresses and detachment of the community from the surface of the substratum. Some cells detach from regions of the biofilm and disperse to form biofilms in new environmental niches

biofilm environment is a rich layer of nutrients that can support rapid growth of the microorganisms within it. Biofilms are thus accumulations of microorganisms, EPS, multivalent cations, inorganic particles, biogenic materials, colloidal, and dissolved compounds. As the microcolonies grow, additional species, the so-called secondary colonizers, are recruited through coaggregation and nonspecific aggregation interactions, increasing the biofilm biomass and species complexity (Katharios-Lanwermeyer et al. 2014). This process allows species that cannot adhere to the noncolonized surfaces to become part of the biofilm (Gorbushina 2007). In the mature biofilm, complex diffusion channels transport nutrients, oxygen, and other necessary elements for cells, and carry away metabolic waste products. The last stage of biofilm lifestyle is the dispersal of cells from the biofilm colony, which enables it to spread and colonize new surfaces (Fig. 1.1).

Exposed heritage surfaces are mostly low-nutrient environments. The formation of biofilms is a survival mechanism adopted by microorganisms. Other survival mechanisms are nutritional versatility, cellular morphological changes, and formation of spores (Mittelmann 2018). Moreover, microbial cells can enter a viable but non-culturable state reducing their respiration rates and densities yet retaining their essential metabolic activities in response to nutrient deficiency (Mittelmann 2018). Associations of phototrophic and heterotrophic microorganisms often compose the biofilms that grow on heritage objects. Photosynthetic microorganisms sustain the development of heterotrophs through the production of EPS such as polysaccharides

(Villa et al. 2015). The heterotrophs, in turn, can promote the growth of autotrophs by supplying key substances and other products.

A **lichen** is a mutualistic symbiotic association between green algae or cyanobacteria and fungi. The mutual life form has properties quite different from those of its component organisms growing separately. The fungal component of the lichen is called mycobiont, while the photosynthesizing organism is called photobiont. Photobionts are green algae and cyanobacteria. Most lichens are associated to green algae. The photobiont produces the nutrients needed by the entire lichen, while the mycobiont produces large numbers of secondary metabolites, the lichen acids, which play a role in the biogeochemical weathering of rocks and in soil formation. The nonreproductive tissue, or vegetative body part, is called thallus. Nearly all lichens have an outer cortex, which is a dense, protective layer of hyphae. Below it, there is the photosynthetic layer. Then, a layer of loose hyphae, the medulla, is present. Foliose lichens have a lower cortex with rhizines, which anchor them to the substratum, while crustose lichens attach to the substratum through the medulla. Lichens come in many colors (white, black, red, orange, brown, yellow, and green), shapes, and forms. Three major growth forms are recognized: crustose that is flattened and adheres tightly to the substratum; foliose with flat leaf-like lobes that lift from the surface; fruticose that has tiny, multiple leafless branches three-dimensional growing.

The mycobiont produces the fruiting bodies, which are spore-producing structures. The most common sexual fruiting bodies are the apothecium and the perithecium. Apothecia typically are structures cup or disc-shaped, often with a distinct rim around the edge. The hymenium, the tissue containing the asci, forms the disc, the upper surface of the apothecium. Two basic types of apothecium are recognized in lichens, differing in their margins and underside (together named "the exciple"). In lecanorine apothecia, the thallus tissue extends up the outside of the apothecium to form the exciple and the rim. This margin generally retains the color of the thallus and normally contains algal cells. In lecideine apothecia, the exciple is part of the true apothecial tissue and does not contain algal cells. Discs that become very contorted and appear as line segments on the surface of the thallus are called lyrellae.

Perithecia are usually flask-shaped fruiting bodies enclosing the asci. At maturity, an apical opening, the ostiole, releases the ascospores. Perithecia are usually partially embedded in the thallus or in the substratum and are relatively small, rarely more than 1 mm in diameter and commonly much less.

# 3   Factors Influencing Microbial Colonization

Inorganic cultural heritage materials differ in surface texture, hardness, porosity, pH, and chemical composition, characteristics that make them favorable or unfavorable to microbial colonization.

The susceptibility of **stones** to hold organisms is called **bioreceptivity**, a term coined by Guillitte (1985). In other words, it is the aptitude of a stone to be vulnerable to organisms' colonization. Guillitte further defined this concept. The primary bioreceptivity indicates the potential of a healthy material to be colonized; the secondary bioreceptivity is the result of the deterioration caused by abiotic and biotic factors; finally, the tertiary bioreceptivity is prompted by nutrients contained in the substratum (e.g. dead biomass, dust particles, animal feces, water repellents and consolidants, biocides, etc.).

There are several reports in the literature on studies conducted either in situ or under laboratory conditions on the assessment of bioreceptivity. Highly porous materials are more susceptible to microbial colonization because of their capacity to absorb more water and retain it in for longer periods of time. Surface roughness affects the trapping of moisture and concentrates it in micro-fissures where growth is usually more abundant. It also enables the accumulation of particles - soot, organic and inorganic debris, pollens, spores, and salts (Jacob et al. 2018). In addition, rougher surfaces can be a preferential site for colonization because they provide shelter from wind, desiccation, and excessive solar radiation (Miller et al. 2012b; Jacob et al. 2018). Evidence of the importance of these physical characteristics is given by the resistance to microbial colonization of smooth and impermeable surfaces of glazed ceramics unless fissures are present.

Studies on primary and secondary bioreceptivity of granites showed that the growth of phototrophic biofilms is strongly enhanced by high open porosity, capillary water content, surface roughness, and abrasion pH rather than by differences in the chemical and mineralogical composition of the rocks (Prieto and Silva 2005; Vázquez-Nion et al. 2018a, b). Abrasion pH, measured after grinding the rock in distilled water, relates to the number of basic cations released by the rock when in contact with aqueous solutions. Open porosity is correlated with void spaces in the rock; thus, it refers to the capacity of rock to absorb water. Capillary water absorption, providing information about the pattern of the pore network, is connected to the time the rocks remain wet. Despite the above-mentioned stone properties are assumed as the most important to determine bioreceptivity of natural and artificial stones, some studies showed that the chemical composition and petrography appear to be crucial factors as well (Miller et al. 2012b). A study (Miller et al. 2006) on limestone, granite, and marble samples artificially inoculated with two photosynthetic microorganisms (the cyanobacterium *Gloeocapsa alpina* and the green microalga *Stichococcus bacillaris*) showed that limestone and marble, which had the highest (>17%) and the lowest (≤1%) porosity, respectively, supported the greatest microorganisms' colonization while granite showed just a limited growth. The result apparently depends mainly on the chemical composition rather than on the physical characteristics of the stones. Similarly, the importance of stone chemical composition emerged from a laboratory experiment (Olsson-Francis et al. 2012) that showed basalts have higher rates of cyanobacterial growth and dissolution than rhyolitic rocks. According to the authors, the difference is due to a higher content of quartz, which has a low rate of weathering, and to lower concentrations of bio-essential elements, such as Ca, Fe, and Mg. A research on the prevention of biological colonization in the archaeological area of Fiesole (Italy) (Pinna et al. 2012, 2018) assessed the secondary bioreceptivity of some stones (sandstone, marble, plaster) where crustose lichens were previously removed. The 8-year-long study showed that the recolonization of the three substrata after the cleaning depended mainly on their bioreceptivity and on climatic conditions. Marble showed a high bioreceptivity as, at the end of the monitoring, fungi and lichens covered the surfaces. On the contrary, sandstone and plaster showed very low bioreceptivity. Unlike marble, the pioneer species on sandstone were lichens that started developing more than 4 years after cleaning. At the end of the monitoring, the lichens grew extensively on sandstone surfaces, but did not cover them completely. Although the position near the ground of the test area of the original Roman plaster would be expected to be favorable to biological growth, it was the least bioreceptive substratum as, at the end of the monitoring, almost no biological colonization was detected on the surfaces. It showed a low water absorption due to the presence of *cocciopesto*,[1] of a natural wax, and of an outer thin whitish calcite layer. These substances likely made it non-hygroscopic and prevented microbes' growth. As the environmental conditions of the tested areas did not differ and the porosity was quite low on all the stones, arguably the different bioreceptivity can relate to their chemical composition, pH, shape and orientation of pores, pore size distribution and surface texture.

Different types of stone show different bioreceptivities toward lichen colonization as it has been shown in a study of gravestones in Jewish urban cemeteries in north-eastern Italy (Caneva and Bartoli 2017). Trachyte mainly hosted the growth of the lichen *Protoparmeliopsis muralis* which covered horizontal surfaces. Verona stone, marble, and Istrian stone were less colonized while Nanto stone, a soft yellow-brown limestone, was the material mostly affected by lichen colonization, and suffered the highest degree of deterioration.

**Substratum pH** influences biological colonization because some organisms prefer specific values or tolerate a narrow pH range. Extreme pH values are not favorable because of the damaging effect of $H^+$ or $OH^-$ ions. Although most microorganisms tend to live in pH neutral conditions, some may colonize cement over a wide pH range (Allsopp et al. 2004; Prieto and Silva 2005; Miller et al. 2012b). Many fungi prefer slightly acidic substrata (e.g. granites, some sandstones), while alkaline conditions (e.g. limestone, marble, lime mortars) may favor cyanobacteria (Caneva and Ceschin 2008). As a fresh concrete has a pH of 12–13, it permits microbial growth when pH is lowered by reaction with atmospheric carbon dioxide (carbonation) (Allsopp et al. 2004). The alkalinity of concrete is crucial to decrease the rate of biodeterioration by inhibiting the microbial activity (Noeiaghaei

---

[1]Cocciopesto is a decorative plaster dating back to Roman times. The Romans utilized it to implement durable floors or as a base to lay their intricate stone mosaics. Made with crushed soft fired brick and lime putty, cocciopesto is a highly resistant finish able to withstand wet or humid environments.

et al. 2017). Thus, the degree of carbonation and pH values play a key role in the receptivity of concrete and mortars to microbial colonization.

Differences in abundance of colonization on stones can be due to tertiary bioreceptivity related to nutrients contained on the surfaces or in the stones (Salvadori and Charola 2011). They can derive from existing biological growth on the surface, bird droppings, organic compounds used in restoration practice, pollutants. Air and rain carry nutrients in the form of dust and soil particles. Soil fertilization leads to an accumulation of nitrates and phosphates (eutrophication) that are contained in bird droppings too. Some lichens (nitrophilic species) have been adapted to high nitrogen levels. Generally, they have an orange thallus, easily observed on roofs, architectural moldings, horizontal surfaces of statues where the eutrophication derived from bird droppings and/or fertilizers transported by air and rain.

Along with substrata' bioreceptivity and capacity to retain water, **the environment surrounding the monument** and the monument itself act as limiting factors of biological growth. Local microclimate, macroclimate, wind-driven rain, pollution, geographical location, architectural design and details of monuments or sculptures are remarkable factors influencing organisms' colonization (Tonon et al. 2019). Tropical climate conditions in Far East, for example, enhance the establishment and widespread colonization of microflora on cultural heritage objects due to high water availability during monsoon season (Zhang et al. 2019).

The importance of wind-driven rain to the abundance of microbial colonization of surfaces emerged from a five-year-long study in the archaeological site of Pompeii, Italy (Traversetti et al. 2018). The extent of microbial growth was much broader on northern and western exposures respect to south-exposed walls. Comparison of climatic parameters, data of the dominant winds and microorganisms' occurrence (cyanobacteria, algae, and lichens) showed that wind-driven rain along with lower temperatures and poor ventilation of northern exposure strongly affected stones' wetness and played a key role in the promotion of microbial development. A similar study documented that the moisture contents in the walls of a brick tower was higher near the edges of the walls than at the center just for wind-driven precipitation, inducing fungal mold colonization especially at this position (Abuku et al. 2009). In these studies, the orientation of the object towards the light was an aspect affecting microbial colonization as well. In wet climates and northern latitudes, north-exposed surfaces get less direct sunlight. Therefore, once wet, they remain damp for much longer than other stones. This slow-drying condition is much more favorable to biological colonization than the hostile condition of rapid wetting and drying cycles experienced on the south faces (Adamson et al. 2013). Similarly, in a different climate and latitude (south of Brazil) painted surfaces showed higher fungal colonization on south-facing sides that received less solar radiation than north-facing ones. If the surface temperature fell below dew point at night, they remained moist for longer periods of time after wetting (Shirakawa et al. 2010).

Temperature and relative humidity are key parameters influencing the development of fungi (Jain et al. 2009). Most of them may be grouped into three categories on a thermal basis. Psychrophiles thrive at temperatures between 0 °C and 5 °C, mesophiles between 20 °C and 45 °C, while thermophiles at or above 55 °C. Members of the *Mucoraceae* and *Deuteromycetes* grow on substrata at relative humidity of 90–100% and not below, while members of the *Aspergillus glaucus* group grow at relative humidity as low as 65% (Jain et al. 2009). Temperature may also affect the stone water content. The influence of surface temperature showed a significant positive correlation with green algae biofilms in a survey of four sandstone heritage structures in central Belfast (Ireland) exposed for around 100 years (Cutler et al. 2013). Areas with lower temperatures were, on average, greener than warmer areas. As reported by the authors, it is possible to model algal greening of sandstones from the scaled-down outputs of regional climate models as it mostly relates to climate and atmospheric particulates.

In hot climates sunny surfaces are more hostile and variable than those shaded, thus the production of extra cellular polymeric substances tends to be higher to protect the cells from the adverse conditions (Scheerer 2008). Moreover, microbial biomass and species diversity are usually much higher in shaded areas (Ramírez et al. 2010).

# 4   Biodeterioration Processes Caused by Biofilms and Lichens

Microorganisms that play a potential role in biodeteriorative processes of inorganic materials are autotrophic and heterotrophic bacteria, fungi, algae and lichens.

Microbial biofilms interact with inorganic materials in several ways: physical deterioration, where materials structure is affected by microbial growth (e.g. physical or mechanical breaking); aesthetic deterioration due to fouling; chemical deterioration, mineral and metal transformations due to excretion of metabolites or other substances such as acids that adversely affect the structural properties of the material (e.g. increasing of porosity, weakening of the mineral matrix, dissolution of minerals and formation of biominerals, biocorrosion of metals and alloys).

The physical and chemical weathering of rocks by lichens—mainly epilithic and endolithic crustose species - encompasses the following mechanisms: hyphae growth in intergranular voids and mineral cleavage planes; expansion and contraction of the thallus by wetting and drying, and freeze–thaw cycles; incorporation of mineral fragments into the thallus; swelling and deteriorative action of organic and inorganic acids; dissolution/etching of mineral grains and precipitation of amorphous and crystalline compounds.

In addition, the dissolution of respiratory $CO_2$ contained in water held by biofilms and lichen thalli can lower the pH at the substratum–thallus interface, accelerating the chemical weathering of the rocks (Di Martino 2016).

Effective physical weathering requires chemical weathering since the growth in the stone of biofilms and lichens is facilitated by the dissolution of minerals along grain boundaries, cleavages, and cracks. The overall effect increases porosity and

**Fig. 1.2** Statue in the park of Nympheburg, Munich, Germany. The statue's surface is almost completely covered by biofilms and lichens that strongly compromise its legibility

permeability. Since these weathering processes interact and enhance each other action, it is impossible to separately quantify their role (Bjelland and Thorseth 2002; de Los Rios et al. 2002).

As for natural and artificial stones, the aesthetic aspects assume an important role because biofilms' and lichens' development is an impediment to the site's presentation to the audience when it compromises the legibility of a monument (Ashbee 2010) (Fig. 1.2).

## 4.1  Stone

The so-called biomineralization or biologically induced mineralization is the process by which biological activity induces the precipitation and accumulation of minerals. It is the result of the metabolism of the organisms and is present in all five kingdoms of life. Minerals produced by biofilms and lichens include iron hydroxides, magnetite, manganese oxides, clays, amorphous silica, carbonates, phosphates, oxalates (Konhauser 2007; Dupraz et al. 2009; Miller et al. 2012a; Urzì et al. 2014). Biomineralization produced by fungi occurs when they modify the local microenvironment creating conditions that favor extracellular precipitation of mineral phases (Fomina et al. 2010; Gadd et al. 2014).

**Stone-inhabiting microorganisms** may grow on the surface (epiliths) or some millimeters or even centimeters in the substratum (endoliths) (Gadd et al. 2014). The endolithic colonization is an adaptive strategy developed by cyanobacteria, algae, fungi, and lichens. The use of the substratum as a shield against external stress proves to be a decisive evolutionary selection advantage (Pohl and Scheider 2002). The endolithic mode of life includes different ecological niches: chasmoendoliths and cryptoendoliths occupy preexisting fissures and structural cavities in the rocks, whereas euendoliths grow in soluble carbonatic and phosphatic substrata dissolving the stone immediately below the surface. The first form of growth leads to a co-responsibility in the detachment of scales of material due to the pressure exerted by increasing biomass. This process can occur repeatedly, involving areas increasingly in depth (Pinna and Salvadori 2008). The light that reaches the bulk of the stones limits the growth of phototrophic microorganisms. The presence of water in micropores, especially those with translucent walls, may enhance light penetration, increasing the light available for photosynthesis in the cryptoendolithic habitats (Cámara et al. 2014). Euendoliths form microcavities of varying morphologies according to the species.

Stones subjected to extreme sun irradiation in hyper-arid Atacama Desert (Chile) are colonized by endolithic **algae and cyanobacteria**. In order to endure the harsh environment, they synthesize carotenoids and scytonemin, respectively. The production of these pigments is interpreted as an adaptation strategy against high doses of solar irradiation as they are passive UV-screening pigments (Vítek et al. 2016).

Cyanobacteria and microalgae are often the first colonizers of stone surfaces where they develop phototrophic biofilms. Algae form colored (green, gray, black, brown, and orange) powdery patinas, and gelatinous layers. They usually dominate surfaces in wet and rainy areas. Cyanobacteria typically form dark brown and black patinas but also pink discolorations. Besides the aesthetic disturbance caused by the colored patinas, algae and cyanobacteria cause water retention and damage due to freeze–thaw cycles. Areas colonized by dark biofilms formed by cyanobacteria may absorb more sunlight. Temperature changes increase mechanical stress by expansion and contraction of the biofilms (Scheerer et al. 2009).

The main groups of **fungi** isolated from stone monuments are Hyphomycetes and black meristematic fungi (Salvadori and Casanova 2016). Hyphomycetes, commonly known as mold, are a class of asexual or imperfect fungi. They lack fruiting bodies, the sexual structures used to classify other fungi. The production of conidia (spores) occurs by fragmentation of vegetative hyphae or from specialized hyphae called conidiophores. Since the attachment of airborne spores to surfaces is the first step of fungal colonization, the species diversity of fungi present on stones is similar to the diversity of common airborne spores (Sterflinger and Piñar 2013). Many Hyphomycetes, notably *Aspergillus*, *Fusarium*, and *Penicillium* genera, produce toxic metabolites (mycotoxins). Several Hyphomycetes growing on stone heritage objects are dematiaceous. The term refers to the characteristic dark appearance of these fungi that form dark gray, brown or black colonies. Fungi excrete organic acids (oxalic, citric, acetic, formic, gluconic, glyoxylic, fumaric, malic, succinic, and pyruvic) that can act as chelators. Particularly, oxalic acid has a high capacity of degrading minerals for its complexing and acid properties. Its importance relates to the ability of the oxalate anion to complex and/or precipitate metals as secondary

**Fig. 1.3** Early development of black meristematic fungi on the wealthy surface of a marble model sample exposed outdoor. Stereomicroscope, 50×

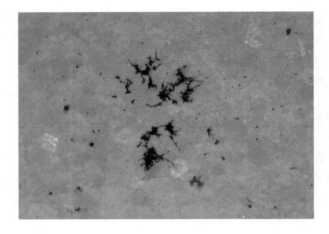

biominerals. The reaction results both in mineral dissolution and in mineral formation (Gadd et al. 2014). Calcium oxalate precipitation has an important influence on biogeochemical processes in soils, acting as a calcium reservoir (Gadd et al. 2014). The precipitation of secondary minerals (carbonates and oxalates) on and within stones, and the mineral dissolution connected to fungal growth lead to the formation of crusts on the surface and around the hyphae that can progressively cement fissures and cracks (Fomina et al. 2010).

An investigation into a thin dark rock coating at the Ngaut Ngaut heritage complex in South Australia showed that it contained a mixture of calcite, quartz, gypsum, and weddellite (Roberts et al. 2015). The dark coating covered petroglyphs engraved in the rock shelter and dated back around 3000 years B.C. According to the authors, the weddellite was likely formed from the reaction of calcite with oxalic acid excreted by surface microflora.

Metal mobilization can also be achieved by chelation ability of siderophores. When living in environments of reduced iron content, fungi produce iron(III)-binding ligands, commonly of a hydroxamate nature, termed siderophores (Salvadori and Casanova 2016).

Black meristematic fungi belong to the genera *Hortaea*, *Sarcinomyces*, *Coniosporium*, *Capnobotryella*, *Exophiala*, *Knufia,* and *Trimmatostroma* (Sterflinger and Piñar 2013). Their cells have thick pigmented walls. They form slowly expanding, cauliflower-like colonies that grow by isodiametric enlargement of the cells (Sterflinger 2010) (Fig. 1.3). In addition to the meristematic growth, many of the black fungi can exhibit a yeast-like growth (De Leo et al. 2003). They abandoned the hyphal phase adopting the microcolonial or yeast phase characterized by an extremely slow growth, in response to the lack of organic nutrients and to stresses of outdoor substrata. The microcolonial phase enhances the survival and persistence of these fungi in the biofilms. They produce various pigments including carotenoids, mycosporines, and melanins that protect them from UV irradiation. Moreover, melanin provides them with extra-mechanical strength making hyphae able to grow better in fissures. Microcolonial fungi, as well as cyanobacteria, algae

**Fig. 1.4** Marble well-curb (Montefiore Conca, Italy) colonized by biofilms (**a**). Peculiar pattern of stone weathering on a sample. Biofilms form little black spot on the surface. Stereomicroscope (**b**). Polished cross section of the sample showing black fungi on the surface and hyaline hyphae in the bulk of the stone (**c**)

and lichens, are poikilohydric microorganisms, which means that they are metabolically active or dormant depending on water availability. They can thrive under extreme conditions including irradiation, temperature, salinity, pH, and moisture. They are stress-tolerant colonizers involved in biodeterioration (de los Rios et al. 2009; Sterflinger 2010; Marvasi et al. 2012; Onofri et al. 2014; Salvadori and Casanova 2016). On marbles, they grow in the inter-crystalline spaces (Fig. 1.4) contributing to the loosening and detachment of crystals and forming biopitting. However, the scarcity of nutrients and the very slow growth limit their contribution to weathering (Hoffland et al. 2004). Biopitting is an alteration caused by euendolithic microorganisms that form little blind holes close together (Golubic et al. 2005) (Fig. 1.5). The stone frequently associated with this kind of deterioration is marble (53%), followed by carbonate rocks (44%), granite, and concrete (3%). Bio pitting is caused mostly by cyanobacteria along with fungi (11%), lichens (10%), and algae (5%) (Lombardozzi et al. 2012).

**Wall paintings** may have a "double nature," inorganic and organic. In the history, they have been made using several techniques. The best known is *fresco* (from the Italian word *affresco* that derives from the adjective fresco "fresh") that uses water-soluble pigments applied on wet plaster or lime mortar. The pigments are absorbed by the wet plaster. When drying, the plaster reacts with the air in a process called carbonation in which calcium hydroxide reacts with carbon dioxide and forms insoluble calcium carbonate; this reaction fixes the pigment particles in the plaster. The technique called *a secco fresco* uses instead dry plaster (secco means "dry" in

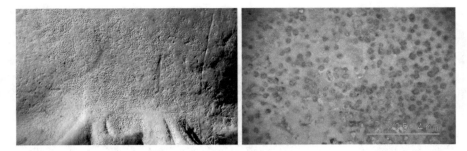

**Fig. 1.5** Biopitting on a stone statue, Uffizi Gallery (Firenze, Italy). Close-up of a detail of the surface (right). The circular micro-holes were likely produced by endolithic lichens that grew when the statue was located outdoors. The fruiting bodies were present inside the holes. After the lichens died, they left just the empty holes

Italian). An organic binding medium, like egg, glue, or oil, is needed to fix the pigment into the plaster. In Classical Greco-Roman times, the encaustic painting technique was in use. Pigments were ground in a molten beeswax binder (or resin binder) and applied to the surface while hot. Today, murals are mostly painted in oils, tempera, or acrylic colors. Microbial colonies may deteriorate either the inorganic component or the organic one causing efflorescence, expansion cracking, peeling and spalling of outer layers, color changes, and stains. Oxalates may be produced from the calcite or from metal-containing pigments of the painted layers (Gadd et al. 2014; Unković et al. 2017).

The deteriorative potential of fungi isolated from wall paintings was evaluated by in vitro studies to assess the risk of deterioration and formulate appropriate conservation treatments. Some biodegradative properties—calcite dissolution, casein hydrolysis, pigment secretion, acid and alkali production—were tested using specific agar culture media (Pangallo et al. 2012; Unković et al. 2018). Many species of the genera *Aspergillus* and *Penicillium* demonstrated the ability to dissolve calcite as they produced and secreted acids; oxalic acid was the main cause of deterioration in most instances (Ortega-Morales et al. 2016). Species of the genus *Cladosporium* hydrolyzed casein showing high proteolytic activity (Unković et al. 2018). They are thought as the primary degraders of protein components of painted layers of wall paintings. Applying the mentioned method, the authors detected also species showing no deterioration capabilities.

Pink patinas and rosy discolorations occurring on stones and wall paintings may be caused by bacteria and algae. In the crypt of the Original Sin (Matera, Italy), the pink color resulted from the production of a ruberin-type carotenoid pigment produced by actinobacteria (Nugari et al. 2009). Similarly, carotenoids caused pink and yellow discolorations on wall paintings of St. Botolph's Church (Hardham, UK) (Kyi 2006). According to the authors, the pigments provide microorganisms with a protective mechanism against damage caused by photooxidation reactions. Rosy-powdered areas on the frescoes of St. Brizio Chapel (Orvieto Cathedral, Italy) contained instead phycoerythrin, a pigment produced by capsulated coccoid cyanobacteria that grew even in the dark conditions of the chapel being able to use the organic compounds present on the fresco surface (Cappitelli et al. 2009). An analogous alteration on wall paintings of two churches in Georgia (Gittins et al. 2002) was due to the bacterium *Micrococcus roseus* while rosy discoloration on walls of Terme del Foro (Pompeii, Italy) were associated to the development of the bacterium *Arthrobacter agilis* which produced the pigment bacterioruberin (Tescari et al. 2018). Green algae belonging to the genus *Trentepohlia* formed red powdery spots on a medieval wall painting in Italy (Zucconi et al. 2012).

**Subterranean environments** like caves and catacombs are ecosystems characterized by unique microhabitats and a very delicate equilibrium of microbial communities living on surfaces and rock paintings (Urzì et al. 2018). The relative humidity is in most cases high (around 90%), air circulation is very low, and the temperature is almost constant through the year (Urzì et al. 2018). The conservation of these sites is challenging; changes of microclimatic parameters (light regime, release of $CO_2$ by visitors, modification of temperature and relative humidity) irreversibly change the whole microclimate resulting in biodeterioration of surfaces and rock paintings (Urzì et al. 2018). The microclimate of the Lascaux cave (France) was strongly affected by visitors (up to 1800 per day) and by the artificial lights installed to allow cave's view. Consequently, a green biofilm developed on the worldwide famous prehistoric drawings. Called "le maladie verte," it was formed by green algae belonging to *Chlorophyta* (Bastian et al. 2010). Then the paintings suffered from a fungal invasion by *Fusarium solani* which formed long white mycelia, and by dematiaceous fungi which produced black stains on the ceiling and passage banks (Bastian et al. 2010). These problems, which generated worldwide interest because the prehistoric drawings represent a priceless cultural heritage for all humankind, led to the closure of the cave in 1963. A similar case was that of the wall paintings of Takamatsuzuka Tumulus (Japan) discovered in 1972 (Miura 2006). The temperature of the tomb was quite stable until 1980, when it began to rise following the outdoor temperature. By 2000, concurrently with the opening to visitors, the temperature reached nearly 20 °C creating an ideal microclimate for fungi that widely colonized the wall paintings.

Several brown spots were spread on the wall paintings of the tomb of King Tutankhamun (Valley of the Kings, Luxor, Egypt) when it was discovered in 1922 (Szczepanowska and Cavaliere 2004). GC/MS analysis of the spots indicated that they contained 16% (by weight) of malic acid, suggesting microbial involvement in their formation (Vasanthakumar et al. 2013). Although located in a desert, over the centuries periodic floods occurred allowing moisture to enter the tomb chambers. The perspiration and breathe of visitors were a further source of moisture. According to a report by the Getty Conservation Institute (www.getty.edu/conservation), six people breathing in the tomb chamber for an hour raised the RH level by 3%.

The paleolithic cave of Altamira in Spain was close to the public in 2002 after photosynthetic bacteria and fungi deteriorated the prehistoric paintings (Jurado et al. 2014).

In subterranean environments phototrophic biofilms composed primarily by cyanobacteria associated with heterotrophic bacteria and microalgae are a relevant cause of damage. The extent to which phototrophic microorganisms colonize the substrata highly depends on light intensity (Urzì et al. 2018). A progressive deepening of their growth in the substrata leads to the mobilization of elements and to higher water retention by polysaccharide sheaths. Mineral crystals were observed on the cells of some cyanobacteria (*Scytonema julianum* and *Loriella* sp.) and of heterotrophic bacteria of the genus *Streptomyces* isolated from caves' walls (Albertano et al. 2005). As reported by the authors, *Streptomyces*, associated to *S. julianum*, promote the precipitation of calcium carbonate on the polysaccharide sheath of cyanobacteria in the form of calcite crystals. A white crust formation is generally associated to the deposition of these crystals and it results in a stromatolitic layer on the stone surface. Calcium carbonate dissolution and precipitation are the main processes in the biotransformation of calcareous substrata in caves (Albertano et al. 2005). Similar $CaCO_3$ crystals precipitation induced by *Actinobacteria, Firmicutes,* and *Proteobacteria* that grew on surfaces of Roman catacombs showed instead a different morphology being in the form of a white fluffy (Urzì et al. 2014). *Actinobacteria* produced extracellular pigments that caused color change of the wall paintings located in the Snu-Sert-Ankh tomb, Egypt (Elhagrassy 2018). Similarly, the development of chemoorganotrophic bacteria harmfully affected the painted surfaces of Hal Saflieni Hypogeum (Paola, Malta) where they extensively formed white alterations on ochre-decorated surfaces dated back to 3300–3000 B.C. (Zammit et al. 2008).

Biodeteriorative effects of **lichens** on stone cultural heritage objects are well documented. However, they varied considerably from species to species. The mechanisms involved in the weathering have been mentioned above. Lichens produce hundreds of carbon-based secondary substances (so-called lichen substances), most of which are unique to the lichen symbiosis. Depending on their chemical structure, they can act as complexing agents. One of the organic acids produced by lichens is oxalic acid, a strong complexing agent, as it has been discussed. It forms chelating bonds producing calcium, magnesium, manganese, and copper oxalate crystals at the rock–epilithic lichen interface, in the thalli, on their surface, and in the bulk of the stones down to the depth reached by fungal hyphae (Edwards et al. 2003; Skinner and Jahren 2003; Konhauser 2007; Dupraz et al. 2009; Viles 2012; Salvadori and Casanova 2016). Oxalates have been implicated in Fe, Si, Mg, Ca, K, and Al mobilization from sandstone, basalt, granite, limestone, and silicates (Gadd et al. 2014; González-Gómez et al. 2018). Calcium oxalate occurs in two crystalline forms—the dihydrate (weddellite $CaC_2O_4 \cdot 2H_2O$) and the more stable monohydrate (whewellite, $CaC_2O_4 \cdot H_2O$).

*Dirina massiliensis* f. *sorediata* is one of the most deteriorating lichens, harmful to stones and wall paintings. Hyphae may develop into calcite up to 20 mm and calcium oxalate is present within the stone (Salvadori and Casanova 2016). Its vegetative reproduction allows a fast colonization of artworks' surfaces in a few years (Salvadori and Casanova 2016). In a medieval cave exposed to external environmental conditions this lichen grew on wall paintings and its gray-green powdery thalli formed a compact and continuous crust that included materials derived from the paintings. Old thalli appeared inflated at the center creating blister-like structures and causing the detachment of painting layers (Nugari et al. 2009). The decay and detachment of the center part of the thallus occur in other lichen species and open the underlying area to further weathering, resulting in cratered mounts on the rock surface (Mottershead and Lucas 2000).

Endolithic lichens develop their biomass in the bulk of the stones, with colonization patterns varying among species. A study on the anatomy of five calcicolous endolithic lichens showed that the photobiont layer is located at the same depth, 100–180 μm, from the surface, but the arrangement and depth of diffusion of hyphae into the stones vary considerably among species (Pinna et al. 1998). A different, atypical pattern of growth is that of *Pyrenodesmia erodens*, an euendolithic lichen able to dissolve limestones to a depth of several millimeters. It forms, with aging, confluent, moniliform depressions (Tretiach et al. 2003). Its thallus is composed by clusters of green algae surrounded by inflated, appressed hyphae. The clusters are arranged in bores formed by dissolution of the rock. This lichen has been found in dry sites of the Mediterranean region and of the southern Alps on exposed, subvertical faces of limestone and dolomite rocks (Tretiach et al. 2003). Its occurrence has been reported on the calcareous rocks of monumental remains in Pasargadae, UNESCO world heritage site in Iran (Sohrabi et al. 2017). The study focused on the lichen colonization and deterioration patterns in the semi-arid conditions of the area. Besides the effects of *Pyrenodesmia erodens*, other epilithic and endolitihic lichens showed damaging action such as pitting and granular disaggregation. Therefore, lichen communities are a potential threat for the conservation of the archaeological site. Moreover, lichen colonization and deterioration patterns do not appear peculiar respect to what has been previously described by the literature about colonized limestones in temperate and semi-arid areas around the Mediterranean basin.

The chemical deterioration of silicate and carbonate rocks by endolithic lichens relates to the secretion of siderophore-like compounds and of carbonic anhydrase (Favero-Longo et al. 2011). This enzyme increases the speed of the reaction $CO_2 + H_2O \leftrightarrow HCO_3^- + H^+$, thus it accelerates the dissolution of calcium carbonate. However, the study of calcium carbonate deterioration by endolithic lichens needs further experimental evidence and it is an interesting goal for further research. The present results, yet relevant, enlighten the complexity of the phenomenon.

Endolithic lichens form biopitting on limestones, with pits' diameter ranging from 0.2 to 1–2 mm depending on the size of fruiting bodies (Salvadori and Casanova 2016) (Fig. 1.5). When lichen death occurs, the empty pits are progressively enlarged by water (rainfall, water runoff, water accumulation). Then they can coalesce forming larger interconnected depressions. Unlike epilithic species, endolithic ones are not characterized by the production of calcium oxalates.

A one-year long laboratory experiment on lichen–rock interactions was conducted using mycobionts and photobionts of the endolithic lichens *Bagliettoa baldensis* and *B. marmorea* (Favero-Longo et al. 2009). They were isolated and then inoculated on marble and limestone samples. The same species growing on limestone outcrops and abandoned marble quarries showed penetration pathways similar to those reproduced in vitro. The study highlighted that erosion processes caused by lichen development increased the availability of hyphae passageways only after long-term colonization. The differences in hyphae growth depended on the mineral composition and structure of the lithotypes.

Different climates affect the endolithic growth of lichens (Pohl and Scheider 2002). In the humid Northern Alps (Austrian glacier), the bulk of the calcareous rock, just under the surface, showed three layers. At a depth of 0–150 µm, the substratum is mostly intact, just partly interlocked with the lichen litho cortex. At 150–300 µm, photobionts (green algae or cyanobacteria) are capable of actively dissolving the substratum, thus creating habitable cavities. Then several mm in depth, there is the mycobiont with hyphae actively solubilizing the substratum and often forming dense networks. Most biomass of endoliths on carbonate rocks in the arid Mediterranean-Maritime Alps (Provence, France) was instead confined just under the surface. The average colonization intensity and growth depth are markedly deeper in the more humid substratum of the Austrian Alps.

## 4.2 Stained-Glass Windows

Glass is a non-crystalline, amorphous inorganic solid. The most familiar types of glass are prepared by melting and rapidly cooling quartz sand and other ingredients - sodium carbonate ($Na_2CO_3$, "soda"), lime (calcium oxide—CaO), magnesium oxide (MgO), and aluminum oxide ($Al_2O_3$). The resulting glass contains about 70–74% silica by weight and is called a soda–lime glass (Na-rich glass) which accounts for about 90% of manufactured glass. To color the glass, powdered metals are added to the mixture while the glass is still molten. The medieval stained-glass windows that decorate many European churches were made using, besides sand, a different ingredient, the so-called potash that is wood ash ($K_2O$, K-rich glass). These glasses are more easily decayed than Na-rich ones.

Biofilms' growth on glass is correlated to tertiary bioreceptivity. Neither the inorganic composition nor the physical features of glass favor microbial colonization, but the organic residues of various origins, as dust deposits, dead microbial material, and bird droppings, can be a source of nutrients. The deteriorative action of microorganisms on glass is a modest, slow, yet continuous process that can accelerate its weathering. Research focused mainly on medieval stained-glass windows of European churches where corrosion, patinas, pitting, cracks, and mineral crusts occur. Microorganisms may contribute to all these decay forms (Carmona et al. 2006; Marvasi et al. 2009; Piñar et al. 2013). Analyses revealed complex bacterial communities consisting of members of the phyla *Proteobacteria*, *Bacteroidetes*, *Firmicutes* and *Actinobacteria* (Piñar et al. 2013). Fungi showed less diversity than bacteria, and species of the genera *Aspergillus*, *Cladosporium,* and *Phoma* were dominant (Carmona et al. 2006; Piñar et al. 2013). Thus, historical glass windows are a habitat in which both fungi and bacteria form complex microbial consortia of high diversity. The bacteria are genetically similar to those that cause mineral precipitation on stones. Regarding the detected fungi, they are ubiquitous airborne species. According to Piñar and coauthors (2013), the pitting present on the surfaces of glass windows could relate to other more specialized fungi that grew in the past and are not detectable now.

The chemical composition of the glass affects microbial development; copper contained in green glasses acts as an inhibitor (Marvasi et al. 2009; Piñar et al. 2013). In a laboratory test, historically accurate reconstructions of glass windows (fifteenth and seventeenth centuries, Sintra, Portugal) were inoculated with fungi of the genera *Cladosporium* and *Penicillium* previously isolated from the original stained-glass windows. The fungi produced clear damage on glass surfaces in form of stains, erosion, pitting, crystals, and leaching (Rodrigues et al. 2014).

Biofilms on modern glasses may alter their transparency (Shirakawa et al. 2016) and smoothness (Corrêa Pinto et al. 2019). However, Na-rich glass samples inoculated with a mixture of four fungal species showed high resistance against the microorganisms' growth (Corrêa Pinto et al. 2019).

## 4.3 Metals

Physicochemical interactions between a **metallic material and its environment** can lead to electrochemical corrosion. It is a chemical reaction that occurs when electrons from metal are transferred to an external electron acceptor, causing release of the metal ions into the surrounding medium and deterioration of the metal (Beech and Sunner 2004). This process consists of a series of oxidation (anodic) and reduction (cathodic) reactions of compounds present in direct contact with or in proximity to the metallic surface (Beech and Sunner 2004). The rate of the metal deterioration decreases gradually with time, because the oxidation products (for example, oxidation of metallic iron to $Fe^{2+}$ ion) adhere to the surface forming a protective layer that provides a diffusion barrier to the reactants.

**Biocorrosion or microbiologically influenced corrosion (MIC)** is caused by the interaction between the metal surface, abiotic corrosion products, and microbial cells and their metabolites. MIC is an electrochemical process in which microorganisms initiate, facilitate, or accelerate a corrosion reaction on a metal surface through the utilization and release of electrons (Kadukova and Pristas 2018). Microbial activity of biofilms present on surfaces of metallic objects may affect the kinetics of cathodic and/or anodic reactions and may also considerably modify the chemistry of any protective layers (Beech and Sunner 2004). Recent evidence indicates that bacteria can transport electrons via organic molecules of the surrounding environment (Gu 2019).

Most investigations have addressed the impact of biofilms on corrosion behavior of iron, copper, aluminum, and their alloys. The effects of microbial activity on iron or steel range from pitting, crevice formation, under-deposit corrosion to stress corrosion cracking (Aramendia et al. 2015).

Microorganisms mostly involved in the biocorrosion of metals are aerobic and anaerobic bacteria (Zanardini et al. 2008). In open air systems, anaerobic corrosion still occurs if a corrosive anaerobic biofilm grows underneath a top aerobic biofilm (Xu et al. 2016). While in oxic environments metal corrosion is a fast process that can be enhanced by microorganisms, under anoxic conditions it would take very long time without the help of microorganisms (Kadukova and Pristas 2018).

Anaerobic sulfate reducing bacteria (SRB) have been considered key culprits in MIC of a wide range of industrial structures because the sulfur cycle is linked to microbial metabolism, affecting the integrity of metals (Beech et al. 2014; Rémazeilles et al. 2010). Indeed, one of the most known phenomena is the production of sulfides induced in anoxic environments by SRB (Aramendia et al. 2015). However, also other types of bacteria have been associated with metals in terrestrial and aquatic habitats—sulfur oxidizing bacteria, iron oxidizing/reducing bacteria, manganese-oxidizing bacteria, nitrate reducing bacteria and bacteria secreting organic acids and slime (Beech and Coutinho 2003; Xu et al. 2016; Kadukova and Pristas 2018). Regarding the influence of SRB in the corrosion of iron, it has been documented that not only different genera of these organisms but also species within the same genus vary in their ability to deteriorate iron (Beech et al. 2014).

Extracellular polymeric substances are able to bind metal ions. This property is important to MIC and depends both on bacterial species and on the type of metal ion (Kinzler et al. 2003). Metal binding by EPS involves interaction between the metal ions and anionic functional groups (e.g. carboxyl, phosphate, sulfate, glycerate, pyruvate, and succinate groups) that are commonly present on the protein and carbohydrate components of EPS. In particular, the affinity of multidentate anionic ligands for multivalent ions, such as $Ca^{2+}$, $Cu^{2+}$, $Mg^{2+}$, and $Fe^{3+}$, can be very strong (Beech and Sunner 2004).

Biocorrosion is recognized as an important category of corrosion leading to important economic loss in many industries and services including oil and gas pipelines, water utilities and the power generation industry (Xu et al. 2016; Eid et al. 2018). Therefore, MIC has been the subject of extensive studies on these systems for the past decades and several models have been proposed to explain mechanisms governing biocorrosion (Beech and Sunner 2004). Regarding cultural heritage objects, many papers documented the role of microorganisms in corrosion of metal artifacts (Brown and Masters 1980; Dhawan 1987; Gilbert 1987; McNeil et al. 1991; Sánchez del Junco et al. 1992; Little et al. 1998; Rémazeilles et al. 2010; Aramendia et al. 2015). MIC is associated to the presence of anaerobic SRBs on the surfaces of archaeological objects recovered from terrestrial and aquatic environments (Sánchez del Junco et al. 1992). Regarding buried metal objects, it is worth mentioning that the degree of preservation of them in soil is specific to the type of metal (Kibblewhite et al. 2015). Ag is less resistant to corrosion than Au but more than Cu while Zn corrodes faster. Cu artefacts may contain As, and this element is also commonly a minor constituent of bronze (an alloy of Cu and Sn, which is more resistant to corrosion than pure Cu). Fe is much more easily corroded than Cu, while Pb is resistant to corrosion in most aqueous environments. Aluminum forms a protective surface oxide coating that gives it some resistance to oxidative corrosion (Kibblewhite et al. 2015). Degradation rates of these materials (e.g. Fe) can be influenced by the soil type. Sulfide and disulfide ions generated by SRB in soils

**Fig. 1.6** Widespread lichen colonization on metals. Ancient cannon, Ioannina, Greece (left). A close-up of the colonized metal surface (right)

enhance the corrosive activity of chloride ions. Moreover, biogenic sulfides may be used as alternative substrates in the cathodic reaction at low oxygen concentration (Sánchez del Junco et al. 1992). Biocorrosion of objects in soil can be markedly accelerated by alternate anaerobic and aerobic conditions. Sulfides formed by SRB in anaerobic conditions are transformed to polysulfides, and sulfur and ferric sulfates when soil pH decreases.

The formation of black corrosion stains containing sulfides on copper alloy buried objects was attributed to SRB, prevalent in those soils, through the production of hydrogen sulfide (Sánchez del Junco et al. 1992).

Some studies have focused on the involvement of fungi in the MIC. The development of *Aspergillus niger* accelerated the corrosion process (severe pitting corrosion and aluminum depletion accompanied with copper enrichment) of aluminum alloy 2024 samples in a laboratory experiment (Dai et al. 2016). The corrosion rate was four times greater than that of samples exposed to 3.5 wt% NaCl solution. Oxalic acid, identified as the dominant metabolite of *Aspergillus niger*, caused the corrosion. The same compound formed iron oxalates and calcium oxalates on the surface of parts of a weathering steel sculpture (Aramendia et al. 2015). Irregularities and discolorations of the object were thus caused by oxalic acid excreted by microorganisms growing on the steel surface.

To the best of my knowledge, no studies have been carried out yet on the interaction of lichens with metallic works of art. As illustrated in Fig. 1.6, crustose lichens completely cover a metal object. Thus, carrying out these studies would be beneficial.

While most studies have documented the negative aspects of the biofilms' development on metallic materials, an emerging topic deals with a positive aspect of this interaction, that is corrosion inhibition. The mechanisms involved are supposed to be the neutralization of the action of environmental corrosive substances (e.g. acidic compounds), and the formation of protective films or the stabilization of preexisting protective films on a metal (e.g. EPS cohesive effects) (Videla and Herrera 2009). Both mechanisms are linked to a marked modification of the environmental conditions at the metal/solution interface due to biological activity. However, the inhibitory action of bacteria may change to the contrary, that is to a corrosive action by other bacteria located within the biofilm (Videla and Herrera 2009). Therefore, the corrosive and the inhibitory actions of bacteria can coexist at the metal surfaces where complex biofilm/protective films interactions occur.

## 5 Bioprotection of Stones by Biofilms and Lichens

The damaging capability of biofilms and lichens on rock art and monuments has been documented, as previously mentioned. Nonetheless, recent research has provided increasing evidence of a minor effect and even of protection. Therefore, the correlation among biofilms, lichens and stone deterioration is no longer being considered an axiom (Pinna 2014). Their presence on works of art does not automatically indicate a change in the physicochemical properties of the materials. A study on petroglyphs colonized by lichens demonstrated, for example, that the organisms were not one of the key factors to define their state (Chiari and Cossio 2004). Lichens neither acted damaging the outermost layer of the sandstones nor protecting them from rain, sun, etc. Lichens filled the micro-fissures between the grains, which were large enough to hold them without exerting relevant pressure. In that way lichens filled the pores so that the porosity of the outer layer was the same as in the bulk of the rock. The deterioration of the sandstone related mainly to the nature of the sandstone itself, particularly to the dimension of the quartz grains: the larger the grains, the greater the porosity, and consequently water absorption, fragility, and decohesion of the rock.

Phototrophic biofilms colonizing marbles of the cathedral of Monza (Italy) did not affect the stones that were instead altered by meteoric waters, wind, solar radiation, and large thermal excursions (Gulotta et al. 2018). The environmental exposure was the main cause of damage of the marbles. Moreover, two types of marble showed the most intense surface decay in the uncolonized areas.

Lithobionts, mainly lichens, have been suggested to temporarily stabilize porous and loosely to moderately cemented stones. Case hardened surfaces that develop over tuff outcrops in Frijoles Canyon (Bandelier National Monument, New Mexico, USA) are a combination of biotic and abiotic crusts that contribute to the preservation of ancient rock-cut caves called cavates (derived from the words "cave" and "excavate"), which were used as dwelling, storage, and special-purpose rooms by the Native American Pueblos (Porter et al. 2016). Lichen cover improved the weather resistance of the tuff and provided some degree of protection to archaeological resources carved in the rock. The lichen cover formed a barrier layer shielding rock surfaces from water flow, wind abrasion, and temperature variation, resulting in reduced weathering rates (Porter et al. 2016). The results were supported by in situ measures of the volume of water absorbed by the rock through time and of surface gas permeability. According to the authors, surface improvements occurred as the result of two different processes: (1) colonization of outcrop surfaces by cyanobacteria, lichens, and other biota cemented wind-blown and water-transported particles to the rock face through secretion of sticky polysaccharides; the formation of dense networks of filaments infiltrated, reinforced, and stabilized the accumulated clay/silt coatings, and shielded the loose ash inside the tuff from wind- and water-driven actions; (2) excretion of organic acids by biotic crust catalyzed biogeochemical reactions that led to the dissolution of fine volcanic glass and cementation of the tuff surface by precipitation of secondary minerals in the network of interconnected pores. Both processes result in occlusion of pores at the surface.

Some studies reported similar benefits conferred by lichen/biofilm cover on various types of rocks. Lichens may provide protection against the natural decay of porous stones either by decreasing the intensity of water exchanges between substrata and environment or by diminishing the damage caused by atmospheric agents (wind, rain, pollutants, salt aerosol) (Ariño et al. 1995; Wendler and Prasartet 1999; Carballal et al. 2001; Bungartz et al. 2004; Gadd and Dyer 2017). More widespread exfoliation, saline efflorescence, flaking, and honeycombing of uncolonized surfaces supported this hypothesis. Lichens may provide thermal blanketing, absorb aggressive chemicals, keep surfaces hot and dry more constantly, reduce boundary layer wind speed, limit erosion and thermal tress (Garcia-Vallès et al. 2003; Carter and Viles 2003; Casanova Municchia et al. 2018).

After the removal of crustose epilithic lichens on marble and sandstone in the archaeological site of Fiesole (Italy), in situ water absorption measurements were carried out (Pinna et al. 2018). Both substrata showed a quite low absorption capacity, which was even close to that of the healthy substrata. The only alteration detected on the stones was that of epilithic crustose lichens forming a uniform and continuous coverage that could have acted as a protective barrier for the substrata, protecting them from abiotic weathering.

Epilithic and endolithic lichens affected hardness and water absorption of Portland and Botticino limestones. Measures of surface hardness and water absorption capacity/infiltration rates were carried out before and after the removal of lichens (Morando et al. 2017). The observed water absorption increase was caused by porosity alteration due to endolithic lichens action. By contrast, the lichen–rock interface revealed unmodified or even increased hardness of the limestones, related to lower weathering rates. According to the authors, these results suggest the opportunity to investigate if and how hardening patterns may also accompany lichen biogeochemical processes on limestones (Morando et al. 2017).

An experimental study showed that endolithic lichens substantially reduced loss of material from a limestone surface (Fiol et al. 1996). Two types of limestone samples, taken from bare natural outcrops and from rocks covered with endolithic lichens, were exposed to artificial conditions in the laboratory, simulating rainstorm using distilled water. The analysis of run-off water showed that the loss of mineral grains and dissolved compounds from the bare limestone surface was substantially higher than that from the lichen-covered surface, suggesting that a process of bioprotection may took place.

According to Hoppert and König (2006), the ecology of succession of biofilms and lichens on stones is the key to explain the phenomenon. Microorganisms characterized by fast-growing ability and short reproductive cycle are more likely to be deteriorative, because they colonize rapidly after disturbance (sporadic events affecting community structure and dynamics, e.g., fire, drought, storms, or erosion) or stress conditions (e.g. increasing temperature and UV levels, decreasing moisture levels), and use a range of strategies to obtain nutrients from the substratum. Such strategies may cause further change of the surface through weathering. In an advanced state of biofilm development, these microorganisms are replaced by colonizers with different life strategies. These "new" colonizers are biofilms and lichens characterized by moderate metabolic and reproductive rates. In order to develop and colonize the rock, they need that the stone is not rapidly decomposed. Therefore, they "protect" the stone forming a compact network of cells and extra-cellular polymers, which surround and enwrap the mineral particles. This biogenic structure temporarily stabilizes the stone and reduces weathering, enabling biofilms and lichens to growth and persist over several years or even decades (Hoppert and König 2006).

Bioprotection by lichens may relate to environmental parameters. Carter and Viles (2005) suggested that a single species may act protectively in one environmental context but can be deteriorative in another. For example, the epilithic crustose lichen *Verrucaria nigrescens* is common in Europe in both wet and dry climates. In wet conditions, the lichen may act as a bioprotector, shielding the stone from the direct action of rainfall, acid attack, wind, runoff, and dissolution. In a hotter and drier context, where there are extreme temperatures, *Verrucaria nigrescens* may have instead a dominant biodeteriorative impact, intensifying temperature fluctuations at the rock surface and possibly leading to thermal stress.

According to McIlroy de la Rosa and coauthors (2013), active bioprotection of a rock surface by crustose lichens is determined primarily by how the thalli are attached to the substratum, by their binding and waterproofing action and whether they provide an effective shield. The production of insoluble substances (usually calcium oxalates) at the lichen–rock interface provides a passive protection of the rock.

The results of the mentioned studies have even led Gadd and Dyer (2017) to suggest actions for promoting the growth of biofilms and lichens on stone surfaces as the biogenic approach is an environmentally friendly method for protection of stones.

A general consideration emerging from the literature so far is that biodeterioration and bioprotection by lichens coexist in a delicate equilibrium that can be destabilized by environmental changes. According to Morando et al. (2017), deterioration caused by the growth of lichen hyphae in limestones is likely counterbalanced by a protective effect on the rock surface by the covering of epilithic and endolithic species. They shelter and protect the substratum from more aggressive abiotic weathering agents limiting erosion processes.

Despite significant research efforts, there are still many general issues that need to be investigated. A homogeneous coverage of crustose lichens on stone surfaces is usually formed by many species growing close each other. Some of them can be overall protective, other can be overall deteriorative and other can have a negligible effect, resulting in differences of the spatial distribution of surface biomodifications. This mélange of situations is difficult to manage from a conservative point of view. In another scenario the surfaces are colonized by few little lichen thalli surrounded by biofilms. What happens in this situation? Are just some little areas protected while others are not? Furthermore, even a single lichen thallus can have either a protective or a deteriorative action within the same thallus (Mottershead and Lucas 2000).

The mentioned studies attribute bioprotective properties to epilithic and endolithic crustose lichens. However, there are crustose lichens like those belonging to *Lepraria* genus that do not form a continuous layer over the substratum. In that case it is unlikely they act as a local shield. In addition, atmospheric factors (e.g. moisture, temperature, and insolation) affect lichen metabolic activity and, consequently, their impact on the substratum, including their protective role. Climate changes alter the structure and function of microbial communities as well. According to Viles and Cutler (2012), areas where increased frequency of climatic disturbances will likely occur, will be subjected to biodeteriorative effects for the same reasons previously discussed about the ecology of succession of biofilms and lichens on stones. On the contrary, areas where increased stresses will occur are likely to be colonized by microorganisms adapted to persist in low-resource environments. In such a context, there will be a strong decrease of biodeterioration.

Another gap in our knowledge of bioprotection effects on stones by biofilms and lichens derives from the fact that the mentioned studies applied different methods to assess the protective properties of lichens, therefore their results are not comparable. The assessment of standard test methods and procedures to evaluate the protective and/or damaging action by biofilms and lichens is crucial for future studies on this subject.

Further research is required to understand the importance of microbial ecology in biodeteriorative vs bioprotective effects (i.e. the difference between actively altering and non-altering microbial communities on stones exposed to the same environment). New lines of research proposed by Villa and coauthors (2016) will help explain these issues. Viewing the biofilms from an ecological perspective, new studies should deal with a system-level approach which is potentially very valuable in understanding the functional capacities and adaptive potentials that drive aggregation and survival of microbial communities on stones (Villa et al. 2016). The study of functional traits (morphological, biochemical, physiological, structural, phenological, or behavioral features) will contribute to shed light on the interaction of microorganisms with their environment and with stones (Villa et al. 2016). These advances will also help explain how such aspects affect spatial and temporal distributions of biotic and abiotic reactions. Ongoing research in understanding the bioprotective vs biodeteriorative role of biofilms and lichens will also lead to make suggestions regarding the actions that should or should not be taken to retard deterioration and enhance protection of monuments (Jacob et al. 2018).

# References

Abuku M, Janssen H, Roels S (2009) Impact of wind-driven rain on historic brick wall buildings in a moderately cold and humid climate: numerical analyses of mould growth risk, indoor climate and energy consumption. Energ Buildings 41:101–110

Adamson C, McCabe S, Warke PA, McAllister D, Snith BJ (2013) The influence of aspect on the biological bolonization of stone in Northern Ireland. Int Biodeterior Biodegradation 84:357–366

Albertano P, Bruno L, Bellezza S (2005) New strategies for the monitoring and control of cyanobacterial films on valuable lithic faces. Plant Biosyst 139:311–322

Allsopp D, Seal K, Gaylarde C (2004) Introduction to biodeterioration. Cambridge University Press, Cambridge

Aramendia J, Gomez-Nubla L, Bellot-Gurlet L, Castro K, Arana G, Madariaga JM (2015) Bioimpact on weathering steel surfaces: oxalates formation and the elucidation of their origin. Int Biodeterior Biodegradation 104:59–66

Ariño X, Ortega-Calvo JJ, Gomez-Bolea A, Saiz-Jimenez C (1995) Lichen colonization of the Roman pavement at Baelo Claudia (Cadiz, Spain): biodeterioration vs bioprotection. Sci Total Environ 67:353–363

Ashbee J (2010) Ivy and the presentation of ancient monuments and building. In: Sternberg T (ed) Ivy on Walls, Seminar Report, English Heritage, pp 9–15

Bastian F, Jurado V, Nováková A, Alabouvette C, Saiz-Jimenez C (2010) The microbiology of Lascaux Cave. Microbiology 156:644–652

Beech IB, Coutinho CLM (2003) Biofilms on corroding materials. In: Lens P, Moran AP, Mahony T, Stoodly P, O'Flaherty V (eds) Biofilms in medicine, industry and environmental biotechnology - characteristics, analysis and control. IWA Publishing of Alliance House, London, pp 115–131

Beech IB, Sunner J (2004) Biocorrosion: towards understanding interactions between biofilms and metals. Curr Opin Biotechnol 15:181–186

Beech IB, Sztyler M, Gaylarde CC, Smith WL, Sunner J (2014) Biofilms and biocorrosion. In: Liengen T, Feron D, Basseguy R, Beech IB (eds) Understanding biocorrosion: fundamentals and applications. Woodhead Publishing, Cambridge, pp 33–50

Bjelland TH, Thorseth IH (2002) Comparative studies of the lichen–rock interface of four lichens in Vingen, Western Norway. Chem Geol 192:81–98

Bridier A, Piard J-C, Pandin C, Labarthe S, Dubois-Brissonnet F, Briandet R (2017) Spatial organization plasticity as an adaptive driver of surface microbial communities. Front Microbiol 8:1364

Brown PW, Masters LW (1980) Factors affecting the corrosion of metals in the atmosphere. In: Ailor WH (ed) Atmospheric corrosion. John Wiley and Sons, New York, pp 31–49

Bungartz F, Garvie LAJ, Nash TH (2004) Anatomy of the endolithic Sonoran Desert lichen Verrucaria rubrocincta Breuss: implications for biodeterioration and biomineralization. Lichenologist 36:55–73

Cámara B, Suzuki S, Nealson KH, Wierzchos J, Ascaso C, Artieda O, de los Ríos A (2014) Ignimbrite textural properties as determinants of endolithic colonization patterns from hyperarid Atacama Desert. Int Microbiol 17:235–247

Caneva G, Bartoli F (2017) Botanical planning and lichen control for the conservation of gravestones in Jewish urban cemeteries in North-Eastern Italy. Israel J Plant Sci 64:1–14

Caneva G, Ceschin S (2008) Ecology of biodeterioration. In: Caneva G, Nugari MP, Salvadori O (eds) Plant biology for cultural heritage: biodeterioration and conservation. Getty Conservation Institute, Los Angeles, pp 35–58

Cappitelli F, Abbruscato P, Foladori P, Zanardini E, Ranalli G, Principi P, Villa F, Polo A, Sorlini C (2009) Detection and elimination of cyanobacteria from frescoes: the case of the St. Brizio Chapel (Orvieto Cathedral, Italy). Microbial Ecol 57:633–639

Carballal R, Paz-Bermúdez G, Sánchez-Biezma MJ, Prieto B (2001) Lichen colonization of coastal churches in Galicia: biodeterioration implications. Int Biodeterior Biodegradation 47:157–163

Carmona N, Laiz L, Gonzalez JM, Garcia-Herasa M, Villegasa MA, Saiz-Jimenez C (2006) Biodeterioration of historic stained glasses from the Cartuja de Miraflores (Spain). Int Biodeterior Biodegradation 58:155–161

Carter NEA, Viles HA (2003) Experimental investigations into the interactions between moisture, rock surface temperatures and an epilithic lichen cover in the bioprotection of limestone. Build Environ 38:1225–1234

Carter NEA, Viles HA (2005) Bioprotection explored: the story of a little known earth surface process. Geomorphology 67:273–281

Casanova Municchia A, Bartoli F, Taniguchi Y, Giordani P, Caneva G (2018) Evaluation of the biodeterioration activity of lichens in the Cave Church of Üzümlü (Cappadocia, Turkey). Int Biodeterior Biodegradation 127:160–169

Chiari G, Cossio R (2004) Lichens of Wyoming sandstone, do they cause damage? In: St. Clair L, Seawards M (eds) Biodeterioration of rock surfaces. Kluwer Academic Publishers, Dordrecht, pp 99–113

Corrêa Pinto AM, Palomar T, Alves LC, da Silva SHM, Monteiro RC, Macedo MF, Vilarigues MG (2019) Fungal biodeterioration of stained-glass windows in monuments from Belém do Pará (Brazil). Int Biodeterior Biodegradation 138:106–113

Cutler NA, Viles HA, Ahmad S, McCabe S, Smith BJ (2013) Algal 'greening' and the conservation of stone heritage structures. Sci Total Environ 442:152–164

Dai X, Wang H, Ju LK, Cheng G, Cong H, Newby BMZ (2016) Corrosion of aluminum alloy 2024 caused by *Aspergillus niger*. Int Biodeterior Biodegradation 115:1–10

De Leo F, Urzì C, de Hoog GSA (2003) New meristematic fungus, *Pseudotaeniolina globosa*. Antonie Van Leeuwenhoek 83:351–360

de Los Rios A, Wierzchos J, Ascaso C (2002) Microhabitats and chemical microenvironments under saxicolous lichens growing on granite. Microb Ecol 43:181–188

de los Rios A, Cámara B, García del Cura MA, Rico VJ, Galván V, Ascaso C (2009) Deteriorating effects of lichen and microbial colonization of carbonate building rocks in the Romanesque churches of Segovia (Spain). Sci Total Environ 407:1123–1134

Dhawan S (1987) Role of microorganisms in corrosion of metals – a literature survey. In: Agrawal OP (ed) Asian regional seminar on conservation of metals in humid climates. ICCROM, Roma, pp 100–102

Di Martino P (2016) What about biofilms on the surface of stone monuments? Open Confer Proc J 6:14–28

Diao M, Taran E, Mahler S, Nguyen AV (2014) A concise review of nanoscopic aspects of bioleaching bacteria–mineral interactions. Adv Colloid Interf Sci 212:45–63

Dupraz C, Reid RP, Braissant O, Decho AW, Norman RS, Visscher PT (2009) Processes of carbonate precipitation in modern microbial mats. Earth-Sci Rev 96:141–162

Edwards HGM, Seaward MRD, Attwood SJ, Little SJ, De Oliveira LFC, Tretiach M (2003) FT-Raman spectroscopy of lichens on dolomitic rocks: an assessment of metal oxalate formation. Analyst 128:1218–1221

Eid MM, Duncan KE, Tanner RS (2018) A semi-continuous system for monitoring microbially influenced corrosion. J Microbiol Methods 150:55–60

Elhagrassy AF (2018) Isolation and characterization of actinomycetes from mural paintings of Snu-Sert-Ankh tomb, their antimicrobial activity, and their biodeterioration. Microbiol Res 216:47–55

Favero-Longo SE, Borghi A, Tretiach M, Piervittori R (2009) In vitro receptivity of carbonate rocks to endolithic lichen-forming aposymbionts. Mycol Res 113:1216–1227

Favero-Longo SE, Gazzano C, Girlanda M, Castelli D, Tretiach M, Baiocchi C, Piervittori R (2011) Physical and chemical deterioration of silicate and carbonate rocks by meristematic microcolonial fungi and endolithic lichens (Chaetothyriomycetidae). Geomicrobiol J 28:732–744

Fiol L, Fornós JJ, Ginés A (1996) Effects of biokarstic processes on the development of solutional rillenkarren in limestone rocks. Earth Surf Process Landf 21:447–452

Flemming HC, Wingender J, Szewzyk U, Steinberg P, Rice SA, Kjelleberg S (2016) Biofilms: an emergent form of bacterial life. Nat Rev Microbiol 14:563–575

Fomina M, Burford EP, Hillier S, Kierans M, Gadd GM (2010) Rock-building fungi. Geomicrobiol J 27:624–629

Gadd GM, Dyer TD (2017) Bioprotection of the built environment and cultural heritage. Microb Biotechnol 10:1152–1156

Gadd GM, Bahri-Esfahani J, Li Q, Rhee YJ, Wei Z, Fomina M, Liang X (2014) Oxalate production by fungi: significance in geomycology, biodeterioration and bioremediation. Fungal Biol Rev 28:36–55

Garcia-Vallès M, Topal T, Vendrell-Saz M (2003) Lichenic growth as a factor in the physical deterioration or protection of Cappadocian monuments. Environ Geol 43:776–781

Gilbert M (1987) Black spots on bronzes. Newsletter ICOM Committee for Conservation, Metal Working Group, 3:12

Gittins M, Vedovello S, Dvalishvili M, Kuprashvili N (2002) Determination of the treatment and restoration needs of medieval frescos in Georgia. In: Proceeding of ICOM-CC 13th triennial meeting, Rio de Janeiro, Brazil, pp 560–564

Golubic S, Pietrini AM, Ricci S (2005) Euendolithic activity of the cyanobacterium *Chroococcus lithophilus* Erc. in biodeterioration of the pyramid of Caius Cestius, Rome, Italy. Int Biodeterior Biodegradation 100:7–16

González-Gómez WS, Quintana P, Gómez-Cornelio S, García-Solis C, Sierra-Fernandez A, Ortega-Morales O, De la Rosa-García SC (2018) Calcium oxalates in biofilms on limestone walls of Maya buildings in Chichén Itzá, Mexico. Environ Earth Sci 77: 230

Gorbushina AA (2007) Life on the rocks. Environ Microbiol 9:1613–1631

Gu JD (2019) Corrosion, microbial. In: Schmidt T (ed) Encyclopedia of microbiology, 4th edn. Elsevier, Amsterdam, pp 762–771

Guillitte O (1985) Bioreceptivity: a new concept for building ecological studies. Sci Total Environ 167:215–220

Gulotta D, Villa F, Cappitelli F, Toniolo L (2018) Biofilm colonization of metamorphic lithotypes of a renaissance cathedral exposed to urban atmosphere. Sci Total Environ 639:1480–1490

Hentzer M, Eberl L, Nielsen J, Givskov M (2003) Quorum sensing: a novel target for the treatment of biofilm infections. BioDrugs 17:241–250

Hoffland E, Kuyper TW, Wallander H, Plassard C, Gorbushina AA, Haselwandter H, Holmstrom S, Landeweert R, Lundstrom US, Rosling A, Sen R, Smits MM, van Hees PAW, van Breemen N (2004) The role of fungi in weathering. Front Ecol Environ 2: 258–264

Hoppert M, König S (2006) The succession of biofilms on building stone and its possible impact on biogenic weathering. In: Fort R, Alvarez de Buergo M, Gomez-Heras M, Vazquez-Calvo C (eds) Heritage, weathering and conservation. Taylor & Francis Group, London, pp 311–315

Jacob JM, Schmull M, Villa F (2018) Biofilms and lichens on eroded marble monuments. APT Bull 49:55–60

Jain A, Bhadauria S, Kumar V, Chauhan RS (2009) Biodeterioration of sandstone under the influence of different humidity levels in laboratory conditions. Build Environ 44:1276–1284

Jurado V, Laiz L, Sanchez-Moral S, Saiz-Jimenez C (2014) Pathogenic microorganisms related to human visits in Altamira Cave, Spain. In: Saiz-Jimenez C (ed) The conservation of subterranean cultural heritage. CRC Press/Balkema, Leiden pp. 229–238

Kadukova J, Pristas P (2018) Biocorrosion – microbial action. In: Wandelt K (ed) Encyclopedia of interfacial chemistry: surface science and electrochemistry. Elsevier, Amsterdam, pp 20–22

Katharios-Lanwermeyer S, Xi C, Jakubovics NS, Rickard AH (2014) Mini-review: microbial coaggregation: ubiquity and implications for biofilm development. Biofouling 30:1235–1251

Kibblewhite M, Tóth G, Tamás Hermann T (2015) Predicting the preservation of cultural artefacts and buried materials in soil. Sci Total Environ 529:249–263

Kinzler K, Gehrke T, Telegdi J, Sand W (2003) Bioleaching - a result of interfacial processes caused by extracellular polymeric substances (EPS). Hydrometallurgy 71:83–88

Konhauser K (2007) Introduction to geomicrobiology. Blackwell, Oxford

Kyi C (2006) The significance of appropriate sampling and cultivation in the effective assessment of biodeterioration. Z Kunsttechnol Konservierung 20:344–351

Little B, Wagner P, Hart K (1998) The role of biomineralization in microbiologically influenced corrosion. Biodegradation 9:1–10

Liu W, Order HL, Madsen JS, Bjarnsholt T, Sorensen SJ, Burmolle M (2016) Interspecific bacterial interactions are reflected in multispecies biofilm spatial organization. Front Microbiol 7:1366

Lombardozzi V, Castrignanò T, D 'Antonio M, Casanova Municchia A, Caneva G (2012) An interactive database for an ecological analysis of stone biopitting. Int Biodeterior Biodegradation 73:8–15

López D, Vlamakis H, Kolte R (2010) Biofilms. Cold Spring Harb Perspect Biol 2:a00398

Marvasi M, Vedovato E, Balsamo C, Macherelli A, Dei L, Mastromei G, Perito B (2009) Bacterial community analysis on the mediaeval stained-glass window 'Nativita' in the Florence cathedral. J Cult Heritage 10:124–133

Marvasi M, Donnarumma F, Frandi A, Mastromei G, Sterflinger K, Tiano P, Perito B (2012) Black microcolonial fungi as deteriogens of two famous marble statues in Florence, Italy. Int Biodeterior Biodegradation 68:36–44

McIlroy de la Rosa JPM, Warke PA, Smith BJ (2013) Lichen-induced biomodification of calcareous surfaces: bioprotection versus biodeterioration. Progr Phys Geogr 37:325–351

McNeil MB, Mohr DW, Little BJ (1991) Correlation of laboratory results with observations on long-term corrosion of iron and copper alloys. In: Vandiver PB, Druzik J, Wheeler GS (eds) 2nd Symposium on materials issues in art and archaeology, San Francisco, USA, pp 753–759

Miller AZ, Laiz L, Dionísio A, Macedo MF (2006) Primary bioreceptivity: a comparative study of different Portuguese lithotypes. Int Biodeterior Biodegradation 57:136–142

Miller AZ, Dionísio A, Sequeira Braga MA, Hernández-Mariné M, Afonso MJ, Muralha VSF, Herrera LK, Raabe J, Fernandez-Cortes A, Cuezva S, Hermosin B, Sanchez-Moral S, Chaminé H, Saiz-Jimenez C (2012a) Biogenic Mn oxide minerals coating in a subsurface granite environment. Chem Geol 322–323:181–191

Miller AZ, Sanmartín P, Pereira-Pardo L, Dionísio A, Sáiz-Jiménez C, Macedo MF, Prieto B (2012b) Bioreceptivity of building stones: a review. Sci Total Environ 426:1–12

Mittelmann MW (2018) The importance of microbial biofilms in the deterioration of heritage materials. In: Mitchell R, Clifford J (eds) Biodeterioration and preservation in art, archaeology and architecture. Archetype Publications, London, pp 3–15

Miura S (2006) Conservation of mural paintings of the Takamatsuzuka tumulus and its current situation. In: Yamauchi K, Taniguchi Y, Uno T (eds) Mural paintings of the silk road: cultural exchanges between east and west. Archetype Publications Ltd, Tokio, pp 127–130

Morando M, Wilhelm K, Matteucci E, Martire L, Piervittori R, Viles HA, Favero-Longo SE (2017) The influence of structural organization of epilithic and endolithic lichens on limestone weathering. Earth Surf Process Landforms 42:1666–1679

Mottershead D, Lucas G (2000) The role of lichens in inhibiting erosion of a soluble rock. Lichenologist 32:601–609

Noeiaghaei T, Mukherjee A, Dhami N, Chae SR (2017) Biogenic deterioration of concrete and its mitigation technologies. Construct Building Mater 149:575–586

Nugari MP, Pietrini AM, Caneva G, Imperi F, Visca P (2009) Biodeterioration of mural paintings in a rocky habitat: the crypt of the Original Sin (Matera, Italy). Int Biodeterior Biodegradation 63:705–711

Olsson-Francis K, Simpson AE, Wolff-Boenisch D (2012) The effect of rock composition on cyanobacterial weathering of crystalline basalt and rhyolite. Geobiology 10:434–444

Onofri S, Zucconi L, Isola D, Selbmann L (2014) Rock-inhabiting fungi and their role in deterioration of stone monuments in the Mediterranean area. Plant Biosyst 148:384–391

Ortega-Morales BO, Narváez-Zapata J, Reyes-Estebanez M, Quintana P, de la Rosa-García SC, Bullen H, Gómez-Cornelio S, Chan-Bacab MJ (2016) Bioweathering potential of cultivable fungi associated with semi-arid surface microhabitats of Mayan buildings. Front Microbiol 7:201

Pangallo D, Kraková L, Chovanová K, Šimonovičová A, de Leo F, Urzì C (2012) Analysis and comparison of the microflora isolated from fresco surface and from surrounding air environment through molecular and biodegradative assays. World J Microbiol Biotech 28:2015–2027

Piñar G, Garcia-Valles M, Gimeno-Torrente D, Fernandez-Turiel JL, Ettenauer J, Sterflinger K (2013) Microscopic, chemical, and molecular-biological investigation of the decayed medieval stained window glasses of two Catalonian churches. Int Biodeterior Biodegradation 84:388–400

Pinna D (2014) Biofilms and lichens on stone monuments: do they damage or protect? Front Microbiol 5:1–3

Pinna D (2017) Coping with biological growth on stone heritage objects. Methods, products, applications, and perspectives. Apple Academic Press, Palm Bay

Pinna D, Salvadori O (2008) Biodeterioration processes in relation to cultural heritage materials. Stone and related materials. In: Caneva G, Nugari MP, Salvadori O (eds) Plant biology for cultural heritage. The Getty Conservation Institute, Los Angeles, pp 128–143

Pinna D, Salvadori O, Tretiach M (1998) An anatomical investigation of calcicolous endolithic lichens from the Trieste karst (NE Italy). Plant Biosyst 132:183–195

Pinna D, Salvadori B, Galeotti M (2012) Monitoring the performance of innovative and traditional biocides mixed with consolidants and water-repellents for the prevention of biological growth on stone. Sci Total Environ 423:132–141

Pinna D, Galeotti M, Perito B, Daly G, Salvadori B (2018) In situ long-term monitoring of recolonization by fungi and lichens after innovative and traditional conservative treatments of archaeological stones in Fiesole (Italy). Int Biodeterior Biodegradation 132:49–58

Pohl W, Scheider J (2002) Impact of endolithic biofilms on carbonate rock surfaces. In: Siegesmund S, Weiss T, Vollbrecht A (eds) Natural stone, weathering phenomena, conservation strategies and case studies, geological society. Special Publications, London 205:177–194

Porter D, Broxton D, Bass A, Neher DA, Weicht TR, Longmire P, Domingue R (2016) The role of case hardening in the preservation of the cavates and petroglyphs of Bandelier. e-δialogos 5:12–31

Prieto B, Silva B (2005) Estimation of the potential bioreceptivity of granitic rocks from their intrinsic properties. Int Biodeterior Biodegradation 56:206–215

Ramírez M, Hernández-Mariné M, Novelo E, Roldán M (2010) Cyanobacteria containing biofilms from a Mayan monument in Palenque. Mexico Biofouling 26:399–409

Rémazeilles C, Saheb M, Neff D, Guilminot E, Tran K, Bourdoiseau JA, Sabot R, Jeannin M, Matthiesen H, Dillmann P, Refait P (2010) Microbiologically influenced corrosion of archaeological artefacts: characterization of iron(II) sulfides by Raman spectroscopy. J Raman Spectrosc 41:1425–1433

Roberts A, Campbell I, Pring A, Bell G, Watchman A, Popelka-Filcoff R, Lenehan C, Gibson C, Franklin N (2015) A multidisciplinary investigation of a rock coating at Ngaut Ngaut (Devon Downs), South Australia. Aust Archaeol 80:32–39

Rodrigues A, Gutierrez-Patricio S, Miller AZ, Saiz-Jimenez C, Wiley R, Nunes N, Vilarigues M, Macedo MF (2014) Fungal biodeterioration of stained-glass windows. Int Biodeterior Biodegradation 90:152–160

Salvadori O, Casanova A (2016) The role of fungi and lichens in the biodeterioration of stone monuments. Open Conf Proc J 7:39–54

Salvadori O, Charola AE (2011) Methods to prevent biocolonization and recolonization: an overview of current research for architectural and archaeological heritage. In: Charola AE, McNamara C, Koestler RJ (eds) Biocolonization of stone: control and preventive methods. Smithsonian Institute Scholarly Press, Washington, pp 37–50

Sánchez del Junco A, Moreno DA, Ranninger C, Ortega-Calvo JJ, Saiz-Jimenez C (1992) Microbial induced corrosion of metallic antiquities and works of art: a critical review. Int Biodeterior Biodegradation 29:367–375

Scheerer S (2008) Microbial biodeterioration of outdoor stone monuments. Assessment methods and control strategies. PhD Thesis, Cardiff University, UK

Scheerer S, Ortega-Morales O, Gaylarde C (2009) Microbial deterioration of stone monuments - an updated overview. Adv Appl Microbiol 66:97–139

Shirakawa MA, Tavares RG, Gaylarde CC, Taqueda ME, Loh K, John VM (2010) Climate as the most important factor determining anti-fungal biocide performance in paint films. Sci Total Environ 408:5878–5886

Shirakawa MA, Vanderley MJ, Mocelin A, Zilles R, Toma SH, Araki K, Toma HE, Thomaz AC, Gaylarde CC (2016) Effect of silver nanoparticle and TiO$_2$ coatings on biofilm formation on four types of modern glass. Int Biodeterior Biodegradation 108:175–180

Siegesmund S, Snethlage RE (2014) Stone in architecture. Properties, durability. Springer, Berlin Heidelberg, Germany

Skinner HCW, Jahren AH (2003) Biomineralization. In: Schlesinger WH (ed) Treatise on geochemistry, vol 8. Elsevier, Amsterdam, pp 117–184

Sohrabi M, Favero-Longo SE, Perez-Ortega S, Ascaso C, Haghighat Z, Talebian MH, Fadaei H, de los Ríos A (2017) Lichen colonization and associated deterioration processes in Pasargadae, UNESCO world heritage site, Iran. Int Biodeterior Biodegradation 117:171–182

Sterflinger K (2010) Fungi: their role in deterioration of cultural heritage. Fungal Biol Rev 24:47–55

Sterflinger K, Piñar G (2013) Microbial deterioration of cultural heritage and works of art - tilting at windmills? Appl Microbiol Biotechnol 97:9637–9646

Szczepanowska HM, Cavaliere AR (2004) Tutankhamun tomb, a closer look at biodeterioration: preliminary report. In: Rauch A, Miklin-Kniefacz S, Harmssen A (eds) Schimmel: Gefahr für Mensch und Kulturgut durch Mikroorganismen. Konrad Theiss Verlag GmbH & Co, Stuttgart, pp 42–47

Tescari M, Frangipani E, Caneva G, Casanova Municchia A, Sodo A, Visca P (2018) *Arthrobacter agilis* and rosy discoloration in "Terme del Foro" (Pompeii, Italy). Int Biodeterior Biodegradation 130:48–54

Tolker-Nielsen T (2015) Biofilm development. Microbiol Spectr 3: MB-0001-2014

Tonon C, Favero-Longo SE, Matteucci E, Piervittori R, Croveri P, Appolonia L, Meirano V, Serino M, Elia D (2019) Microenvironmental features drive the distribution of lichens in the house of the Ancient Hunt, Pompeii, Italy. Int Biodeterior Biodegradation 136:71–81

Traversetti L, Bartoli F, Caneva G (2018) Wind-driven rain as a bioclimatic factor affecting the biological colonization at the archaeological site of Pompeii, Italy. Int Biodeterior Biodegradation 134:31–38

Tretiach M, Pinna D, Grube M (2003) *Caloplaca erodens* [sect. *Pyrenodesmia*], a new lichen species from Italy with an unusual thallus type. Mycolog Prog 2:127–136

Unković N, Erić S, Šarić K, Stupar M, Savković Ž, Stanković S, Stanojević O, Dimkić I, Vukojević J, Ljaljević Grbić M (2017) Biogenesis of secondary mycogenic minerals related to wall paintings deterioration process. Micron 100:1–9

Unković N, Dimkić I, Stupar M, Stanković S, Vukojević J, Ljaljević Grbić M (2018) Biodegradative potential of fungal isolates from sacral ambient: in vitro study as risk assessment implication for the conservation of wall paintings. PLoS One 13:e0190922

Urzì C, De Leo F, Bruno L, Pangallo D (2014) New species description, biomineralization processes and biocleaning applications of Roman catacombs-living bacteria. In: Saiz-Jimenez C (ed) The conservation of subterranean cultural heritage. CRC Press/Balkema, Leiden/Netherlands, pp 65–72

Urzì C, Bruno L, De Leo F (2018) Biodeterioration of paintings in caves, catacombs and other hypogean sites. In: Mitchell R, Clifford J (eds) Biodeterioration and preservation in art, archaeology and architecture. Archetype Publications Ltd, London, pp 114–129

Vasanthakumar A, DeAraujo A, Mazurek J, Schilling M, Mitchell R (2013) Microbiological survey for analysis of the brown spots on the walls of the tomb of King Tutankhamun. Int Biodeterior Biodegradation 79:56–63

Vázquez-Nion D, Silva B, Prieto B (2018a) Influence of the properties of granitic rocks on their bioreceptivity to subaerial phototrophic biofilms. Sci Total Environ 610–611:44–54

Vázquez-Nion D, Troiano F, Sanmartín P, Valagussa C, Cappitelli F, Prieto B (2018b) Secondary bioreceptivity of granite: effect of salt weathering on subaerial biofilm growth. Mater Struct 51:158

Vega LM, Mathieu J, Yang Y, Pyle BH, McLean RJC, Alvarez PJJ (2014) Nickel and cadmium ions inhibit quorum sensing and biofilm formation without affecting viability in *Burkholderia multivorans*. Int Biodeterior Biodegradation 91:82–87

Videla HA, Herrera LK (2009) Understanding microbial inhibition of corrosion. A comprehensive overview. Int Biodeterior Biodegradation 63:896–900

Viles HA (2012) Microbial geomorphology: a neglected link between life and landscape. Geomorphology 157–158:6–16

Viles HA, Cutler NA (2012) Global environmental change and the biology of heritage structures. Glob Chang Biol 18:2406–2418

Villa F, Pitts B, Lauchnor E, Cappitelli F, Stewart PS (2015) Development of a laboratory model of a phototroph-heterotroph mixed-species biofilm at the stone/air interface. Front Microbiol 6:1251

Villa F, Stewart PS, Klapper I, Jacob JM, Cappitelli F (2016) Subaerial biofilms on outdoor stone monuments: changing the perspective towards an ecological framework. Bioscience 66:285–294

Vítek P, Ascaso C, Artieda O, Wierzchos J (2016) Raman imaging in geomicrobiology: endolithic phototrophic microorganisms in gypsum from the extreme sun irradiation area in the Atacama Desert. Anal Bioanal Chem 408:4083–4092

Wendler E, Prasartet C (1999) Lichen growth on old Khmer-style sandstone monuments in Thailand: damage factor or shelter? In: Bridgland J, Brown J (eds) Proceedings of the 12th triennial meeting of the ICOM Committee for Conservation. James and James, London, pp 750–754

Xu D, Li Y, Gu T (2016) Mechanistic modeling of biocorrosion caused by biofilms of sulfate reducing bacteria and acid producing bacteria. Bioelectrochemistry 110:52–58

Zammit G, De Leo F, Albertano P, Urzì C (2008) A preliminary study of microbial communities colonizing ochre-decorated chambers at the Hal Saflieni Hypogeum at Paola, Malta. In: Lukaszewicz JW, Niemcewicz P (eds) Proceedings of the 11th international congress on deterioration and conservation of stone. Torun, Poland, pp 555–562

Zanardini E, Cappitelli F, Ranalli G, Sorlini C (2008) Biodeterioration processes in relation to cultural heritage materials - metals. In: Caneva G, Nugari MP, Salvadori O (eds) Plant biology for cultural heritage: biodeterioration and conservation. Getty Conservation Institute, Los Angeles, pp 153–156

Zhang G, Gong C, Gu J, Katayama Y, Someya T, Gu JD (2019) Biochemical reactions and mechanisms involved in the biodeterioration of stone world cultural heritage under the tropical climate conditions. Int Biodeterior Biodegradation 143:104723

Zucconi L, Gagliardi M, Isola D, Onofri S, Andaloro MC, Pelosi C, Pogliani P, Selbmann L (2012) Biodeterioration agents dwelling in or on the wall paintings of the Holy Saviour's cave (Vallerano, Italy). Int Biodeterior Biodegradation 70:40–46

# Chapter 2
# Microbiota and Biochemical Processes Involved in Biodeterioration of Cultural Heritage and Protection

**Ji-Dong Gu and Yoko Katayama**

**Abstract** The world cultural heritage sites face new challenges for an effective protection and management because of destruction and damage initiated by both natural and anthropogenic causes. Fresh rock and sandstone surfaces of buildings are quickly colonized and covered by a layer of microorganisms, including phototrophs, lithotrophs, and heterotrophs to form a biofilm that alters the local conditions of the stone surfaces, especially under the favorable tropical climate conditions for autotrophic microorganisms and plants. Biofilms had been studied with indigenous or pure cultures of isolated microorganisms, but the selective ones that contribute to deterioration of the cultural heritage cannot be confirmed easily. Currently, high-throughput sequencing and metegenomics analyses are capable of obtaining microbial community and composition in great depth, but they also suffer from similar weakness unable to identify the culprits in the community. With these as background, this article presents a different approach by focusing on the biochemical processes and the responsible microorganisms involved to reveal the destruction processes for management and protection. Among these different functional groups of microorganisms, lichens are known as pioneering rock-decomposing microorganisms, and both sulfur-oxidizing bacteria and fungi participate in the decomposition of sandstone via sulfur cycling and initiation of salt attack of the stone afterward, resulting in defoliation and cracking of stone. Other microorganisms including ammonia-oxidizing bacteria and archaea, especially the latter, have been recently detected on sandstone monuments providing evidence on the new organisms involved in the deterioration of cultural heritage and buildings. In addition, fungi can colonize the surfaces of the matured biofilms and play a new role in the removal

J.-D. Gu (✉)
Environmental Engineering, Guangdong Technion Israel Institute of Technology, Shantou, Guangdong, People's Republic of China

Laboratory of Environmental Microbiology and Toxicology, School of Biological Sciences, The University of Hong Kong, Hong Kong SAR, People's Republic of China
e-mail: jidong.gu@gtiit.edu.cn

Y. Katayama
Tokyo National Research Institute for Cultural Properties, Tokyo, Japan

© The Author(s) 2021
E. Joseph (ed.), *Microorganisms in the Deterioration and Preservation of Cultural Heritage*, https://doi.org/10.1007/978-3-030-69411-1_2

of them, which has a potential biotechnological application in conservation of cultural heritage. The new proposed approach by focusing the microorganisms with identified biochemical function is more productive than a description of the community composition and assembly when assessing cultural heritage biodeterioration, and this provides basic and useful information for effective protection strategies and management.

**Keywords** Biodeterioration · Sandstone · Biofilms · Defoliation · Monuments · Sulfur-oxidizing · Ammonia-oxidizing · Salt attack

# 1 Introduction

Deterioration of the world cultural heritage and historic buildings is a result of both natural and/or anthropogenic contributors involving flora, fauna, and microorganisms as well as pollutants. Early investigations on cultural heritage biodeterioration were mainly focused on the isolation and identification of the cultural microorganisms for a description of them without specific function to the biodeterioration (May 2000). Such practice had been persisted before the application of polymerase chain reaction (PCR) to cultural heritage microbiology research to reveal the complex microbial community without culturing and isolation (Rölleke et al. 1998). The latest high-throughput sequencing and metagenomics provide a much deeper description of the microbial community and composition without culturing or isolation (May 2000; McNamara et al. 2006; Sterflinger and Pinar 2013; Zhang et al. 2019; Ma et al. 2020; Meng et al. 2020). This new approach also has some major weakness in its inability to identify the active microbes from the genomic DNA-based community analysis and, in addition, the active deteriorating ones are not identified from the community. Because of these, the advances made on knowing the community better have not translated to useful results on the deteriorating processes for a more effective management and prevention (Liu et al. 2020).

Microorganisms colonizing surfaces of cultural heritage can destruct the underlying materials through their influences on the physical, chemical, and bioreceptibility of the substratum materials, especially metabolic activities and biochemical reaction. Biofilms as a physical layer on the outer surface of cultural heritage alter the thermal property and moisture contents, which in turn have their impact on the biofilms and the activities of the microorganisms. The mechanisms involved in biodeterioration of natural sandstone and man-made materials include both abiotic and biological ones (Mansch and Bock 1998; Mitchell and Gu 2000; Warscheid and Braams 2000; Gu 2003; Liu et al. 2018a; Zhang et al. 2019; Ding et al. 2020; Liu et al. 2020). On the basis of microbiology and biochemistry, the initial colonization by the pioneering microorganisms to form biofilms on rock and stone surfaces can initiate a number of subsequent physical and chemical changes of the substratum materials, including the discoloration in appearance, alteration of porosity and vapor/moisture diffusivity in and out of the stone, accumulation of organic substances, acid production, solubilization and mobility of ions and salts, and mineral crystallization within the sandstone when interacting with the surrounding environments, e.g., drying (Ariño and Saiz-Jimenez 1996; Mitchell and Gu 2000; Warscheid and Braams 2000; Gaylarde et al. 2003; Perry et al. 2005; Gu and Mitchell 2013; Zhang et al. 2019). The tropical South Asia Cambodia, as an example, has a large collection of historical monuments and temples which are the most important existing records of the ancient civilizations, history, and humanity of people in the region of the world, including the Khmer (Freeman and Jacques 1999). Unfortunately, similar to many other cultural heritages around the world, these cultural heritage monuments are showing irreversible deterioration and destruction under the tropical climate conditions which support a colonizing microbial community and flora, and, more importantly, the active growth of various microorganisms under favorable conditions of the seasonality (Flores et al. 1997; Tayler and May 2000; Videla et al. 2000; Piñar et al. 2009; Motti and Stinca 2011; Gu 2012; Adamson et al. 2013; Keshari and Adhikary 2014; Meng et al. 2016, 2017; Liu et al. 2018b; Meng et al. 2020). In addition, a human dimension is increasing its contribution to the deterioration due to anthropogenic influences of pollutants and their deposition (Liu et al. 2018c, 2020; Meng et al. 2020). Microbial biofilms on stone surfaces alter the physical properties, thermal and water uptake and loss from the materials, and also the dynamics of soluble salt uptake and transport, into and out of the stone block, to initiate any physical stress and damage internally, e.g., crystallization of minerals to induce the pressure and delamination (Liu et al. 2018a; Zhang et al. 2019). Apart from the physical and chemical contributions to the damages, microbial biofilms on the stone monuments have been widely recognized for their negative impact mostly as the dominant views (Dornieden et al. 2000; Papida et al. 2000; Saiz-Jimenez and Laiz 2000; Warscheid and Braams 2000; Sterflinger and Prillinger 2001; Crispim and Gaylarde 2005; Li et al. 2007, Lan et al. 2010; Li et al. 2010; Meng et al. 2016; Sterflinger and Pinar 2013; Essa and Khallaf 2014; Meng et al. 2020).

Microbial community of Angkor monuments was analyzed initially with 16S rRNA gene-based polymerase chain reaction (PCR) and clone library (Lan et al. 2010; Kusumi et al. 2013) and, at the same time, isolation and identification of bacteria and fungi of the sulfur cycle and capable of acid production were focused at the beginning to obtain a general overview of the microbial assembly and their role on the potential destruction of these sandstone monuments under the tropical climate (Meng et al. 2020; Mitchell and Gu 2000; Hosono et al. 2006; Li et al. 2010; Kusumi et al. 2011; Hu et al. 2013; Ding et al. 2020). With the non-invasive sampling technique developed at the Japan Space Agency, a spatial distribution of the different microbial groups was also visualized on selective sandstone bas relief of Bayon temple of Angkor Thom after taking sampling on the same selective locations for more than 40 times to construct the microbial community spatially by extraction of DNA and then PCR to obtain enough DNA for further community characterization to show the microbial communities over the sampling sequences as spatial information (Kusumi et al. 2013).

**Fig. 2.1** Photographs of the UNESCO World Cultural Heritage Angkor Wat in Cambodia. A distance view of the central tower with surrounding ones (**a**); a corner section of the wall of the central tower (**b**); and a close up view of the carving of an Apsara on the sandstone wall (**c**)

Sulfur-oxidizing bacteria and fungi are detected at a number of Angkorian temples and they actively oxidize elemental S ($S^0$) to produce acidity to contribute to the sandstone deterioration. Both mycobacteria and fungi were initially isolated and identified from Angkor monuments, and population dynamics were also monitored at a number of sites to show an active population of these microorganisms over a period of multiple years (Li et al. 2007, 2010; Kusumi et al. 2011). Lately, an accumulation of $NO_3^-$ was initially observed at several locations of the sandstone temples in Cambodia and this has lead us to investigate the microbiological contributors for this phenomenon and also the possible causes to it (Meng et al. 2017; Ding et al. 2020). After several years of sampling and analyses, it then becomes clear that ammonia-oxidizing archaea (AOA) are much more abundant and also biologically active than ammonia-oxidizing bacteria (AOB) on these Angkor monuments in Cambodia using both genomic DNA first and then reverse transcripts of RNA in samples for *amoA* gene abundance quantification (Meng et al. 2016, 2017; Liu et al. 2018b; Meng et al. 2020). The analyses over several years at a number of monuments in the main cluster of Angkor Wat, Angkor Thom and also smaller ones at relatively short distance from the city Siem Reap, namely Phnom Krom and Wat Athvea, allow further confirmation on the observation for the ubiquity of the phenomenon (Fig. 2.1). The dominance of AOA at these temples is a wider occurring phenomenon, not restricted to a few selective cultural heritage sites (Meng et al.

2017). Microorganisms on sandstone, especially those with specific biochemical function and capabilities, still reveal new information on the microbiology and biochemical reaction mechanisms for attack of sandstone differing from the community-based analysis without any specific biochemical function confirmed. A new analysis reveals that Comammox bacteria, oxidizing ammonia directly to $NO_3^-$ in a one-step reaction instead of the conventional two-step processes via nitrite as an intermediate, are also detected on Preah Vihear Temple (Ding et al. 2020).

# 2  Microbial Colonization

## 2.1  Pioneering Colonizers and Colors

Phototrophic microorganisms and lichens have major advantages over heterotrophic ones to initiate their colonization on fresh stone surfaces initially under the tropical climate conditions with plenty of sunlight and relatively high humidity. The description of these microorganisms and on different materials and niches is widely available (Sterflinger 2000; Lisci et al. 2003; Crispim et al. 2003; Liu et al. 2018c; Zhang et al. 2019; Liu et al. 2020). Due to the frequent rainfalls in monsoon season and plenty of sunlight, the sandstone under the tropical climate faces different challenges from those under temperate climate, which experience harsh winter and possibly hot summer. Under the tropical conditions, microbial activity is much more active and the life span of the sandstone materials for these monuments is consequently shortened compared to temperate climate (Meng et al. 2017, 2020). It is due to this fact that Angkor monuments are covered extensively with a wide range of microorganisms today showing in different colors (Fig. 2.1).

Natural sandstone contains very little organic carbon to support the active growth of heterotrophic population initially, but, after the initial colonization by the pioneering autotrophic microbial population on stones and rocks, the dead biomass and metabolites of them on the stone as a conditioning layer or crust become a readily available source of organic carbon for heterotrophic microorganisms to grow subsequently. Microbial colonization on inorganic stone and rock follows a general pattern and a series of sequences under the open environmental conditions (May 2000). Cyanobacteria, algae, and lichens are pioneering colonizers before active development of heterotrophic bacteria and then fungi (Gaylarde and Gaylarde 2000; Tayler and May 2000; Li et al. 2007; Kusumi et al. 2011, 2013; Liu et al. 2020). Lichens are mainly on stone of the open conditions under direct sunlight, but cyanobacteria and algae are abundant and active inside temples and monuments without direct sunlight but the moisture or water supply is constant.

Surface properties of the underlying sandstone can be altered in responses to thermal and hydrological processes after the colonization by microorganisms as biofilm, which in turn affect the physical and chemical characteristics of the sandstone during the succession and dynamics of different physiological and functional microbes (Gu et al. 1996; Gu and Mitchell 2013; Liu et al. 2016; Zhang et al. 2018;

Ding et al. 2020). The main sequence of events after the initial microbial colonization on stone includes: (1) improvement of water retention in sandstone, immobilization of organic carbon onto surfaces and into sandstone through photosynthesis by the pioneering surface colonizers and (2) development of a population consisting of heterotrophic bacteria and fungi to form a more complex community composition of both autotrophic and heterotrophic microorganisms. Besides the biology, the surface properties are modified after microbial colonization for porosity, water regime (permeability, evaporation, and availability of water), and adhesion and trapping of atmospheric pollutants onto cultural heritage (Ariño and Saiz-Jimenez 1996; Mitchell and Gu 2000; Gu 2012; Liu et al. 2018a). Angkor Wat and the nearby temples constructed of sandstone are particularly susceptible to the colonization by different microorganisms over the past near one thousand years due to material and also the tropical conditions with plenty of sunlight and an active water cycle (Liu et al. 2018b). The phototrophic microorganisms including cyanobacteria and microalgae occur widely on the inside gallery walls of Angkor Wat and temples without direct sunlight, especially where moisture can be retained for a long period of time (Li et al. 2007; Lan et al. 2010; Kusumi et al. 2011; Meng et al. 2017; Zhang et al. 2019; Meng et al. 2020). Similar events are also observed on historic buildings frequently elsewhere (Crispim et al. 2003).

The different colors appearing on these monuments and buildings are closely associated with the material types, locations, dynamics of the microbial community development, seasonality, and also the ambient climate conditions. At any given selective location, the differences in color appearance are largely a result of the different microbial groups at different stages of their physiological growth and development on the specific sandstone type (Kusumi et al. 2013; Liu et al. 2018c; Zhang et al. 2019). Cultural heritage monuments and historic buildings often show different colors by microorganisms and also visible plants under natural conditions (Fig. 2.2). Cyanobacteria change its intensity of the green color from light to bright and then dark from early to maturity during the community development. The biofilms on Bayon temple of Angkor Thom show a change from the initial cyanobacteria and algae to the black fungi of the pigment-producing Ascomycetes at a late stage (Lan et al. 2010; Hu et al. 2013; Zhang et al. 2019). Many of the sandstone monuments in Southeast Asia and South Americas appear stained black or darker on the surface due to the fungi. At Angkor monuments, orange-pink pigmentation is associated with *Lecythophora* sp., dark brown with *Cladosporium cladosporioides* melanin, and black with *Coniosporium apollinis* melanin (Lan et al. 2010; Kusumi et al. 2013; Liu et al. 2018a). The pigmentation and deposition on sandstone surfaces have visual and aesthetical effects, but production of inorganic and organic acids by the colonizers has a much greater harm to the underlying sandstone and rock (May et al. 2000; Etienne and Dupont 2002; Liu et al. 2018b).

**Fig. 2.2** Photographs of the sandstone foundation and current conditions at Angkor Wat (**a**) and Bayon temple of Angkor Thom from a distance (**b**) and a close up (**c**)

## 2.2  *Halophilic*

Sandstone and rock are unique niches and the microbial community and composition on sandstone are actually dominated by halophiles. *Haloarchaea* species are found on Angkor temples and the deteriorated sandstone walls of Bayon temple contain carbonates, chlorides, nitrates, sulfates, etc. as the main constituents (Uchida et al. 2000; Zhang et al. 2018). This information reminds of the early isolation and identification studies yielding a large number of common and known microorganisms, and such results shall be taken carefully for assessment of any damage to the sandstone because those microorganisms are not the dominant members of the indigenous community. The local conditions with sufficiently high concentration of salts enrich halotolerant and halophilic microorganisms to be the dominant members of the microbiome on sandstone, especially during the dry seasons. For example, the *Haloarchaea* species are a group of *Euryarchaeota* which are known for their preference for high salt concentrations (Kates 1978; May 2000).

## 2.3  Archaea

*Archaea* as an independent domain of life is a significant milestone discovery in understanding of the biological world (Woese 1978). The non-methanogenic archaea have been expanded rapidly due to the many new phyla and superphyla of archaea discovered with the available high-throughput sequencing technology, metagenomics, and bioinformatics tools available (Liu et al. 2018a, b; Zhou et al. 2018, 2019). The information has illuminated on new lineages of the archaea, but almost all of these new archaea are discovered mainly based on 16S rRNA gene sequences and metagenomes available. To further advance the physiology and biochemistry of these archaea, enrichment and isolation in pure cultures are the key step to advance the understanding on their ecological contribution and significance. As an example, AOA has only been recently detected as the dominant group responsible for oxidation of ammonia on sandstone monuments in Cambodia after observing accumulation of $NO_3^-$ in sandstone (Meng et al. 2016, 2017). After analyzing samples from three different temples in Siem Reap of Cambodia, namely Bayon temple of Angkor Thom, Wat Athevea, and Phnom Krom, AOA was the dominant ones in the genomic DNA and also transcribed cDNA of the same set of samples, while AOB could only be detected at lower abundance and in a small fraction of the samples (Meng et al. 2016, 2017). Three PCR amplified sequences of Bayon temple were closely related to *Nitrosopumilus maritimus,* the first pure culture isolated from a tropical marine aquarium (Könneke et al. 2005; Walker et al. 2010), which is designated as Group I.1a. Major sequences of Phnom Krom samples fell into Group I.1a-associated closely affiliated with  *Ca.* Nitrosotalea devanaterra (Tourna et al. 2008, 2011) and into Group I.1b with *Ca.* Nitrososphaera (Meng et al. 2016, 2017). The occurrence of these two groups is in good agreement with the fact that they are tolerant to low pH values (Hatzenpichler et al. 2008; Hatzenpichler 2012). *Nitrosotalea* and *Nitrososphaera* can grow when urea is available under acidic conditions (Kowalchuk et al. 2000; Laverman et al. 2001; Lu et al. 2012). In contrast, AOB were less abundant significantly compared with AOA on cultural heritage monuments under the tropical climate (Meng et al. 2016, 2017).

## 3  Key Biochemical Processes of Biodeterioration

## 3.1  Carbon Sequestration

As discussed above, phototrophic microorganisms are pioneer colonizers on fresh sandstone surfaces of cultural heritage under the tropical climate conditions and then the subsequent impacts from the microorganisms can be also variable due to the different species and the specific biochemical processes carried out by them (Gu and Mitchell 2013; Liu et al. 2018a). The most significant event is to sequestrate carbon dioxide from atmosphere initially onto sandstone to alter the exterior surface properties of the sandstone materials for an active dynamic advance of microorganisms with different physiological capabilities to provide chances for heterotrophic microorganisms to colonize over time. In addition to the natural cycle of carbon from inorganic to organic, anthropogenic pollution and pollutants cannot be ignored for their contribution to the development of a different microbial community that utilize the deposition to grow and also cause damage of the underlying materials (Mitchell and Gu 2000).

## 3.2  Nitrogen Transformation

Among the different biochemical reactions of N cycle, $N_2$-fixation and nitrification are two of the most significant ones on sandstone monuments of cultural heritage and historic buildings. At the same time as $CO_2$ is sequestrated onto sandstone in biomass, $N_2$ fixation is also taking place when cyanobacteria and algae are the dominant ones in the biofilm community (Zhang et al. 2019; Liu et al. 2020). For the nitrification or ammonia oxidation, the *amoA* genes of bacteria and archaea are phylogenetically distinct possessing different evolution and phenotypic characteristics (De Boer and Kowalchuk 2001; Nicol and Schleper 2006; Dodsworth et al. 2011). Abundance of the AOA *amoA* gene is much higher than that of AOB in selective habitat, including open ocean (Wuchter et al. 2006; Nakagawa et al. 2007; Beman et al. 2008; Cao et al. 2013), oil reservoirs (Li et al. 2011), acidic soils (Leininger et al. 2006; Nicol and Schleper 2006; Adair and Schwartz 2008; Nicol et al. 2008; Onodera et al. 2010), and estuaries and wetlands (Caffrey et al. 2007; Wang and Gu 2013).

Deteriorated sandstone samples are mainly comprised of decomposition products of minerals and clays from sandstone and components of biofilms at different development stages. Intrinsic sandstone properties dictating the colonization of microorganisms to form biofilms include mineral composition, porosity and permeability and retaining of water by the materials and susceptibility to dissolution (Ariño and Saiz-Jimenez 1996; Essa and Khallaf 2014). Ammonium released from decomposition of organic biomass and deposition from atmosphere is oxidized through microbial metabolism and AOA is the dominant group over AOB at several temples including Bayon temple of Angkor Thom, Wat Athevea, and Phnom Krom in Cambodia (Meng et al. 2016). AOA are easily detected at all sandstone monuments and the pH values of these samples are also generally low. Coupling the relative high concentration of $NO_3^-$ detected with the low pH condition and the more abundant AOA than AOB, an active biochemical process of ammonia oxidation or nitrification is operative on these cultural heritage to result in an accumulation of biogenic nitric acid, which contributes to the biodeterioration of sandstone. At the same time, the accumulation of $NO_3^-$ in sandstone of Angkor temples is also related to the inactive removal of the $NO_3^-$ by denitrification and anammox of the relevant microorganisms (Ding et al. 2020).

## 3.3 Sulfur Transformation

The four most important elements to life are carbon, nitrogen, phosphorus, and sulfur. Both nitrogen and sulfur metabolizing microorganisms are responsible for production of inorganic acids that can erode the sandstone and rock of cultural heritage. Previously, emphasis has been put more on corrosion initiated by sulfate/sulfur-reducing bacteria and sulfur-oxidizing bacteria (Li et al. 2007; Li et al. 2010; Kusumi et al. 2013) and AOB (Sand and Bock 1991; Mansch and Bock 1998). After the discovery of ecological role of AOA, this group of archaea has also been detected on sandstone monuments using both DNA- and also RNA-based quantification of the *amoA* gene of this functional group of organisms (Meng et al. 2016, 2017). The abundance of AOA was higher than that of AOB, implying that AOA is a major contributor to the deterioration of sandstone (discussed above). In addition, sulfur-oxidizing bacteria and fungi are also detected on sandstone cultural heritage and their population is proportional to the sampling site where the damage is the most severe (Li et al. 2010; Kusumi et al. 2011). Based on these results, special attention shall be given to the microorganisms involved in sulfur cycling to effectively protect cultural heritage. Sulfate was detected in the liquid medium inoculated with deteriorated sandstone samples, and the chemolithoautotrophic *Mycobacterium* spp. (Kusumi et al. 2011) and 19 sulfur-oxidizing fungal strains (Li et al. 2010) capable of sulfur oxidation were isolated from the deteriorated sandstones in Angkor Wat, Bayon, and Phnom Krom temples to indicate their contribution.

Sources of sulfur may be from the biofilm and preformed biomass and also fecal materials of bats and wild animals living in these temples. In the early part of the previous century, heterotrophic fungi were confirmed for sulfur oxidation and various fungi can oxidize $S^0$ for energy to grow, including *Trichoderma harzianum* and *Aspergillus niger*. A number of fungal genera are capable of oxidizing inorganic sulfur compounds, but the biochemical sulfur oxidation pathway and enzymes involved are still not known clearly. *Fusarium solani* strain THIF01, isolated from Angkor monuments, is capable of utilizing $S^0$ chemolithoautotrophically as the energy source (Li et al. 2010). This fungus contains an endobacterium *Bradyrhizobium* sp. which has sulfur oxidation capability, but the bacterium-free culture (strain THIF01BF) that was obtained by treatments with antibiotics of strain THIF01 showed chemolithoautotrophic growth with $S^0$ (Xu et al. 2018).

## 3.4 Other Elements

Other elements are also important to life and the cycling of nutrients by microorganisms in different ways. They can serve in dissimilatory and assimilatory biochemical processes depending on the element types and the specific microorganisms involved producing transformation products and biomass, respectively (Ehrlich 2002). As an example, iron in the environment can play a role to participate in the corrosion and biodeterioration of stone materials (Warscheid and Braams 2000), but the relative significance of this biochemical process compared with major elements C, N, and S shall be much smaller or insignificant. Because of this, involvement and contribution of microorganisms utilizing Fe and Mn and other elements to deterioration of stone are not further discussed here, but their significance in revealing new biochemical processes shall not be underestimated in future research as a focus area.

# 4 Interactions Among Sandstone, Microbiota, and the Local Environment

## 4.1 Material Types

Sandstone is a porous material of natural origin and is not strong enough compared to other stone and rock from igneous formation. Such property of this material makes it to be very susceptible to attack by weathering through water absorption and freeze-and-thaw cycle. Colonization by microflora and plants adds another dimension to the stability challenge to sandstone in open climate conditions (Liu et al. 2018b, 2020). Under tropical conditions of Southeast Asia and South America, both sunlight and water are plenty to provide the basic requirements for phototrophic life forms of plants and microorganisms to colonize and grow on surfaces. As soon as the initial colonization on surfaces is established, e.g., lichens either the initial protection or further damage of the materials can take place through the specific microorganisms actively metabolizing on surfaces of stone (Liu et al. 2018c, 2019). Both autotrophic and heterotrophic ones can actively grow in the surface colonizing community and alter the materials through their metabolic processes and the metabolites. Such progress and development of the community over time on sandstone surfaces maintain an active and dynamic community on the sandstone surfaces and allow sequestration of inorganic $CO_2$ from atmosphere and cycling of nutrients through the life cycle of different microorganisms and their biomass coupling with the seasonality.

Sandstone is interacting with the surrounding environment, including water, air, microbiota, and anthropogenic contributors. Among all of them, temperature and availability of water play a crucial role in limiting or promoting microbial colonization and biofilm formation on sandstone under favorable conditions of the tropics (Liu et al. 2018a). Because of the porosity of sandstone and biofilm formation to prevent more quick exchanges of water between the surrounding and those retained inside sandstone, increasing retention of water holding in sandstone supports more active microbial colonization and growth to result in more damage of stone from biological activities (Zhang et al. 2019). At the same time, movement of water containing soluble salts by siphon effects promotes formation of mineral crystals upon drying, which initiates external surface defoliation progressively to lead to

**Fig. 2.3** Photographs of Devatas carved on sandstone columns showing differences in conditions and destructions with more red pigment deposition (**a**) and more severe damage (**b**) at Bayon temple of Angkor Thom

significant weakening of the sandstone mechanical strength to collapse over time. Such phenomena are widely observed on monuments of cultural heritage and buildings showing the progression to defoliation and destabilization of sandstone structure (Fig. 2.3). This kind of damage and destruction is widespread at stone monuments in Southeast Asia and, at Angkor Wat in particular, the sandstone columns for structural support at the gallery section show much severe damage in the basal portion closer to the floor level where rainwater can be retained for longer period of time. This suggests that water management is an important issue in the protection of world cultural heritage for an effective and long-term plan (Liu et al. 2019).

## 4.2 Available Nutrients and Pollutants

World cultural heritage in tropical regions is much more susceptible to deterioration mainly due to the availability of sunlight and more importantly the rainwater onto sandstone which contains abundant minerals to allow microbial colonization and growth with the available water. In the tropical regions after the initial microbial settlement and colonization on sandstone and rock, a bioactive layer covers the exterior surface of sandstone and temples/monuments to buffer the underling materials with the surrounding environment from protection and damage at different stages of the development (Liu et al. 2018b). Initially oligotrophic niche on sandstone becomes enriched through phototrophic growth to sequestrate organic substances onto and into the surface layer of sandstone, in doing so the local conditions are altered from the oligotrophic to organic carbon rich one. After this transformation, heterotrophic microorganisms become the dominant population including both bacteria and fungi to change the color of sandstone from its original gray or light brown to dark and black. Fresh microbial biofilm consisting of algae and cyanobacteria is reported (Lan et al. 2010) and the different colors and pigmentations on sandstone wall are associated with different microbial groups at Bayon and Angkor Wat (Kusumi et al. 2013). These microorganisms need to be further evaluated for their specific role in the bioprotection and/or biodeterioration at different development stages considered to advance our comprehensive understanding of their specific role over time to provide more accurate information for future management of cultural heritage effectively. Subsequent results indicate the acid attack initiated by microbial oxidation of $S°$ from reduction product of $H_2S$ or sulfate (Li et al. 2007, 2010). More recent research information showed that ammonia oxidation is an active biochemical reaction mechanism carried out mainly by AOA at several temples in Cambodia (Meng et al. 2016, 2017). This is a new development for further assessment of its flux and activities so that future management can be made considering such results in the conservation plan. It is certain that more contributors are involved in the biodeterioration of sandstone monuments and destruction of valuable galleries and bas relief, key factors shall be identified and ranked according to the extent of destruction by each of them so that management can be focused on a few most important ones first to monitor the effectiveness over time to verify the conservation results.

Cultural heritage of monuments and temples is in constant exchange with the surrounding environmental conditions, both water and biological attacks may take place simultaneously. In addition, the increasing anthropogenic pollution cannot be ignored in the current and future investigation and protection. With the pre-existing biofilms and modified surfaces of sandstone to increase the roughness of the surfaces and porosity, the color change enhances thermal property of the underlying sandstone to absorb solar radiation, but emits less due to the biofilm layer especially when water contents in the surface layer are high. In addition, the pollutants in the air can be deposited and trapped by the porous biofilms on surfaces to be potential sources of organic carbon to selective biofilm microorganisms capable of utilizing them to grow. The active surface layer coated on sandstone is of significant research interest to investigate the physical and biochemical processes actively involved and also the dynamics of them spatially and temporally to gain further insights of this micro-niche and system for new discoveries.

## 5 Protection Strategies

As cultural heritage and historic buildings are unique in their design, cultural identity, Angkor monuments as an example have a very important role in the understanding of Khmer culture and civilization history. Without them, it is a great loss to the world humanity for the future generations to appreciate and comprehend the world history for an understanding of the rich culture and civilization in Southeast Asia. It is apparent that the deterioration and damage of these monuments and temples as world cultural heritage are constant threats to the future protection of these monuments (Fig. 2.4), therefore effective protective measures, based on scientific data, shall be formulated, tested, and implemented to preserve them for the future generations.

Deterioration of stone monuments is a result of natural decomposition process involving physical, chemical, and biological ones, protection cannot stop these processes (Liu et al. 2020), but delay their progression under specific conditions. With such information in mind, effective management and protection strategies shall be the focus on identification of the key processes involved for destruction and rank them in order for management strategy than an overall community analysis without identification of the detrimental processes. The mechanisms of the most active processes together with the responsible microorganisms shall be delineated to

**Fig. 2.4** Photographs of Apsaras of Angkor Wat (**a**) and Devatas at Bayon temple (**b**) showing different conditions on the same sandstone of different quality initially

advance the knowledge for protection and management. Biofilms on sandstone have different roles at different stages of the colonization and biofilm development, such understanding is vitally important to a deeper understanding of biofilm dynamics and its relationship with the integrity of sandstone (Warscheid and Braams 2000; Gu et al. 2011; Liu et al. 2018c; Zhang et al. 2019). As a matter of fact, physical removal of biofilms is not going to solve the deterioration issue from microbial activity, but accelerate the damage because of removing the protective biofilms to expose the underlying surfaces for another cycle of colonization and biofilm development. Microorganisms can also play a role in control of biofilms. Fungi appearing on matured biofilms are correlated with the peeling off of old biofilms, which has potential to be further developed into biotechnology for biological control and cleaning of colonized surfaces (Gleeson et al. 2005; Hu et al. 2013).

The biochemical processes should be identified when evaluating the susceptibility of sandstone to microorganisms for colonization and biodeterioration (Meng et al. 2017; Zhang et al. 2019). Phototrophic microorganisms initially establish on the sandstone surfaces to be the pioneering biofilms when water and mineral nutrients are readily available from the substratum. Apart from the external factors of the environment, sandstone has reasonable porosity to allow water retention and also promotion of microbial colonization. At the same time, the small pores also trap nutrients from atmospheric deposition containing dust and microbes to promote biosusceptibility and attack of sandstone. Currently, major efforts have been given to the analysis of the microbial community with the latest available sequencing technology independent of culturing and isolation, but the biochemical functions by the selective members of the community need to be delineated so that initiation for damage of the cultural heritage can be prevented and limited with means available. The interfaces between sandstone surface and biofilms remain a hotspot for biological activity and biochemical processes when analysis of cultural heritage is the target (Ma et al. 2015; Wu et al. 2017; Kakakhel et al. 2019). Biofilm and plants can establish the cultural heritage over time when routine maintenance is not carried out effectively. A close monitoring and clearance can achieve a good protection from weeds and plants, which otherwise would grow into the sandstone to cause dislocation of the stone blocks and collapse of the structure (Liu et al. 2018a; 2019). To protect them, maintenance work shall be conducted to remove newly grown plants and weeds regularly so that no large plants can be established and remain on these structures.

Maintenance work also involves repair of the damaged and destructed structure and stone pieces. This may involve the use of chemical adhesive and consolidants, so that the repaired stone may last longer when exposed to both the ambient climate conditions and invasive growth of microorganisms, particularly fungi (Gu et al. 2011; Wu et al. 2017; Zhang et al. 2018). At the same time, water repellants are also used to seal the stone from natural rainwater and keep dry to avoid water-associated damage. But such practice has a serious potential drawback in that trapped water inside the sandstone is more destructive than those without such protections used so nature regulates the water movement (Sterflinger and Pinar 2013; Liu et al. 2018b; Sterflinger et al. 2018). Considering these factors, sandstone monuments and temples are better kept away from water, especially stagnant water, if possible or let natural rainwater to uptake and drain freely from the structures when raining or drying accordingly. However, after the formation of biofilm on surfaces, these exchange processes for water are being slowed down for both into and out of the stone, resulting in less efficient movement of water to initiate damage of the stone through both biofilm development and also salting effects.

Polymeric materials are increasingly used in applications including conservation and protection of cultural heritage. Some of the successful cases are usually restricted to specific environmental conditions of museums, but such information is not widely reported for the specificity of the materials to the selective outdoor environmental conditions. Misuse of such information has resulted in disaster in conservation work due to outbreaks of microbial growth after application of consolidants or even antimicrobials to cultural heritage exposed to outdoor conditions (Gu 2003; Gu et al. 2011; Wu et al. 2017; Liu et al. 2018c; Kakakhel et al. 2019). Polymeric materials may perform well under harsh conditions including high humidity and temperature, but such polymeric materials are generally very difficult to apply as consolidants because of their high viscosity, insolubility in even solvents, and associated colors. In contrast, those polymers that have been used successfully in museums are generally not resistant to metabolism by indigenous microorganisms because fungi are airborne and particularly capable of attacking high strength and engineering polymers due to fillers, additives, and plasticizers (Gu 2003; Mitchell and Gu 2000; Gu et al. 2011; Gu 2012).

Inorganic materials have many good applications in construction and protection of building. Such practice has been proven successful when the candidate materials are used to appropriate niche and conditions. For example, lime is a common material to apply to surfaces of building to protect the biomaterials from insect attack. Such materials are also available now on the market, but the composition of the commercially available products contains not only lime but also additives, including polymeric materials, cellulose, and cellulose esters, which are biodegradable and cause serious biofilm establishment and growth on surfaces of structure (Gu and Mitchell 2013).

New materials and technologies are also becoming available through research and innovation, and some of the promising ones include nanomaterials and technology for effective protection against fast growth of colonized biofilm microorganisms (Liu et al. 2018a). At the same time, corroding chemical and biochemical processes may be inhibited through intervention of chemical treatment or approaches to advance the protection more effectively.

Antimicrobial chemicals are commercially available for inhibiting the growth of microorganisms under selective or specific conditions, e.g., disease pathogen control and eradication, but the available information indicates that successive applications promote the development of resistance in microorganisms, which is a major public health concern when the resistant genes are enclosed on plasmids that are more readily transmitted among the same species and also between different species. It is unlikely to achieve an effective long-term control using a single or few chemicals. Conversely, when selecting candidate chemicals, the current testing protocols for screening are problematic because the testing conditions are mainly based on inhibitory testing in liquid cultures, not the application surface or the environmental niche conditions. The very little resemblance between the target conditions and the laboratory testing ones results in non-effectiveness after applications in situ on cultural heritage (Gu 2003; Essa and Khallaf 2014; Liu et al. 2018b, 2020).

# 6 Summary

Stone cultural heritage and historic buildings are susceptible to colonization by microorganisms of a wide range of physiological characteristics depending on the environmental conditions and also the stage of the development. Both the specific active microorganisms and also their physiological functions shall be identified and elucidated for a better and more specific understanding on the deterioration biochemical processes and mechanisms involved for effective protection of the sandstone temples/monuments. Microorganisms involved in carbon, nitrogen, and sulfur cycles are the most important ones, especially the biochemical processes that generate acidic or corrosive products to erode sandstone and dissolve the minerals. On Angkor temples/monuments, sulfur-oxidizing bacteria and fungi, ammonia-oxidizing archaea, and also phototrophic cyanobacteria and algae are identified as priority members for investigations of their biochemical contribution to the deterioration. Related processes, e.g., denitrification, anammox, and comammox, are also important for the comprehensive knowledge on the accumulation of nitrate in many sandstone monuments.

Physical and environmental factors of the materials and hydrological dynamics shall not be ignored from the biology for protection and management of cultural heritage sites because they affect the biology on stone fundamentally and then the stability of the stone. Equipped with the information on the materials of cultural heritage and buildings, biology, and biochemistry, preservation strategies can be formulated for protection of sandstone by taking advantage of the latest development and technologies in chemistry, nanomaterials, and molecular biology.

**Acknowledgements** This project was supported by Safeguarding of Bayon Temple of Angkor Thom (JASA, Japan), APSARA Authority of Cambodian Government, the UNESCO/Japanese Funds-in-Trust for the Preservation of the World Cultural Heritage, and partially by a Hong Kong RGC GRF Grant (No. 17302119).

**Ethical Statement   Conflict of Interest:** All authors declare that they have no conflict of interest in this study.

**Ethical Approval:** This article does not contain any studies with human participants or animals performed by any of the authors of this investigation.

Sampling at temples was granted by APSARA Authority of Cambodian Government, the Kingdom of Cambodia.

# References

Aair KL, Schwartz E (2008) Evidence that ammonia-oxidizing archaea are more abundant than ammonia-oxidizing bacteria in semiarid soils of northern Arizona, USA. Microbiol Ecol 56:420–426

Adamson C, McCabe S, Warke PA, McAllister D, Smith BJ (2013) The influence of aspect on the biological colonization of stone in Northern Ireland. Int Biodeterior Biodegradation 84:357–366

Ariño X, Saiz-Jimenez C (1996) Factors affecting the colonization and distribution of cyanobacteria, algae and lichens in ancient mortars. In: Riederer J (ed) 8th International Congress on Deterioration and Conservation of Stone. Möller Druck und Verlag, Berlin, pp 725–731

Beman JM, Popp BN, Francis CA (2008) Molecular and biogeochemical evidence for ammonia oxidation by marine Crenarchaeota in the Gulf of California. ISME J 2:429–441

Caffrey JM, Bano N, Kalanetra K, Hollibaugh JT (2007) Ammonia oxidation and ammonia-oxidizing bacteria and archaea from estuaries with differing histories of hypoxia. ISME J 1:660–662

Cao H, Auguet JC, Gu J-D (2013) Global ecological pattern of ammonia-oxidizing archaea. PLoS One 8:e52853

Crispim C, Gaylarde C (2005) Cyanobacteria and biodeterioration of cultural heritage: a review. Microbiol Ecol 49:1–9

Crispim CA, Gaylarde CC, Gaylarde PM, Copp J, Neilan BA (2003) Molecular biology for investigation of cyanobacterial populations on historic buildings in Brazil. In: Saiz-Jimenez C (ed) Molecular biology and cultural heritage. Balkema, Lisse, pp 141–143

De Boer W, Kowalchuk G (2001) Nitrification in acid soils: micro-organisms and mechanisms. Soil Biol Biochem 33:853–866

Ding X, Lan W, Li Y, Wu J, Hong Y, Urzi C, Katayama Y, Ge Q, Gu J-D (2020) Microbiome and nitrate removal by denitrifying and anammox on the sandstone Preah Vihear temple in Cambodia revealed by metagenomics and N-15 isotope. Appl Microbiol Biotechnol 104(22):9823–9837. https://doi.org/10.1007/s00253-020-10886-4

Dodsworth JA, Hungate BA, Hedlund BP (2011) Ammonia oxidation, denitrification and dissimilatory nitrate reduction to ammonium in two US Great Basin hot springs with abundant ammonia-oxidizing archaea. Environ Microbiol 13:2371–2386

Dornieden T, Gorbushina AA, Krumbein WE (2000) Biodecay of cultural heritage as a space/time-related ecological situation – an evaluation of a series of studies. Int Biodeterior Biodegradation 46:261–270

Ehrlich HL (2002) Geomicrobiology. Marcel Dekker, New York, p 768

Essa AM, Khallaf MK (2014) Biological nanosilver particles for the protection of archaeological stones against microbial colonization. Int Biodeterior Biodegradation 94:31–37

Etienne S, Dupont J (2002) Fungal weathering of basaltic rocks in a cold oceanic environment (Iceland): comparison between experimental and field observations. Earth Surf Proc Land 27:737–748

Flores M, Lorenzo J, Gómez-Alarcón G (1997) Algae and bacteria on historic monuments at Alcala de Henares, Spain. Int Biodeterior Biodegradation 40:241–246

Freeman M, Jacques C (1999) Ancient Angkor. River Books, Ltd., Bangkok

Gaylarde C, Silva MR, Warscheid T (2003) Microbial impact on building materials: an overview. Mater Struct 36:342–352

Gaylarde PM, Gaylarde CC (2000) Algae and cyanobacteria on painted buildings in Latin America. Int Biodeterior Biodegradation 46:93–97

Gleeson DB, Clipson N, Melville K, Gadd GM, McDermott FP (2005) Characterization of fungal community structure on a weathered pegmatitic granite. Microb Ecol 50:360–368

Gu J-D (2003) Microbiological deterioration and degradation of synthetic polymeric materials: recent research advances. Int Biodeterior Biodegrad 52:69–91. https://doi.org/10.1016/S0964-8305(02)00177-4

Gu J-D (2012) Biofouling and prevention: corrosion, biodeterioration and biodegradation of materials. In: Kultz M (ed) Handbook of environmental degradation of materials, 2nd edn. Elsevier, Waltham, pp 243–282

Gu J-D, Mitchell R (2013) Biodeterioration. In: Rosenberg E, DeLong EF, Lory S, Stackebrandt E, Thompson F (eds) The prokaryotes: applied bacteriology and biotechnology. Springer, New York, pp 309–341

Gu J-D, Ford TE, Mitchell R (1996) Fungal degradation of concrete. In: Sand W (ed) DECHEMA monographs. Biodeterioration and biodegradation. VCH, Frankfurt, pp 135–142

Gu J-D, Ford TE, Mitton B, Mitchell R (2011) Microbial degradation of materials: general processes. In: Revie W (ed) The Uhlig corrosion handbook, 3rd edn. John Wiley & Sons, New York, pp 351–363

Hatzenpichler R (2012) Diversity, physiology, and niche differentiation of ammonia-oxidizing archaea. Appl Environ Microbiol 78:7501–7510

Hatzenpichler R, Lebedeva EV, Spieck E, Stoecker K, Richter A, Daims H, Wagner M (2008) A moderately thermophilic ammonia-oxidizing crenarchaeote from a hot spring. Proc Natl Acad Sci U S A 105:2134–2139

Hosono T, Uchida E, Suda C, Ueno A, Nakagawa T (2006) Salt weathering of sandstone at the Angkor monuments, Cambodia: identification of the origins of salts using sulfur and strontium isotopes. J Archaeol Sci 33:1541–1551

Hu H, Katayama Y, Kusumi A, Li SX, Wang J, de Vries RP, Gu J-D (2013) Occurrence of *Aspergillus allahabadii* on sandstone at Bayon Temple, Angkor Thom, Cambodia. Int Biodeterior Biodegradation 76:112–117

Kakakhel MA, Wu F, Gu J-D, Feng H, Shah K, Wang W (2019) Controlling biodeterioration of cultural heritage objects with biocides: a review. Int Biodeterior Biodegrad 143:104721. https://doi.org/10.1016/j.ibiod.2019.104721

Kates M (1978) The phytanyl ether-linked polar lipids and isoprenoid neutral lipids of extremely halophilic bacteria. Prog Chem Fats Other Lipids 15:301–342

Keshari N, Adhikary SP (2014) Diversity of cyanobacteria on stone monuments and building facades of India and their phylogenetic analysis. Int Biodeterior Biodegradation 90:45–51

Könneke M, Bernhard AE, José R, Walker CB, Waterbury JB, Stahl DA (2005) Isolation of an autotrophic ammonia-oxidizing marine archaeon. Nature 437:543–546

Kowalchuk GA, Stienstra AW, Heilig GHJ, Stephen JR, Woldendorp JW (2000) Molecular analysis of ammonia-oxidising bacteria in soil of successional grasslands of the Drentsche A (The Netherlands). FEMS Microbiol Ecol 31:207–215

Kusumi A, Li XS, Katayama Y (2011) Mycobacteria isolated from Angkor monument sandstones grow chemolithoautotrophically by oxidizing elemental sulfur. Front Microbiol 2:104

Kusumi A, Li X, Osuga Y, Kawashima A, Gu J-D, Nasu M, Katayama Y (2013) Bacterial communities in pigmented biofilms formed on the sandstone bas-relief walls of the Bayon Temple, Angkor Thom, Cambodia. Microbes Environ 28:422

Lan W, Li H, Wang WD, Katayama Y, Gu J-D (2010) Microbial community analysis of fresh and old microbial biofilms on Bayon temple sandstone of Angkor Thom, Cambodia. Microb Ecol 60:105–115

Laverman A, Speksnijder A, Braster M, Kowalchuk G, Verhoef H, van Verseveld H (2001) Spatiotemporal stability of an ammonia-oxidizing community in a nitrogen-saturated forest soil. Microb Ecol 42:35–45

Leininger S, Urich T, Schloter M, Schwark L, Qi J, Nicol G, Prosser J, Schuster S, Schleper C (2006) Archaea predominate among ammonia-oxidizing prokaryotes in soils. Nature 442:806–809

Li H, Mu B-Z, Jiang Y, Gu J-D (2011) Production processes affected prokaryotic *amoA* gene abundance and distribution in high-temperature petroleum reservoirs. Geomicrobiol J 28:692–704

Li X, Arai H, Shimoda I, Kuraishi H, Katayama Y (2007) Enumeration of sulfur-oxidizing microorganisms on deteriorating stone of the Angkor monuments, Cambodia. Microbes Environ 23:293–298

Li XS, Sato T, Ooiwa Y, Kusumi A, Gu J-D, Katayama Y (2010) Oxidation of elemental sulfur by *Fusarium solani* strain THIF01 harboring endobacterium *Bradyrhizobium* sp. Microb Ecol 60:96–104

Lisci M, Monte M, Pacini E (2003) Lichens and higher plants on stone: a review. Int Biodeterior Biodegradation 51:1–17

Liu J-F, Wu W-L, Yao F, Wang B, Zhang B-L, Mbadinga SM, Gu J-D, Mu B-Z (2016) A thermophilic nitrate-reducing bacterium isolated from production water of a high temperature oil reservoir and its inhibition on sulfate-reducing bacteria. Appl Environ Biotechnol 1(2):35–42. https://doi.org/10.18063/AEB.2016.02.004

Liu X, Meng H, Wang Y, Katayama Y, Gu J-D (2018a) Water is the critical factor to establishment biological and stability of Angkor temple sandstone in Southeast Asia. Int Biodeterior Biodegrad 133:9–16. https://doi.org/10.1016/j.ibiod.2018.05.011

Liu X, Li M, Castelle CJ, Probst AJ, Zhou Z, Pan J, Liu Y, Banfield JF, Gu J-D (2018b) Insights into the ecology, evolution and metabolism of the widespread Wosearchaeotal lineage. Microbiome 6:102

Liu X, Pan J, Liu Y, Li M, Gu J-D (2018c) Diversity and distribution of Archaea in global estuarine ecosystems. Sci Total Environ 637:349–358

Liu Y-F, Qi Z-Z, Shou L-B, Liu J-F, Yang S-Z, Gu J-D, Mu B-Z (2019) Anaerobic hydrocarbon degradation in candidate phylum 'Atribacteria' (JS1) inferred from genomics. ISME J 13:2377–2390. https://doi.org/10.1038/s41396-019-0448-2

Liu X, Koestler RJ, Warscheid T, Katayama Y, Gu J-D (2020) Microbial biodeterioration and sustainable conservation of stone monuments and buildings. Nature Sustainability (*accepted*)

Lu L, Han W, Zhang J, Wu Y, Wang B, Lin X, Zhu J, Cai Z, Jia Z (2012) Nitrification of archaeal ammonia oxidizers in acid soils is supported by hydrolysis of urea. ISME J 6:1978–1984

Ma Y, Zhang H, Du Y, Tian T, Xiang T, Liu X, Wu F, An L, Wang W, Gu J-D (2015) The community distribution of bacteria and fungi on ancient wall paintings of the Mogao Grottoes. Sci Rep 5:7752

Ma W, Wu F, Tian T, He D, Zhang Q, Gu J-D, Duan Y, Wang W, Feng H (2020) Fungal diversity and potential biodeterioration of mural paintings on bricks in two 1700-year-old tombs of China. Int Biodeterior Biodegrad 152:104972. https://doi.org/10.1016/j.ibiod.2020.104972

Mansch R, Bock E (1998) Biodeterioration of natural stone with special reference to nitrifying bacteria. Biodegradation 9:47–64

May E (2000) Stone biodeterioration. In: Mitchell R, McNamara CJ (eds) Cultural heritage microbiology: fundamental studies in conservation science. American Society for Microbiology, Washington, DC, pp 221–234

May E, Papida S, Abdulla H, Tayler S, Dewedar A (2000) Comparative studies of microbial communities on stone monuments in temperate and semi-arid climates. In: Ciferri O, Tiano P, Mastromei G (eds) Of microbes and art: the role of microbial communities in the degradation and protection of cultural heritage. Kluwer Academic Publishers, Dordrecht, pp 49–62

McNamara CJ, Perry TD, Bearce KA, Hernandez-Duque G, Mitchell R (2006) Epilithic and endolithic bacterial communities in limestone from a Maya archaeological site. Microb Ecol 51:51–64

Meng H, Luo L, Chan HW, Katayama Y, Gu J-D (2016) Higher diversity and abundance of ammonia-oxidizing archaea than bacteria detected at the Bayon Temple of Angkor Thom in Cambodia. Int Biodeterior Biodegradation 115:234–243

Meng H, Katayama Y, Gu J-D (2017) Wide occurrence and dominance of ammonia-oxidizing archaea than bacteria at three Angkor sandstone temples Bayon, Phnom Krom and wat Athvea in Cambodia. Int Biodeterior Biodegradation 117:78–88

Meng H, Zhang X, Katayama Y, Ge Q, Gu J-D (2020) Microbial diversity and composition of the Preah Vihear temple in Cambodia by high-throughput sequencing based on both genomic DNA and RNA. Int Biodeterior Biodegrad 149:104936

Mitchell R, Gu J-D (2000) Changes in the biofilm microflora of limestone caused by atmospheric pollutants. Int Biodeterior Biodegrad 46:299–303

Motti R, Stinca A (2011) Analysis of the biodeteriogenic vascular flora at the Royal Palace of Portici in southern Italy. Int Biodeterior Biodegradation 65:1256–1265

Nakagawa T, Mori K, Kato C, Takahashi R, Tokuyama T (2007) Distribution of cold-adapted ammonia-oxidizing microorganisms in the deep-ocean of the northeastern Japan Sea. Microbes Environ 22:365–372

Nicol GW, Schleper C (2006) Ammonia-oxidising Crenarchaeota: important players in the nitrogen cycle? Trends Microbiol 14:207–212

Nicol GW, Leininger S, Schleper C, Prosser JI (2008) The influence of soil pH on the diversity, abundance and transcriptional activity of ammonia oxidizing archaea and bacteria. Environ Microbiol 10:2966–2978

Onodera Y, Nakagawa T, Takahashi R, Tokuyama T (2010) Seasonal change in vertical distribution of ammonia-oxidizing archaea and bacteria and their nitrification in temperate forest soil. Microbes Environ 25:28–35

Papida S, Murphy W, May E (2000) Enhancement of physical weathering of building stone by microbial population. Int Biodeterior Biodegradation 64:305–317

Perry IV TD, McNamara CJ, Mitchell R (2005) Biodeterioration of stone, scientific examination of art: modern techniques in conservation and analysis. Sackler National Academy of Sciences Colloquium, pp 72–86

Piñar G, Ripka K, Weber J, Sterflinger K (2009) The microbiota of a sub-surface monument the medieval chapel of St. Virgil (Vienna, Austria). Int Biodeterior Biodegradation 63:851–859

Rölleke S, Witte A, Wanner G, Lubitz W (1998) Medieval wall paintings-a habitat for archaea: identification of archaea by denaturing gradient gel electrophoresis (DGGE) of PCR-amplified gene fragments coding for 16S rRNA in a medieval wall painting. Int Biodeterior Biodegradation 41:85–92

Saiz-Jimenez C, Laiz L (2000) Occurrence of halotolerant/halophilic bacterial communities in deteriorated monuments. Int Biodeterior Biodegradation 46:319–326

Sand W, Bock E (1991) Biodeterioration of mineral materials by microorganisms – biogenic sulfuric and nitric acid corrosion of concrete and natural stone. Geomicrobiol J 9:129–138

Sterflinger K (2000) Fungi as geologic agents. Geomicrobiol J 17:97–124

Sterflinger K, Pinar G (2013) Microbial deterioration of cultural heritage and works of art – tilting at windmills? Appl Microbiol Biotechnol 97:9637–9646

Sterflinger K, Prillinger H (2001) Molecular taxonomy and biodiversity of rock fungal communities in an urban environment (Vienna, Austria). Anton Leeuw 80:275–286

Sterflinger K, Little B, Pinar G, Pinzari F, de los Rios A, Gu J-D (2018) Future directions and challenges in biodeterioration research on historic materials and cultural properties. Int Biodeterior Biodegradation 129:10–12

Tayler S, May E (2000) Investigations of the localisation of bacterial activity on sandstone from ancient monuments. Int Biodeterior Biodegradation 46:327–333

Tourna M, Freitag TE, Nicol GW, Prosser JI (2008) Growth, activity and temperature responses of ammonia-oxidizing archaea and bacteria in soil microcosms. Environ Microbiol 10:1357–1364

Tourna M, Stieglmeier M, Spang A, Könneke M, Schintlmeister A, Urich T, Engel M, Schloter M, Wagner M, Richter A (2011) *Nitrososphaera viennensis*, an ammonia oxidizing archaeon from soil. Proc Natl Acad Sci U S A 108:8420–8425

Uchida E, Ogawa Y, Maeda N, Nakagawa T (2000) Deterioration of stone materials in the Angkor monuments, Cambodia. Eng Geol 55:101–112

Videla HA, Guiamet PS, Gomez de Saravia S (2000) Biodeterioration of Mayan archaeological sites in the Yucatan peninsula, Mexico. Int Biodeterior Biodegradation 46:335–341

Walker C, de La Torre J, Klotz M, Urakawa H, Pinel N, Arp D, Brochier-Armanet C, Chain P, Chan P, Gollabgir A (2010) *Nitrosopumilus maritimus* genome reveals unique mechanisms for nitrification and autotrophy in globally distributed marine crenarchaea. Proc Natl Acad Sci U S A 107:8818–8823

Wang YF, Gu J-D (2013) Higher diversity of ammonia/ammonium-oxidizing prokaryotes in constructed freshwater wetland than natural coastal marine wetland. Appl Microbiol Biotechnol 97:7015–7033

Warscheid T, Braams J (2000) Biodeterioration of stone: a review. Int Biodeterior Biodegrad 46:343–368

Woese CR (1978) Bacterial evolution. Microbiol Rev 51:221–271

Wu F, Duan Y, Wang W, He D, Gu J-D, Feng H, Chen T, Liu G, An L (2017) The microbial community characteristics and biodeterioration assessment of ancient wall paintings in Maijishan Grottoes, China. PLoS One 5:12, e0179718

Wuchter C, Abbas B, Coolen MJL, Herfort L, van Bleijswijk J, Timmers P, Strous M, Teira E, Herndl GJ, Middelburg JJ, Schouten S, Damste JSS (2006) Archaeal nitrification in the ocean. Proc Natl Acad Sci U S A 103:12317–12322

Xu H-B, Tsukuda M, Takahara Y, Sato T, Gu J-D, Katayama Y (2018) Lithoautotrophical oxidation of elemental sulfur by fungi including *Fusarium solani* isolated from sandstone Angkor temples. Int Biodeterior Biodegrad 126:95–102. https://doi.org/10.1016/j.ibiod.2017.10.005

Zhang G, Gong C, Gu J, Katamaya Y, Ji-Dong G (2019) Biodeterioration and the mechanisms involved of sandstone monuments of World Cultural Heritage sites in tropical regions. Int Biodeter Biodegr 143:104723

Zhang X, Ge Q, Zhu X, Deng M, Gu J-D (2018) Microbiological community analysis of the Royal Palace in Angkor Thom and Beng Mealea of Cambodia by Illumina sequencing based on 16S rRNA gene. Int Biodeterior Biodegradation 134:127–135

Zhou Z, Liu Y, Li M, Gu J-D (2018) Two or three domains: a new view of the tree of life in the genomics era. Appl Microbiol Biotechnol 102:3049–3058

Zhou Z, Liu Y, Lloyd KG, Pan J, Yang Y, Gu J-D, Li M (2019) Genomic and transcriptomic insights into the ecology and metabolism of benthic archael cosmopolitan, *Thermoprofundales* (MBG-D archaea). ISME J 13:885–901

# Chapter 3
# Molecular-Based Techniques for the Study of Microbial Communities in Artworks

**Katja Sterflinger and Guadalupe Piñar**

**Abstract** Thanks to the revolutionary invention of the polymerase chain reaction and the sequencing of DNA and RNA by means of "Sanger sequencing" in the 1970th and 1980th, it became possible to detect microorganisms in art and cultural assets that do not grow on culture media or that are non-viable. The following generation of sequencing systems (next generation sequencing, NGS) already allowed the detection of microbial communities on objects without the intermediate step of cloning, but still most of the NGS technologies used for the study of microbial communities in objects of art rely on "target sequencing" linked to the selectivity of the primers used for amplification. Today, with the third generation of sequencing technology, whole genome and metagenome sequencing is possible, allowing the detection of taxonomic units of all domains and kingdoms as well as functional genes in the produced metagenome. Currently, Nanopore sequencing technology is a good, affordable, and simple way to characterize microbial communities, especially in the field of Heritage Science. It also has the advantage that a bioinformatic analysis can be performed automatically. In addition to genomics and metagenomics, other "-omics" techniques such as transcriptomics, proteomics, and metabolomics have a great potential for the study of processes in art and cultural heritage, but are still in their infancy as far as their application in this field is concerned.

**Keywords** Biodeterioration · Cultural heritage · Microbiology · Next generation sequencing · Metagenomics · Microbial pedigree · -Omic technologies · Nanopore sequencing

K. Sterflinger (✉) · G. Piñar
Institute of Natural Sciences and Technology in the Arts, Academy of Fine Arts Vienna, Vienna, Austria
e-mail: k.sterflinger@akbild.ac.at; g.pinarlarrubia@akbild.ac.at

© The Author(s) 2021
E. Joseph (ed.), *Microorganisms in the Deterioration and Preservation of Cultural Heritage*, https://doi.org/10.1007/978-3-030-69411-1_3

59

# 1    Introduction

Works of art, including modern and historical paintings, murals, sculptures made of various materials, architectural surfaces, books, and depot material are colonized and destroyed by microorganisms under suitable conditions (Koestler et al. 2003). Object surfaces in combination with their corresponding environment are in most cases an extreme habitat for microorganisms: In the open air, the microbial community is influenced by UV radiation, changing humidity, and dehydration as well as nutrient deficiency. Indoors, the main problem for the organisms is the low availability of water. On wall surfaces, osmotic stress is caused by salt contamination, and layers of paint can even be toxic due to lead or zinc additives. Accordingly, works of art contain microbial communities that can be classified as stress-tolerant— extremotolerant—or even stress-loving—extremophilic. Thus, osmotic sites are often dominated by halophilic bacteria and archaea, which are recognizable as an ecological group by their characteristic pink coloration, but for the most part are unable to grow on culture media (Ettenauer et al. 2014); halotolerant and xerotolerant fungi also occur and are often overgrown by rapidly growing transients on culture media and thus not noticed (Sterflinger et al. 2018a). Contemporary materials, e.g. PVC, PE, PLA are colonized and decomposed by highly specialized bacteria and fungi, which in turn do not grow easily on ordinary laboratory media (Fig. 3.1).

Extremotolerant and extremophilic organisms have very specific growth requirements, which can often only be simulated to a limited extent or not at all in the laboratory and with conventional culture media. What has long been known and accepted in general environmental microbiology, therefore, also applies to microorganisms in artifacts and cultural objects: only a small part of the microorganisms can be detected by sampling and enrichment on culture media, the majority of the microflora remains undetected with these classical microbiological enrichment techniques (Amann et al. 1995). Accordingly, until a few decades ago, information on

**Fig. 3.1** Seiko Mikami, The World Memorable: Suitcases (1993); "Radioactive Waste Bags," all the suitcase shows growth of the xerophilic fungus *Aspergillus restrictus*, which is often overseen in culture media approaches due to its slow growth (the installation consists of three suitcase "Suitcase Accident Air Noxious Particles Waste Containers" and "Biohazard Autoclave Bags" that are not shown here)

microbial biodiversity was limited, but also the knowledge about the ecology of organisms and the processes of damage they cause.

Only the routine use of PCR in the 1980s and 1990s made it possible to use microbial DNA for research on biogenic damage processes on art and cultural assets in order to overcome the hurdle of cultivation and enrichment and to complete and expand the data obtained from this. Initially, in addition to cloning and Sanger sequencing (which is considered a first-generation sequencing technology developed by Frederick Sanger and colleagues in 1977), DNA-based fingerprinting techniques were successfully used to study the biological agents that cause the biodeterioration of cultural artifacts.

Denaturing gradient gel electrophoresis (DGGE) in combination with clone libraries and first-generation sequencing of amplified DNA fragments has been successfully used to characterize microbial communities on walls and murals, paper, parchment, glass, stone, textiles, human remains, and other materials of cultural heritage (Rölleke et al. 1999; Schabereiter-Gurtner et al. 2001, 2002a, b; Laiz et al. 2003; Di Bonaventura et al. 2003; Saiz-Jimenez 2003). The application of this molecular strategy, together with imaging and chemical analyses (Pinzari et al. 2010; Wiesinger 2018), has provided relevant information on the microbial communities associated with different materials and the microbiological risk they present on the surface of the substrates studied (Sterflinger and Piñar 2013). In addition, such studies have enabled the implementation of preservative and disinfectant treatments as well as the monitoring during and after the conservation treatment of valuable objects (Jroundi et al. 2015).

Although first-generation sequencing technology provided only limited data, the first studies opened a new door into the world of biodiversity of artifacts and cultural assets and into the phylogeny of the organisms involved. These studies also led to the description of new species involved in biodeterioration (Sterflinger et al. 1997; Wieser et al. 2005). The studies were also an important basis—and justification— for the further development and application of technologies in the field of Heritage Science. Nevertheless, the work based on DNA-fingerprints, clone libraries, and first-generation sequencing was rather time-consuming and not free of errors, since the selection of suitable primers for DNA amplification already resulted in a strong selection of the initial data. Since Sanger sequencing was only able to sequence short parts of the total DNA contained in a cell in a reasonable time and with reasonable effort, the studies were limited to taxonomically and phylogenetically relevant DNA sections such as 16S-, 18S-rDNA, and ITS sequences. This provided information on biodiversity, but not on the potential of functional genes in potentially harmful organisms. To establish a link between the identity of an organism and its function was, once again, only possible by means of experiments on pure cultures, if available.

## 2   High-Throughput Sequencing Methods: Next Generation Sequencing (NGS)

*"The commercial launch of the first massively parallel pyrosequencing platform in 2005 ushered in the new era of high-throughput genomic analysis now referred to as next-generation sequencing (NGS)"* (Völkerding et al. 2009). Although these technologies generate shorter reads (between 25 and 500 bp), they are capable of delivering many hundreds of thousands or millions of reads in a relatively short time (a few hours). This leads to a high coverage, but the assembly process is much more computationally intensive. These technologies are far superior to Sanger sequencing due to the high data volume and the relatively short time required to sequence an entire genome (Metzker 2010). In recent years, the so-called next generation sequencing technologies (NGS) have been developed and have been a revolutionary innovation for deciphering the complexity of genomes and metagenomes. NGS methods have enabled significant advances, such as the preparation of NGS libraries in a cell-free system without cloning; the thousands to many millions of sequencing reactions produced in parallel; and direct sequencing output that is performed cyclically and in parallel. The large number of reads generated by NGS enables the sequencing of entire genomes and metagenomes at an unprecedented speed. Today, there are several sequencing platforms based on different technologies (Goodwin et al. 2016). Therefore, the choice of the sequencing platform to be used for a particular study depends on the requirements and resources, which vary depending on the type of sample, study finality, cost/runtime, data output, tolerated amplicon size, data storage capacities, and error rates.

Different methods available on the market until 2009 are explained in detail by Völkerding et al. (2009); the most relevant ones are only briefly described here:

– The Roche 454 technology was derived from the technological convergence of pyrosequencing and emulsion PCR. The system has been efficiently used both for sequencing the entire genome and for amplicon sequencing in environmental and medical studies. However, the technology was no longer competitive and was discontinued in 2013.
– Illumina/Solexa: In 2006, the Solexa Genome Analyzer, the first "short-read" sequencing platform, was commercially launched. Acquired by Illumina in 2006, the Genome Analyzer uses a flow-through cell consisting of an optically transparent slide with individual lanes on whose surface oligonucleotide anchors are bound. The template DNA is fragmented into lengths of several hundred base pairs and finally repaired to produce 5'-phosphorylated blunt ends. The polymerase activity of the Klenow fragment is used to attach a single A base to the 3' end of the blunt phosphorylated DNA fragments. This addition prepares the DNA fragments for ligation to oligonucleotide adapters that have a single T-base overhang at their 3' end to increase ligation efficiency. The adapter oligonucleotides are complementary to the flow cell anchors. Under boundary dilution conditions, adapter-modified single-stranded template DNA is added to the

flow cell and immobilized on the anchors by hybridization. Currently, Illumina is one of the most widely used sequencing systems both in science and clinical applications.

- The SOLiD (Supported Oligonucleotide Ligation and Detection) System 2.0 platform, marketed by Applied Biosystems, is a ligation-based short-read sequencing technology. Applied Biosystems has refined the technology and in 2007 launched the SOLiD instrumentation. Sample preparation is similar to the 454 technology in that DNA fragments are ligated to oligonucleotide adapters, bound to beads, and clonally amplified using emulsion PCR.

Recent technologies developed during the last decade are:

- The next generation sequencing technology of Ion Torrent (ThermoFisher) is not based on an optical detection system. The technology was first published in Nature in 2011 (Rothberg et al. 2011) and was launched on the market in 2012. It exploits the fact that the addition of a dNTP to a DNA polymer releases a hydrogen ion. The pH change resulting from these hydrogen ions is measured using semiconductors, with the simultaneous measurement of millions of such changes determining the sequence of each fragment.
- Single Molecule, Real-Time (SMRT) sequencing (Pacific Biosciences) is the core technology that enables sequencing long reads with resulting reads of several tens of kilobases in length. Due to the long fragments, the technology is particularly useful for de novo sequencing and the assembly of entire genomes (Eid et al. 2009).
- Nanopore sequencing technology (Oxford Nanopore Technologies) is based on reading the inserted nucleotides while the DNA sequence passes through a nanopore with an inner diameter of 1 nm (Branton et al. 2008). The electrical conductivity of the protein nanopore is changed as the DNA passes through, and a signal is detected in this way. Sequencing with this innovative technology does not necessarily require intermediate PCR amplification or chemical labeling. In addition, this technology offers a pocket-sized sensor device (Fig. 3.2), the MinION, which was made commercially available in 2015 and whose performance has been further optimized in recent years, resulting in an increasing number of publications in the field of metagenomics.

# 3   Metagenomics in Cultural Heritage

Metagenomics can be performed by the "Sequence it all" approach (simple reading of all DNA sequences present in a sample) or by the "Target sequencing" approach, concentrating on certain conserved sequences, such as only ribosomal RNA genes. The latter procedure has several advantages, such as reducing the complexity of the data obtained and the possibility of assigning more sequences to a particular organism or a group of related organisms. This approach facilitates some

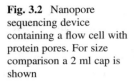

**Fig. 3.2** Nanopore sequencing device containing a flow cell with protein pores. For size comparison a 2 ml cap is shown

semi-quantitative analyses, which is much more difficult in the analysis of all sequences composed of many organisms with largely different genome sizes.

The 16S rRNA gene and ITS regions are the most frequently sequenced regions when using NGS for the examination of art objects (Piñar et al. 2019). This approach is useful in the field of cultural heritage, as DNA of poor quality or low concentration can be amplified by using degenerate primers and PCR. Therefore, metagenomics using NGS analysis has been widely used to analyze the microbial colonization of ancient monuments and valuable objects. Several publications refer to NGS analyses used for biological diagnosis of cultural heritage. Some examples focus on stone monuments (Cutler et al. 2013), murals (Rosado et al. 2014), construction materials of historic buildings (Adamiak et al. 2018), and tombs (Huang et al. 2017). These methods are powerful, but as with any PCR-based method, the primers selection and/or the exponential amplification will distort the results. Furthermore, the quantification of the relative abundance of all different domains of life present in a sample is not possible.

The technological development of molecular methods in recent years has overcome many technical obstacles that have existed up to now. Third-generation sequencing technologies are still under development, but already offer some advantages over second-generation sequencing platforms. The third generation of high-throughput sequencing technologies offers an intelligent solution to the limitations of previous technologies (Schadt et al. 2010). Instead of sequencing a clonally amplified template, a single DNA template is sequenced, leading to a reduction in the use of reagents and chemicals as well as to simplified protocols and miniaturization of the entire technical process and equipment.

Although NGS technologies are not entirely without flaws and limitations (Sterflinger et al. 2018b), the amazing development and establishment of NGS techniques in the field of cultural heritage offers a powerful DNA-based approach for the analysis of valuable artifacts. In addition, the generated data provide information to answer many interesting questions that can arise when handling valuable art objects and might help to:

– provide insight into the selection of materials at the time of manufacture, such as the composition of the materials,
– gain knowledge about the storage conditions of materials and their geographical origin,
– trace the history of use of the object, and
– group objects of unknown origin on the basis of similarities in their microbiomes.

The information enables to understand many open questions in a variety of fields, such as archeology, history, restoration, philology, or criminology, but also can contribute to give a historical benefit to the investigated objects (Piñar et al. 2019).

As described above, there are different sequencing platforms based on different technologies. All of them present advantages and disadvantages; therefore, the choice of the sequencing platform should be decided carefully depending on the purpose of the study. Also, the bioinformatics tools and pipelines necessary for data analysis are manifold and have to be chosen according to the respective data sets and the scientific question to be answered. In recent years, different methods have been tested, their protocols optimized and used to perform metagenomic analyses on artworks at the first time. The following three case studies describe the application of three different sequencing technologies for the study of cultural assets, but with different finalities: (1) Ion Torrent technology, (2) Illumina technology, and (3) Nanopore technology.

(1) Ion Torrent technology. In a metagenome study performed on three ancient marble statues, the main question was whether they might have been smuggled and/or stored together and whether biological traces could help to elucidate their geographical origin (Piñar et al. 2019). Two statues were human torsi, one female and one male, and the third one was a representation of a small girl's head (Fig. 3.3). The finality of this study was to collect as much biological information as possible from the surface of the statues to answer the aforementioned questions. To this end, metagenomic analyses were performed using the Ion Torrent sequencing technology by means of the Ion Personal Genome Machine (PGM), which allows a very good target barcoding sequencing of different life domains. Especially for the screening of prokaryotes, this technology offers a very smart solution thanks to the commercialization of the Ion 16S rDNA Metagenomics Kit that enables the simultaneous targeting of seven of the nine variable regions of the 16S rDNA. This eliminates the bias committed by other NGS technologies that tend to focus only on the V4 region of the 16S rDNA, which even if the most widely used does not cover all phylogenetic groups (Piñar et al. 2019). To approach this study, the deposits and dirt visible in the spaces (e.g., folds, curls of hair) defined by the shape of the sculptures were removed with the help of sterile scalpels and needles without damaging the rock itself. The material contained was mainly powder, dust, dirt, and even some textile fibers.

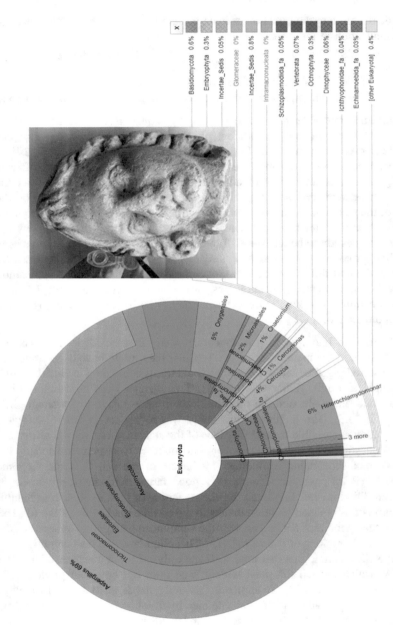

**Fig. 3.3** Krona charts (Ondov et al. 2011) displaying the relative abundance of Eukaryota on the small marble head; data were generated by Ion Torrent sequencing technology

The samples taken from the same statue were later pooled for DNA extraction analysis to perform a single DNA extraction per statue. Since the DNA quantities extracted from the samples were not high enough to directly produce a library (the Ion Torrent technology requires a quantity of 10 ng/μl DNA as input), a genetic strategy, comprising analyses of the 16S ribosomal DNA (rDNA) of prokaryotes (Ion 16S Metagenomics Kit: primer set V2-4-8 and primer set V3-6,7-9), the 18S rDNA of eukaryotes (targeting the V4 region with primers 528F/706R) as well as the internal transcribed spacer regions of fungi (targeting the ITS1 region) was performed. The resulting DNA libraries were barcoded and pooled onto two DNA chips, whose sequencing runs provided an average of more than 4,250,000 total readings with an average unique reading length of 200–300 bp. It is important to note that in the case of the Ion Torren system, the output is a FastQ file and the sequence processing and data mining must be carried out by a bioinformatician. In this study, the raw data were trimmed, filtered, and compared with different databases using a variety of bioinformatics pipelines. After trimming (Martin 2011) only reads longer than 100 nt and containing at least one primer were retained. The prokaryotic reads were initially grouped based on their hypervariable region of origin (HVR). The HVR was derived by a cmscan (Nawrocki and Eddy 2013) with the prokaryotic 16S rRNA model of Rfam (Kalvari et al. 2018) at each read and the start and end position of agreement was compared with the correlating HVR position in the Rfam model. Measured values that were mapped outside the HVR model were discarded, while the other measured values were cut and filtered with Cutadapt similar to the eukaryotic and fungal measured values. The taxonomy was assigned with DADA based on the protocol published in Callahan et al. (2016). For the prokaryotic values, the protocol was applied to each group of IHR separately and to all HVR simultaneously. The results of this study showed to be suitable to draw some conclusions about a possible relationship between the two marble torsi; their microbiomes showed similarities involving many soil-dwelling organisms, which may indicate storage or burial in agricultural land or even pasture soil. For the male torso, it was even possible to deduce a geographical origin, as DNA traces of *Taiwania*, a tree of the cypress family found only in Taiwan and the south of the PRC, were present. In contrast, the statue of the young girl's head showed a more specific microbiome with an abundance of organisms of marine origin, which gives rise to the suspicion that this statue was in contact with the sea. Although the amount of data derived from this method is very large, one must be aware that the amplification step at the beginning of the procedure, targeting specific DNA regions, is a possible bias of the method as it is selective both in terms of taxonomic units and their quantification. Nevertheless, in this study and in the case of the analysis of prokaryotes, this bias was minimized by targeting most of the variable regions of the 16S rDNA simultaneously.

(2) Illumina Technology: Another important application of molecular analysis in cultural heritage research is biocodicology (Fiddyment et al. 2019). A study carried out by Piñar et al. on ancient parchment writings provides an example of

the emerging field of research in biocodicology and shows how metagenomics can help to answer relevant questions that can contribute to a better understanding of the history of ancient manuscripts (Miklas et al. 2019; Piñar et al. 2020c). For this purpose, two Slavonic codices from the eleventh century were investigated by shotgun metagenomics using the Illumina sequencing platform.

The analysis of nucleic acids contained in historical parchments is a difficult matter due to the poor preservation of ancient DNA (aDNA) and the frequent external contamination with modern DNA (Vuissoz et al. 2007). A crucial step in such studies is the extraction of DNA from very little sample material (a few fibers or 1 to 2 millimeters of material) and even from eraser fragments loosened after rubbing the parchment surface (Fiddyment et al. 2019). In the case-study presented here, a combination of two commercial kits was used for the successful extraction of total DNA (ancient and modern), which aim at a silica membrane-based purification of very short DNA fragments (~ 70 bp), i.e. the expected size of the ancient DNA. This approach further enables the elution in very small volumes allowing the concentration of very low DNA yield, as it was the case of the DNA extracted from these ancient objects. The DNA extracts were used to create DNA libraries with the NEBNext Ultra II DNA Library Prep Kit for Illumina and sequenced by means of the Illumina HiSeq 2500 platform using 100 bp single reads. The total number of reads processed ranged from 43,150,589 to 50,416,907, with reads being significantly shorter than when using Ion Torrent or Nanopore. However, the strategy followed in this study showed to be adequate for analyzing the ancient DNA of the parchments. The application of Illumina shotgun sequencing proved to be very effective for the sequencing of short DNA fragments (aDNA) with a high sensitivity, generating a huge amount of data and being less susceptible to chimeric artifacts that can result from PCR-based analyses. Quality control of the reads was done with fastqc (Andrews 2010). Cutadapt was used to trim adapter sequences (Martin 2011). FastQ Screen was used to assess the origin of the parchment probes (Wingett and Andrews 2018). The parchment metagenomes were assessed by first removing human and animal reads for each dataset. This was done for each experiment by filtering out the reads to the human and corresponding animal genomes with STAR (Dobin et al. 2013). The remaining reads were then processed with MetaphlAn2 in order to describe and represent the microbiomes of the parchment probes (Truong et al. 2015). The endogenous DNA enabled to infer the animal origin of the skins used in the production of the two codices, while the DNA obtained from viruses provided insights into the plant origin of the inks used in one of the two codices. In addition, the microbiomes colonizing the surface of the parchments helped to determine their state of preservation and their latent risk of deterioration. The advantage of using this molecular strategy was the detection of extremely small DNA fragments that originate from already strongly degraded ancient DNA, which could not be detected by technologies with longer reading lengths, such as Ion Torrent or Nanopore. Similar to the Nanopore amplification of the entire genome described below, shotgun sequencing enables to obtain the entire genomic information in a given sample: taxonomic tracers from all areas and realms of life as well as functional genes.

(3) Nanopore sequencing technology: Nanopore has been little used in the field of cultural heritage and only two published studies have reported on the benefits of applying this state-of-the-art technology to valuable works of art. In the first study (Šoltys et al. 2019), Nanopore sequencing technology was used to study the microbiome of an eighteenth century wax seal stained with minium. The authors reported on the advantages of sequencing long DNA fragments, which allows sequencing of the almost complete 16S rRNA gene as well as relatively long fragments (more than 600 bp) from the fungal ITS regions and 28S rRNA. However, the authors used target sequencing technology that did not allow for the relative proportion of different phylogenetic groups. In the second study (Piñar et al. 2020a), Nanopore sequencing technology was used for the first time together with a whole genome amplification (WGA) protocol to provide a rapid diagnosis of the biological infection in cultural artifacts. All advantages that this new technology can offer for metagenomic analyses starting from very low DNA concentrations were combined. To test the feasibility of this protocol on cultural assets, samples were taken from two eighteenth and nineteenth century oil paintings on canvas, one of which showed active fungal colonization, while the other had a cracked varnish surface (Fig. 3.4). The total number of reads obtained with Nanopore exceeded the 2,000,000 reads with an average read length of more than 4 Kb in both analyzed paintings. The resulting fast 5 data files were basecalled using the Nanopore GPU basecalling with GUPPY 3.0.3 on UBUNTU 16.04 (Nanopore Community Platform). Once the Fastq files were generated, the data were compared with databases using one of the available pipelines for data analyses of the Nanopore Community Platform, following the steps recommended by manufacturers. The selected workflow chosen was "What's in my pot" (WIMP). It is important to note that this platform offers a series of bioinformatics pipelines that are accessible to the end user and are user-friendly, allowing basic data analysis without the need for a bioinformatics expert. The results obtained with the WGA metagenomic approach proved to be successful for a rapid diagnosis of the microbial colonization of the two tested paintings showing different deterioration phenomena. Their microbiomes showed to be directly related to the conservation status of the paintings as well as with their restoration history (Piñar et al. 2020a).

Until now, the power of metagenomics has been largely tied to expensive and room-sized sequencing technologies, a large number of laboratory ancillary equipment and time-consuming workflows, making analysis far too complex and expensive for routine analysis by museums, collections, and restorers. Nanopore technology opens up new possibilities in this area of research and applied restoration:

– The basic equipment for this technology is the pocket-sized sequencer, combined with a laptop.

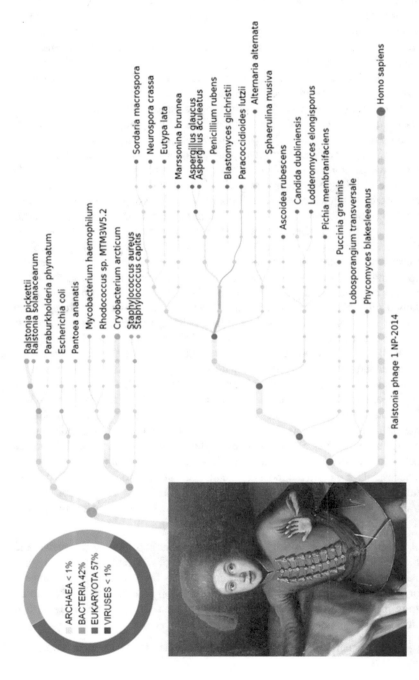

**Fig. 3.4** Part of the phylogenetic tree created by the Oxford Nanopore software "What is in my pot" after WGS from the oil painting "Boy holding a dog" shown on the left

- The protocol, which uses whole genome amplification (WGA), is very efficient and allows the amplification of all phylogenetic groups at once, thus allowing the real metagenome and the relative proportions of the different phylogenetic kingdoms to be derived.
- As mentioned above, for data processing and the creation of phylogenetic trees—including a quantitative representation of these—from the derived data, the company offers a toolbox that is open to all customers of the system. This enables the analysis and interpretation of the data without the involvement of a bioinformatician. However, it should be mentioned here that this only applies to the analysis of taxonomic units. If a targeted search for functional genes or other data mining is to be carried out, a bioinformatics approach is also required here.

As with any other sequencing technology, the user of Nanopore needs to have a background in microbial ecology and the materials under investigation in order to interpret the data correctly and summarize the microbial family tree for each individual sample as a reference data set for current and future comparisons.

In summary, the Nanopore sequencing technology offers several advantages in the field of cultural heritage compared to other NGS technologies, such as the length of the generated sequences, the taxonomic diversity reflecting the real proportions of all domains of life (Piñar et al. 2020a), the acceleration and simplification of the technical process, and the reduction of costs. In particular, the MinION device offers the possibility to perform *in situ* analyses. Instead of collecting samples and sequencing them in the laboratory, the miniature sequencers can be taken into the field and sequencing can be performed on site (Johnson et al. 2017). This last point can of course offer some additional advantages for practical reasons in cultural heritage research, such as *in situ* analysis in museums and depots, at archeological sites or in church interiors, if required. The Nanopore MinION technology offers a fast and affordable way to create microbial pedigrees of artifacts and is now available as a service for museums, restorers, art collections, auction houses, and even criminologists.

Based on the proven success in applying the Nanopore sequencing technology to valuable objects of cultural importance, a project has been launched to study the microbiomes of some of Leonardo Da Vinci's most emblematic drawings, as an example the Codex of the Flight (dated 1505-6). Depending on the available material, samples were represented by swabs, membranes, small particles of glues or impurities removed by conservators from the surfaces. The aim was to generate a specific microbiome from each investigated drawing in order to create a bio-archive to monitor the state of preservation of the drawings and thus be able to use them for possible comparisons in the future (Piñar et al. 2020b, in press).

To give the reader an impression of the steps required for WGA metagenome Nanopore sequencing, we have summarized them in Table 3.1. Details can be found from point 2 onwards (whole genome amplification) in the working instructions of the provider.

**Table 3.1** Basic protocol for whole genome amplification (WGA) and sequencing on objects of cultural heritage using the Nanopore technology[a]

| |
| --- |
| **1. DNA extraction** |
| –Extract DNA from samples (e.g., swabs, membranes, dust) using the FastDNA SPIN Kit for soil (MP Biomedicals, Illkrich, France) as recommended by the manufacturers. |
| –The DNA extracted from each single swab was pooled per sample to obtain a single microbiome from each object. |
| –Assess DNA concentrations by using the Qubit 2.0 fluorometer (Invitrogen Corporation), with the Qubit dsDNA HS Assay Kit. |
| **2. Whole Genome Amplification** |
| –Use the REPLI-g Midi Kit (Qiagen) using the innovative Multiple Displacement Amplification (MDA) technology. Perform reactions in a BioRad C1000 Thermal Cycler as follows: |
| –Load 5 µl template DNA (template DNA was >10 pg) and 5 µl Reconstituted Buffer DLB into a microcentrifuge tube and mix it by vortexing. |
| –Incubate the reaction tubes at room temperature (15–25 °C) for 3 min. |
| –Add 10 µl Stop solution to each reaction tube and mix by vortexing, resulting in 20 µl denatured DNA. |
| –Prepare a 30 µl master mix, containing per reaction 29 µl REPLI-g Midi Reaction Buffer and 1 µl REPLI-g Midi DNA polymerase, and add this to the 20 µl denatured DNA prepared in the previous step. |
| –Incubate the reaction tubes at 30 °C for 16 h. |
| –Finally inactivate the REPLI-g Midi DNA Polymerase by heating the samples at 65 °C for 3 min. |
| –After the WGA, assess the concentration of the amplified DNA with the Qubit 2.0 Fluorometer (Invitrogen Corporation) using the Qubit dsDNA BR Assay Kit. |
| **3. Library construction and quantification, template preparation and sequencing** |
| –Construct DNA libraries following the "Premium whole genome amplification protocol" available in the Oxford Nanopore community using the Ligation Sequencing kit 1D SQK-LSK109 and the Flow cell Priming kit EXP-FLP001 (Oxford Nanopore Technologies). All steps performed for library preparation comprise: (a) the digestion of the amplified DNA with T7 Endonuclease I (New England Biolabs) to remove DNA branching; (b) the preparation of the DNA ends for adapter attachment, using the NEBNext FFPE DNA repair mix (M6630) and the NEBNext End repair/dA-tailing Module (E7546) (New England Biolabs), and (c) the attachment of sequencing adapters supplied in the kit to the DNA ends, by using the NEBNext Quick Ligation Module (E6056) (New England Biolabs), were performed following the specifications of the manufacturers. |
| –After each step of library preparation, perform a clean-up step, using the AMPure XP beads (Agentcourt, Beckman) following the recommendations of the protocol. |
| –The DNA concentration was measured with the Qubit 2.0 Fluorometer (Invitrogen Corporation) using the Qubit dsDNA BR Assay Kit following each clean up step to assure the quality of the library. |
| –Once the library is finished, perform a quality control of the flow cell (SpotOn Flow cell Mk I R9 Version, FLO-MIN 106D) prior to starting the sequencing. The MinKNOW™ software was used to check the number of active pores in the flow cell. |
| –Finally, do the priming and the loading of the DNA library into the flow cell following exactly the recommendations of the manual supplied by the manufacturers! |
| –Start the sequencing run by connecting the Nanopore device (MinION) to a portable computer. |
| –Start the software MinKNOW completing the steps recommended in the protocol, which comprises the selection of the local MinION, entering the Sample ID and the Flow cell ID to be used, and finally, the selection of the appropriate protocol script, allowing the script to run to completion. Runs were performed for 48 h. |

[a]The reader is kindly invited to follow the protocol given here. Please cite the following publication if you use the protocol for your scientific work or services: Piñar G, Poyntner C, Lopandic K,

(continued)

Tafer H, Sterflinger K. 2020. Rapid diagnosis of biological colonization in cultural artefacts using the MinION nanopore sequencing technology. International Biodeterioration and Biodegradation 148: 104908

## 4 Transcriptomics and More -Omic and Meta-Omic Techniques for the Analysis of Cultural Heritage

The DNA-based metagenomics and diversity studies described above have in common that they do not allow conclusions to be drawn about gene expression, so that the potential and actually expressed microbial functions cannot be identified. However, understanding the relationship between microbial diversity and functional processes is a fundamental question. Nowadays, there is a wealth of information on microbial diversity on cultural heritage objects (Sterflinger and Piñar 2013), which often reveal highly specialized microorganisms that can thrive in extreme environments. Nevertheless, the explicit functional and ecological role of individual taxa remains uncertain. Therefore, although the fundamental question: "Who is out there?" is already answered in part, the question "What do organisms do on and with the object?" is still largely unanswered.

The latter question could be answered mainly by using transcriptomics and metatranscriptomics. For the analysis of the transcriptome of a strain or a mixed microbial community, total ribosomal RNA-depleted or poly(A)+ RNA is isolated and converted into cDNA. A typical protocol involves the generation of cDNA of the first strand by random hexamer-primed reverse transcription and subsequent generation of cDNA of the second strand using RNase H and DNA polymerase. The cDNA is then fragmented and ligated to NGS adapters (McGettigan 2013).

In principle, this technology, if applied to whole communities, could be used to clarify what the microorganisms on the object do physiologically under certain conditions—such as temperature, humidity, exposure to light and weather—and whether they are capable of degrading a certain component of the object's material. This would be an absolute breakthrough on the way to the development of suitable measures that "inactivate" the potentially harmful microflora by controlling the environmental parameters without the need for toxic substances or expensive cleaning measures. The biggest challenge at present, however, is the quantity of sample material: the non-destructive methods used—simple removal using swabs or membranes—do not yet allow sufficient amounts of RNA to be obtained for existing technologies. There is a need for development here.

However, transcriptomic studies of individual biodeteriorative strains incubated under defined environmental conditions in the laboratory can provide initial insights into the ecology of individual organisms associated with a particular artifact. For example, comparative genomics and transcriptomics have been used to describe the adaptation strategies of the true halophilic fungus *Aspergillus salisburgensis* (Tafer et al. 2019). The fungus was originally isolated from a 3500-year-old wooden staircase in the salt mines of Hallstatt (Austria), a world cultural heritage site in Austria (Piñar et al. 2016).

The development of high-throughput "omics" methods enables system approaches in which mixed microbial communities are studied as a single unit. Metagenomics, metatranscriptomics, metaproteomics, and metabolomics are used to determine the DNA sequences of the mixed community under investigation, the collectively transcribed RNA, the translated proteins, and the metabolites resulting from cellular processes. All data obtained can be used to identify metabolic pathways and cellular processes within an ecosystem (Kulski 2016). Therefore, the application of meta-omics analyses not only allows the identification of microbial groups in complex communities, but also allows the correlation of taxonomic groups with their ecological function and the observation of structural and functional shifts in communities caused by environmental changes. Meta-omics have the potential to perform in-depth studies on cultural assets in order to gain a deep understanding of the interactions between microorganisms and their substrates. In contrast to the successful application of metagenomics to cultural heritage, there are very few studies to date that demonstrate the success of applying proteomics in this field (Marvasi et al. 2019 and bibliography therein). However, proteomics has been successfully applied to cultural heritage to characterize protein-containing materials in ancient samples (Fiddyment et al. 2015; Vinciguerra et al. 2016). In the future, proteomics could help to analyze degrading enzymes such as proteases, cellulases, lipases, and other proteins with respect to their biodeterioration potential directly from the object of study. Also, the metabolome can provide information about destruction processes (Gutarowska et al. 2015). Conversely, the metabolome could possibly also be a key to the biological control of harmful microbes, since among the numerous primary and secondary metabolites, some of which are volatile organic compounds, there may also be those with antimicrobial effects (Campos and Zampieri 2019).

In conclusion, this review shows the potential of the molecular techniques available to date and the need for development to adapt many of these methodologies to the delicate field of cultural heritage. The further development of technological processes in molecular biology applied to heritage sciences will provide an in-depth understanding of the biodiversity and metabolic functions of the microbial communities that colonize culturally valuable objects. This will help in the first place to understand and mitigate biodeterioration, but also to answer many other interesting questions that may arise when handling valuable art objects.

# References

Adamiak J, Otlewska A, Tafer H, Lopandic K, Gutarowska B, Sterflinger K, Piñar G (2018) First evaluation of the microbiome of built cultural heritage by using the Ion Torrent next generation sequencing platform. Int Biodeter Biodegr 131:11–18

Amann RI, Ludwig W, Schleifer KH (1995) Phylogenetic identification and in situ detection of individual microbial cells without cultivation. Microbiol Rev 59:143–169

Andrews S (2010) FastQC: a quality control tool for high throughput sequence data. Available online at: http://www.bioinformatics.babraham.ac.uk/projects/fastqc

Branton D et al (2008) The potential and challenges of nanopore sequencing. Nat Biotechnol 26:1146–1153

Callahan BJ, Sankaran K, Fukuyama JA, McMurdie P, Holems SP (2016) Bioconductor workflow for microbiome data analysis: from raw reads to community analyses. F1000 Res 5:1492

Campos AI, Zampieri M (2019) Metabolomics-driven exploration of the chemical drug space to predict combination antimicrobial therapies. Mol Cell 74:1291–1303

Cutler NA, Oliver AE, Viles HA, Ahmad S, Whiteley AS (2013) The characterization of eukaryotic microbial communities on sandstone buildings in Belfast, UK, using TRFLP and 454 pyrosequencing. Int Biodeter Biodegr 82:124–133

Di Bonaventura MP, De Salle R, Bonacum J, Koestler RJ (2003) Tiffany's drawings, fungal spots and phylogenetic trees. In: Saiz-Jimenez C (ed) Molecular biology and cultural heritage. Balkema, Lisse, pp 131–148

Dobin A et al (2013) STAR: ultrafast universal RNA-seq aligner. Bioinformatics 29:15–21

Eid J et al (2009) Real-time DNA sequencing from single polymerase molecules. Science 323:133–138

Ettenauer JD, Jurado V, Pinar G, Miller AZ, Santner M, Saiz-Jimenez C, Sterflinger K (2014) Halophilic microorganisms are responsible for the rosy discolouration of saline environments in three historical buildings with mural paintings. PLoS One 9:1–12

Fiddyment S et al (2015) Animal origin of 13th-century uterine vellum revealed using noninvasive peptide fingerprinting. Proc Natl Acad Sci USA 112:15066–15071

Fiddyment S, Teasdale MD, Vnouček J, Lévêque E, Binois A, Collins MJ (2019) So you want to do biocodicology? A field guide to the biological analysis of parchment. Herit Sci 7:35

Goodwin S, McPherson JD, McCombie WR (2016) Coming of age: ten years of next generation sequencing technologies. Nat Rev Genet 17:333–351

Gutarowska B et al (2015) Metabolomic and high-throughput sequencing analysis-modern approach for the assessment of biodeterioration of materials from historic buildings. Front Microbiol 6:979

Huang Z, Zhao F, Li Y, Zhang J, Feng Y (2017) Variations in the bacterial community compositions at different sites in the tomb of Emperor Yang of the Sui Dynasty. Microbiol Res 196:26–33

Johnson SS, Zaikova E, Goerlitz DS, Bai Y, Tighe SW (2017) Real-time DNA sequencing in the Antarctic Dry Valleys using the Oxford Nanopore sequencer. J Biomol Technol 28:2–7

Jroundi F, Gonzalez-Muñoz MT, Sterflinger K, Piñar G (2015) Molecular tools for monitoring the ecological sustainability of a stone bio-consolidation treatment at the Royal Chapel, Granada. PLoS One 10:e0132465

Kalvari I, Argasinska J, Quinones-Olvera N, Nawrocki EP, Rivas E, Eddy SR, Bateman A, Finn RD, Petrov AI (2018) Rfam 13.0: shifting to a genome-centric resource for non-coding RNA families. Nucl Acids Res 46:D335–D342

Koestler RJ, Koestler VH, Charola AE, Nieto-Fernandez FE (2003) Art, biology, and conservation: biodeterioration of works of art. The Metropolitan Museum of Art, New York

Kulski JK (2016) Next-generation sequencing — an overview of the history, tools, and "omic" applications. In: Kulski JK (ed) Next generation sequencing - advances, applications and challenges. INTECH, London, eBook (PDF) ISBN: 978-953-51-5419-8

Laiz L, Piñar G, Lubitz W, Saiz-Jimenez C (2003) Monitoring the colonization of monuments by bacteria: cultivation versus molecular methods. Environ Microbiol 5:72–74

Martin M (2011) Cutadapt removes adapter sequences from high-throughput sequencing reads. EMBnet J 17:10–12

Marvasi M, Cavalieri D, Mastromei G, Casaccia A, Perito B (2019) Omics technologies for an in-depth investigation of biodeterioration of cultural heritage. Int Biodeter Biodegrad 144:104736

McGettigan PA (2013) Transcriptomics in the RNA-seq era. Curr Opi Chem Biol 17:4–11

Metzker ML (2010) Sequencing technologies - the next generation. Nat Rev Genet 11:31–46

Miklas H, Sablatnig R, Brenner S, Schreiner M, Cappa F, Piñar G, Sterflinger K (2019) The Vienna glagolotic projects: past and present. In: Stankovska P, Derganc A, Sivic-Dular A, Rajkko Nahtigal in 100 let slavistike na Univerzi v Ljubljani - Monografija ob 100. oblenici nastanka Oddelka za slavistiko Filozofske fakultete UL, 63-80; Slavica Slovenica, Ljubljana, pp 63–80. Please, clarify this reference. Editor, publisher, city, etc

Nawrocki EP, Eddy SR (2013) Infernal 1.1: 100-fold faster RNA homology searches. Bioinformatics 29:2933–2935

Ondov BD, Bergman NH, Phillippy AM (2011) Interactive metagenomic visualization in a Web browser. BMC Bioinformatics 12:385

Piñar G, Dalnodar D, Voitl C, Reschreiter H, Sterflinger K (2016) Biodeterioration risk threatens the 3100 year old staircase of Hallstatt (Austria): possible involvement of halophilic microorganisms. PLoS One 11:e0148279

Piñar G, Poyntner C, Tafer H, Sterflinger K (2019) A time travel story: metagenomic analyses decipher the unknown geographical shift and the storage history of possibly smuggled antique marble statues. Ann Microbiol 69:1001–1021

Piñar G, Poyntner C, Lopandic K, Tafer H, Sterflinger K (2020a) Rapid diagnosis of biological colonization in cultural artefacts using the MinION nanopore sequencing technology. Int Biodeter Biodegr 148:104908

Piñar G, Sclocchi MC, Sebastiani ML, Sterflinger K (2020b) The "Biological Pedigrees" of some emblematic Leonardo Da Vinci's objects. Gangemi Editore, Rome. in press

Piñar G, Tafer H, Schreiner M, Miklas H, Sterflinger K (2020c) Decoding the biological information contained in two ancient Slavonic parchment codices: an added historical value. Environ Microbiol. In press. https://doi.org/10.1111/1462-2920.15064

Pinzari F, Montanari M, Michaelsen A, Piñar G (2010) Analytical protocols for the assessment of biological dam-age in historical documents. Coalit Newslett 19:6–13

Rölleke S, Muyzer WG, Wanner C, Lubitz W (1999) Identification of bacteria in a biodegraded wall painting by denaturing gradient gel electrophoresis. Appl Environ Microbiol 62:2059–2065

Rosado T, Mirão J, Candeias A, Caldeira AT (2014) Microbial communities analysis assessed by pyrosequening—a new approach applied to conservation state studies of mural paintings. Anal Bioanal Chem 406:887–895

Rothberg J et al (2011) An integrated semiconductor device enabling non-optical genome sequencing. Nature 475:348–352

Saiz-Jimenez C (ed) (2003) Molecular biology and cultural heritage. Balkema, Lisse

Schabereiter-Gurtner C, Piñar G, Lubitz W, Rölleke S (2001) An advanced molecular strategy to identify bacterial communities on art objects. J Microbiol Methods 45:77–87

Schabereiter-Gurtner C, Saiz-Jimenez C, Piñar G, Lubitz W, Rölleke S (2002a) Phylogenetic 16S rRNA analysis reveals the presence of complex and partly unknown bacterial communities in Tito Bustillo cave, Spain, and on its Palaeolithic paintings. Environ Microbiol 4:392–400

Schabereiter-Gurtner C, Saiz-Jimenez C, Piñar G, Lubitz W, Rölleke S (2002b) Altamira cave Paleolithic paintings harbor partly unknown bacterial communities. FEMS Microbiol Lett 211:7–11

Schadt EE, Turner S, Kasarskskis A (2010) A window into third-generation sequencing. Human Mol Genet 19:227–240

Šoltys K, Planý M, Biocca P, Vianello V, Bučková M, Puškárov A, Sclocchi MC, Colaizzi P, Bicchieri M, Pangallo D, Pinzari F (2019) Lead soaps formation and biodiversity in a XVIII Century wax seal coloured with minium. Environ Microbiol 22(4):1517–1534

Sterflinger K, Piñar G (2013) Microbial deterioration of cultural heritage and works of art — tilting at windmills? Appl Microbiol Biotechnol 97:9637–9646

Sterflinger K, De Baere R, de Hoog GS, De Wachter R, Krumbein WE, Haase G (1997) *Coniosporium perforans* and *C. apollinis*, two new rock-inhabiting fungi isolated from marble in the Sanctuary of Delos (Cyclades, Greece). Anton Leeuw 2: 349-363

Sterflinger K, Little B, Piñar G, Pinzari F, De los Rios A, Gu JD (2018a) Future directions and challenges in biodeterioration research on historic materials and cultural properties. Int Biodeter Biodegr 129:10–12

Sterflinger K, Voitl C, Lopandic K, Piñar G, Tafer H (2018b) Big sound and extreme fungi-xerophilic, halotolerant Aspergilli and Penicillia with low optimal temperature as invaders of historic pipe organs. Life 8:22

Tafer H, Poyntner C, Lopandic K, Sterflinger K, Piñar G (2019) Back to the salt mines: Genome and transcriptome comparisons of the halophilic fungus *Aspergillus salisburgensis* and its halotolerant relative *Aspergillus sclerotialis*. Genes 10:381

Truong DT, Franzosa E, Tickle TL, Scholz M, Weingart G, Pasolli E et al (2015) MetaPhlAn2 for enhanced metagenomic taxonomic profiling. Nat Methods 12:902–903

Vinciguerra R, De Chiaro A, Pucci P, Marino G, Biroloa L (2016) Proteomic strategies for cultural heritage: From bones to paintings. Microchem J 126:341–348

Völkerding KV, Dames SA, Durtschi JD (2009) Next-generation sequencing: from basic research to diagnostics. Clin Chem 55:641–658

Vuissoz A, Worobeya M, Odegaard N, Bunce M, Machadoa CA, Lynner N, Peacocke E, Thomas M, Gilbert P (2007) The survival of PCR-amplifiable DNA in cow leather. J Archaeol Sci 34:823–829

Wieser M, Worliczek HL, Kämpfer P, Busse HJ (2005) *Bacillus herbersteinensis* sp. nov. Int J Syst Evol Microbiol 55:2119–2123

Wiesinger R (2018) Material stability of art and cultural heritage, Habilitation Thesis, Institut für Naturwissenschaften und Technologie in der Kunst, Akademie der Bildenden Künste Wien

Wingett SW, Andrews S (2018) FastQ Screen: a tool for multi-genome mapping and quality control. F1000 Res 7:1338

# Chapter 4
# Extreme Colonizers and Rapid Profiteers: The Challenging World of Microorganisms That Attack Paper and Parchment

**Flavia Pinzari and Beata Gutarowska**

**Abstract** Microorganisms form the backbone of life on Earth. Over billions of years, they have colonized and shaped every possible niche on the planet. Microbes have modelled both the land and the sea, and have created favourable conditions for multicellular organisms to thrive in. Our understanding of how microbial diversity is distributed across natural environments and how microbes affect ecosystems is constantly evolving as public databases are set up and new techniques based on massive sequencing are developed. The microbiome found in a particular anthropogenic environment is generally much less complex than those found in natural ones: there is less competition and the main actors are often linked to survival mechanisms regulated by a few limiting factors. Despite this simplicity, it is very difficult to link cause and effect when seeking to identify the role of individual organisms. In the case of biodeterioration of paper and parchment, even when analysing the individual components of a simple phenomenon, it is not always easy to understand the mechanisms at play. Works of art are unique objects and the elements that determine the arrival and establishment of one or more microorganisms and the direction that the biodeterioration process takes are always different. In some cases, however, there are common denominators and predictable mechanisms. The variables that come into play are examined below.

**Keywords** Ecology · Metabolomics · Niche · Colonization · Diagnostics · Mechanisms · Biodeterioration · Substrate · Succession · Metabolites · Enzymes · Indoor environment

F. Pinzari (✉)
Institute for Biological Systems, National Research Council of Italy (CNR), Rome, Italy

Life Sciences Department, Natural History Museum, London, UK
e-mail: flavia.pinzari@cnr.it; f.pinzari@nhm.ac.uk

B. Gutarowska
Department of Environmental Biotechnology, Lodz University of Technology, Łódź, Poland
e-mail: beata.gutarowska@p.lodz.pl

© The Author(s) 2021
E. Joseph (ed.), *Microorganisms in the Deterioration and Preservation of Cultural Heritage*, https://doi.org/10.1007/978-3-030-69411-1_4

# 1 Paper and Parchment as Food Sources for Microbes

## 1.1 Paper

Paper serves as an excellent substrate for various heterotrophic organisms. It is, in fact, a source of organic carbon for many microorganisms, and when it is used in the manufacture of objects such as books, documents, prints, etc., it also contains several organic substances such as animal or vegetable glues, inks, pigments and fillers that can augment its biodegradability. These compounds can represent a nutritional source for many non-specialized microorganisms which, as they grow, prepare the substrate for a more specific and therefore more harmful attack by typically cellulolytic strains—that is, those capable of degrading the cellulose that constitutes the chief component of most papers.

The biodegradability of paper varies depending on different factors, such as the degree of polymerization of cellulose, the percentage ratio between the two types of cellulose (crystalline or amorphous), and the presence of chemical or mechanical pulp (lignin, for example, makes paper unattractive for many fungal species). Additionally, the type of glue used in the pulp or on the paper's surface, which can be of animal or plant origin (starch, gelatin, casein, rosin, etc.), has a significant effect because it can to a greater or lesser extent be either palatable or toxic to microorganisms. The nature of the fillers (e.g. talc, kaolin, calcium carbonate, etc.) used in the manufacture of paper, and possible surface treatments (e.g. colouration, satin-effects, etc.), can have a substantial effect on the outcome of an infection caused by microorganisms (Florian 2002; Pinzari et al. 2006).

Cellulose itself is a long linear chain of glucose. The polymer consists of anhydrous glucopyranose units joined by ß-1,4 glycosidic bonds. A cellulose chain contains approximately 10,000 units of glucose with a molecular weight of approximately 1.5–2 million. The polymer chains are arranged parallel to each other and bind together by means of hydrogen bonds, forming fibrils. These fibrils can form a hydrophobic and very resistant crystalline structure. The enzymatic hydrolysis of crystalline cellulose typically requires the participation of three different enzymes: (1) an endo-1,4-ß-glucanase (EC3.2.1.4) which breaks the molecule, randomly producing chains with free ends; (2) an eso-1,4-ß-glucanase (EC3.2.1.91) that detaches cellobiose units, ß-glucose dimer, from the non-reducing ends of the linear chains; and (3) a ß-glucosidase (EC3.2.1.21) that hydrolyses cellobiose to form glucose (Jayasekara and Ratnayake 2019).

The complete biodegradation of cellulose is carried out by microorganisms that possess all the above-mentioned enzymes; the enzymes operate in succession as a "cellulase complex" capable of breaking down the cellulose polymer into glucose monomers. There are, however, only a few organisms that show real cellulolytic activity: most of the species use simple, easily hydrolysable sugars. The various cellulolytic species produce a wide variety of isoenzymes that attack cellulose in different ways (Sterflinger and Pinzari 2012; El Bergadi et al. 2014; Oetari et al. 2016; Coronado-Ruiz et al. 2018; Jayasekara and Ratnayake 2019). For example, some fungal species do not possess all the enzymes that form the cellulase complex, and therefore exert an incomplete depolymerization effect on cellulose. Many fungi do not possess exoglucanase and act only on amorphous cellulose, leaving the crystalline component intact (Jayasekara and Ratnayake 2019). Bacteria possess cellulolytic enzyme systems that are just as effective and even more diversified than fungi.

The superiority of fungi in cellulose breakdown in comparison to unicellular organisms derives from their growth mode and ecology. Mycelium makes it possible for fungi to colonize a substrate more rapidly and to penetrate it by branching out within. In addition, fungal hyphae can readily move and mobilize the products of polymer degradation over considerable distances through the mycelial network (Szczepanowska et al. 2015). An important factor that influences the degree of cellulose decomposition is a relative lack of nitrogen. Natural cellulosic materials often have a C:N ratio of 100:1 or greater, so the addition of nitrogen is necessary for their decomposition. Consequently, the presence of glues containing proteins, such as gelatin or other nitrogen-rich molecules, increases the biodegradability of paper (Florian 2002).

Fungi that damage paper belong to a few genera that share some ecological traits which characterize them as species that are well suited to colonizing "indoor" environments. Such environments, although artificial, offer sufficient trophic niches to accommodate real trophic successions (Zyska 1997; Pinzari 2011, 2018; Mallo et al. 2017; Karakasidou et al. 2018; Koul and Upadhyay 2018; Lax et al. 2019). The species that attack books and archives are fungi that, in natural settings, are mainly found in soil and on decomposing plant and animal matter (saprophytic fungi) (Treseder and Lennon 2015; Hyde et al. 2019). They primarily feed on cellulose, starch and sugars, but some can also degrade chemically more complex substances such as lignin, keratin, glues and synthetic polymers (Ayu and Teja 2016; Brunner et al. 2018; Hyde et al. 2019). They are organisms that disperse through conidia and spores, which are easily transportable through the air or by insects.

Fungi can damage paper in different ways depending on the species concerned, growth conditions and the type of paper involved. The fungal hyphae can also penetrate cellulose fibrils, fragmenting them through mechanical action (Figs. 4.1, 4.2, 4.3). They can also exploit cellulose as a substrate and cause sheets of paper to become fragile and cracked. Sometimes they release coloured metabolic substances of various kinds between the fibres that can stain sheets or induce chemical alterations, such as oxidation.

Melo et al. (2019) reviewed the known mechanisms behind the formation of stains on paper, the colourants commonly present in the stains and the main fungal species associated with both the production of coloured compounds and their ability to develop in paper. Among the most widespread and staining fungal pigments, melanins are particularly detrimental to books and objects of heritage. Melanins are often contained within the walls of fungi and typically accompany the growth of the hyphae across the cellulose network, infiltrating the structure of paper and causing dark stains that can remain deep in the body of the material (Szczepanowska et al.

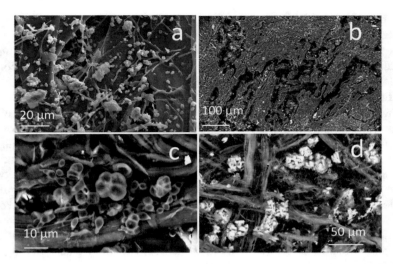

**Fig. 4.1** Scanning Electron Microscopy images of parchment and paper samples affected by biodeterioration: (**a**) surface of a parchment leaf attacked by a fungus. The collagen fibres still appear intact, although fungal hyphae are attached to them. The fungus, in this case, is probably not using collagen as a carbon source, but it is feeding on other components of the material (i.e. waxes, glues, oils); (**b**) the surface of a parchment sample attacked by bacteria. Here there is visible evidence of the formation of cracks between the bundles of fibrils, and disintegration of the matrix that holds the fibrils together; (**c**) surface of a paper sample attacked by a fungus. Despite the fact that the organism has developed among the cellulose fibres producing abundant structures, the fibres still appear intact, indicating that the fungus is not using cellulose as a carbon source; (**d**) surface of a paper sample deeply infiltrated by a fungus that produced abundant acidic compounds, probably oxalic acids. The cellulose fibres appear broken and dissolved. These acidic compounds precipitate in the presence of calcium used in the manufacture of the paper; the precipitated crystals of calcium oxalate are visible as lighter structures in the image (they are brighter than cellulose because of the higher atomic mass of calcium, as revealed by a backscattered electron detector)

2015). Melanins are natural polymers characterized by a strong negative charge, high molecular weight and hydrophobic nature (Nosanchuk et al. 2015). They are produced through the oxidation and polymerization of phenolic/indolic precursors and in fungi they form insoluble pigments (eumelanins) that confer dark colourations and have several functional roles (for example: photoprotection, energy harvesting, heat gain, thermoregulation, and metal scavenging) (Cordero and Casadevall 2017). Colour and concentration of melanins can differ in one and the same microorganism depending on growth conditions, the availability of nutrients and the properties of the substrate. Accordingly, the same fungus can produce very different stains on different paper stocks (Pinzari et al. 2006). This wide variability in colour production also occurs in pigments other than melanin. According to Melo et al. (2019), the colourants predominantly produced by paper-colonizing fungi are hydroxyanthraquinoid (HAQN) pigments. These can be yellow and red or orange, and when widely distributed in paper can change in shade and intensity depending on the local pH and the presence of salts (Figs. 4.1, 4.2, 4.3). In the literature, the

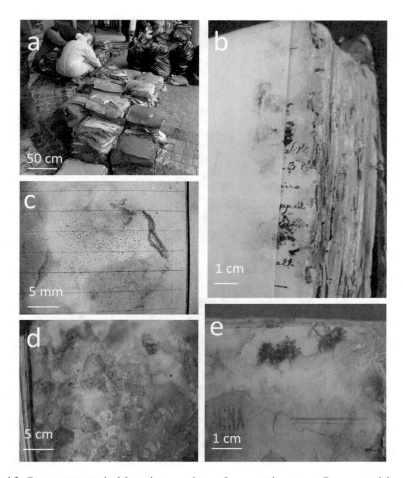

**Fig. 4.2** Documents attacked by microorganisms after a wetting event. Paper materials, once soaked with water, are highly susceptible to degradation by different species of microorganisms whose spores normally inhabit the air and dust of indoor and outdoor environments; (**a**) the salvage of bibliographic material from a library following a flood (Aulla 2011 flood, MS, Italy, photo by Piero Colaizzi, ICRCPAL, Rome); (**b, c, d, e**) books and documents made of paper attacked by microorganisms (fungi and bacteria). The species of fungi and actinomycetes (filamentous bacteria) that take advantage of a sudden availability of water are usually very competitive, able to secrete enzymes, organic acids and pigments, and to develop rapidly at the expense of the substratum, which, in the case of paper, becomes fragile, cracked, stained and mottled with abundant fruiting bodies (which carry the spores or the conidia). In particular, figure (**d**) shows a pattern of stains that form a sort of grid, where the sides of each block are the sign of competition between two different species of fungi, which at the point of contact produce more pigments and metabolic substances aimed at repelling the opponent in order to occupy more space in the material

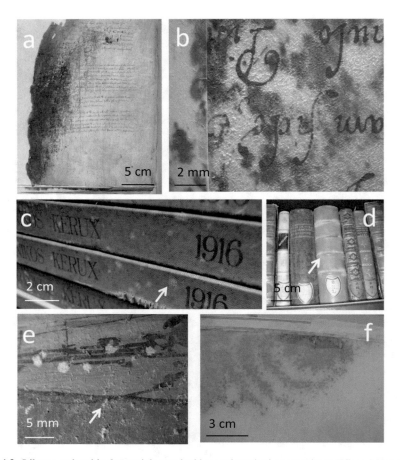

**Fig. 4.3** Library and archival materials attacked by s-selected microorganisms. When the environmental conditions are permanently hostile to the development of most of the microorganisms, some species that are not very competitive but are highly evolved to withstand certain types of stress and limitations, establish themselves on the materials, sometimes giving rise to massive colonization phenomena; (**a, b**) purple stains on parchment caused by halophilic microorganisms, which are associated with the salt used in the manufacturing process, which can remain dormant for many years among collagen fibres; (**c, d**) the typical manifestation of a widespread infection of the fungus *Eurotium halophilicum* in a library. This fungus develops slowly in environments where conservation conditions are good, but takes advantage of the absence of ventilation or other microenvironmental conditions that favour water condensation on some materials; (**e**) fungal efflorescence on the gypsum-based preparatory layer and the mineral pigments that make up the illuminations on a parchment. Many microorganisms find an ideal niche in materials that are typically hostile to other species; (**f**) a peculiar oxidation pattern on a paper sheet affected by biodeterioration. The shape of the stains suggests the periodic growth of microorganisms probably capable of re-starting their growth from the hyphal tips in the presence of a recurring event (i.e. re-wetting or temperature-related)

production of these colourants is often associated with the genera *Aspergillus* and *Penicillium* (Melo et al. 2019), which are also frequently found on stained paper.


## 1.2   Parchment

Parchment was the most commonly used writing material from the second century BC to the end of the Middle Ages, when it was substituted by paper in Europe. It is a thin material made from the skin of animals such as sheep, goats, lambs, pigs and calves. It differs from leather in that it is mainly limed rather than tanned. There were considerable differences in parchment manufacturing methods in different countries and historical periods. Even so, the preparation of parchment involved always the cleaning and mechanical removal of hair by scraping the flayed skin, followed by desiccation of any remaining flesh using sodium or potassium chloride, adjustment of the pH by treatment with ammonium chloride or sulphate with lime, and the application of potash alum (Larsen 2002). Parchment books, depending on their date of manufacture and country of origin, were written using iron gall or carbon pigmented inks, or printed using resinous printing inks; they were frequently illuminated with a wide range of mineral and organic pigments, and embellished with metals such as gold or silver and their alloys. In maedieval times, egg white, natural oils and other materials were applied to the parchment to form a very smooth, highly polished surface for writing and illumination. Parchments were also used in the making of official and private documents (e.g. contracts, wills, genealogies, maps, etc.), often collected in composite volumes bound together by sewing with thread or ribbons, and sometimes with adhesives. The main chemical component of parchment is collagen, a natural biopolymer with a relative molecular mass of 350 kDa. Collagen is formed from a triple helix with random coil telo-peptides. Degradation of parchment is a complex process which involves the oxidative chemical deterioration of the amino acid chains and hydrolytic cleavage of the peptide backbone (Florian 2007).

   Parchment is usually a very robust and durable material; it is particularly resistant to abrasion, tearing and other mechanical damage. Parchment is more resistant than paper to corrosive inks and acid degradation in general, owing to the alkaline agents used in the manufacturing process, such as lime and chalk. However, parchment is extremely hygroscopic, which makes it highly susceptible to infection by bacteria and moulds when exposed to elevated relative humidity.

   Microorganisms with a proteolytic capacity include bacteria and fungi, even if the former possess more specific collagenases capable of digesting the proteins of the parchment through hydrolysis (Figs. 4.1, 4.3). Fungi, on the other hand, develop particularly at the expense of oils and waxes added during the manufacturing process (to make the parchment more flexible), although some species of proteolytic fungi can also utilize collagen as their sole carbon source (Florian 2002, 2007; Pinzari et al. 2012a, b; Lech 2016; Pinzari 2018).

The microorganism species involved in the degradation of parchment can vary considerably with the chemical characteristics of the material, which can be very different depending on the origin of the animal, the part of it used and the manufacturing processes employed. The presence of fats lends parchment a certain degree of water repellency which largely speaking slows down biodeterioration processes. Lipid substances are rather resistant to biological degradation but become more susceptible if they are found as water supported emulsions or in the presence of additional nutritional sources that can easily be metabolized, such as simple sugars (Florian 2002, 2007; Pinzari et al. 2012a, b; Lech 2016; Pinzari 2018).

It has been observed that biodegradation typically starts on the side of the parchment corresponding to the "flesh" side of the skin. In the more advanced stages of degradation, the side bearing the grain (i.e. the "hair" side) is also attacked, where the collagen and elastin fibres form a denser network.

Initial studies on parchment deterioration indicated the Actinomycetes as the bacterial group that plays a major role in the deterioration of many kinds of historical documents and books supported on parchment. Actinomycetes are known to produce many types of enzyme, collagenases (proteases) in particular that are capable of destroying collagen through their hydrolytic activity (Blyskal et al. 2017). Many species are alkaliphiles and therefore develop on parchments (prepared with chalk). However, some more recent studies (Piñar et al. 2015a, b; Migliore et al. 2017, 2019) have highlighted some rather complex mechanisms driving parchment biodegradation, in which actinomycetes only represent some of the actors.

## 2 The Fungal and Bacterial Species That Attack Paper and Parchment

During their storage and utilization, paper and parchment can be colonized by numerous microorganisms, especially fungi and bacteria, including actinomycetes. Examples of microorganisms isolated from paper and parchment are presented in Tables 4.1 and 4.2.

In total, 85 fungal genera and 71 bacterial genera were identified in the book's constituent materials. Paper is a material that is susceptible to a larger number of colonizing microorganisms (80 genera of fungi and 59 genera of bacteria) than parchment (29 genera of fungi, 23 genera of bacteria). The development of modern molecular methods, chiefly NGS (next-generation sequencing), is greatly contributing to the discovery of new genera of microorganisms that colonize these materials. Molecular methods have identified an additional 40 genera of fungi (out of the 85 recognized so far) that inhabit paper but were not previously detected, since they cannot be cultured under laboratory conditions, and 58 genera of bacteria out of the 71 known up to the present. In the case of parchment, thanks to the use of molecular methods, an additional 15 genera of fungi and several taxa of unculturable bacteria have been discovered as frequent colonizers and potential spoilers of the material. It

**Table 4.1** Book materials (paper and parchment) contaminated with **fungi**, identified by means of culture-dependent and independent methods

| Isolated from book materials identified by cultural dependent[a] and culture independent[b] methods | | | |
|---|---|---|---|
| Genera of fungi | Paper | Parchment | References |
| *Acremonium* | a, b | a, b | Arai (2000), Borrego et al. (2012), El Bergadi et al. (2014), Florian and Manning (2000), Gutarowska (2016), Karakasidou et al. (2018), Karbowska-Berent et al. (2014), Kraková et al. (2012, 2018), Lech (2016), Michaelsen et al. (2009, 2010), Nol et al. (2001), Nunes et al. (2015), Oetari et al. (2016), Paiva de Carvalho et al. (2016), Piñar et al. (2015a, b), Principi et al. (2011), Rakotonirainy et al. (2007), Saada et al. (2018), Sato et al. (2014), Szczepanowska et al. (2014), Szulc et al. (2018), Tanney et al. (2015), Zotti et al. (2008, 2011), Zyska (1997) |
| *Acrothecium* | a | | |
| *Alternaria* | a, b | a, b | |
| *Aspergillus* | a, b | a, b | |
| *Aureobasidium* | a, b | b | |
| *Beauveria* | b | | |
| *Bjerkandera* | b | | |
| *Blumeria* | | b | |
| *Botryotrichum* | a | | |
| *Botrytis* | a | | |
| *Byssochlamys* | b | | |
| *Capnobotryella* | b | | |
| *Ceratocystis* | a | | |
| *Chaetomium* | a, b | a, b | |
| *Chalastospora* | a, b | | |
| *Cladonia* | b | | |
| *Chloridium* | a | | |
| *Chrysosporium* | a | | |
| *Cladobotryum* | a | | |
| *Cladosporium* | a, b | a, b | |
| *Coniosporium* | a | b | |
| *Curvularia* | a | | |
| *Diaporthe* | b | | |
| *Diploöspora* | | a, b | |
| *Doratomyces* | a | | |
| *Engyodontium* | | b | |
| *Epicoccum* | a, b | a, b | |
| *Erysiphales* | b | | |
| *Eurotium* | a, b | b | |
| *Fusarium* | a | a | |
| *Geomyces* | a | | |
| *Geosmithia* | a | | |
| *Geotrichum* | a | | |
| *Gleotinia* | b | | |
| *Gliocladium* | a | | |
| *Graphiopsis* | b | | |
| *Gymnoascus* | a | | |
| *Haplographium* | a | | |
| *Helicostylum* | a | | |

(continued)

**Table 4.1** (continued)

| Isolated from book materials identified by cultural dependent[a] and culture independent[b] methods | | | |
|---|---|---|---|
| Genera of fungi | Paper | Parchment | References |
| *Helminthosporium* | a | | |
| *Humicola* | b | | |
| *Hypocrea* | b | | |
| *Isaria* | b | | |
| *Melanospora* | a | | |
| *Monilia* | a | a | |
| *Mucor* | a, b | a, b | |
| *Mycogone* | a | | |
| *Myrothecium* | a | | |
| *Myxotrichum* | a, b | | |
| *Paecilomyces* | a | a | |
| *Panaeolus* | b | | |
| *Phaeomarasmius* | b | | |
| *Pellicularia* | a | | |
| *Penicillium* | a, b | a, b | |
| *Pestalotia* | a | | |
| *Physospora* | a | | |
| *Phoma* | a, b | | |
| *Polyporus* | b | | |
| *Pseudocercospora* | a | | |
| *Ramichloridium* | b | | |
| *Rhodotorula* | a, b | | |
| *Rhizoctonia* | a | | |
| *Rhizopus* | a, b | a, b | |
| *Saccharomyces* | a | | |
| *Saccharicola* | b | | |
| *Scopulariopsis* | a | a | |
| *Sepedonium* | a | a | |
| *Sordaris* | a | | |
| *Spicaria* | a | | |
| *Sporendonema* | | a | |
| *Sporidesmium* | | a | |
| *Sporotrichum* | | a | |
| *Stachybotrys* | b | a, b | |
| *Stemphylium* | a | a | |
| *Sydowia* | b | | |
| *Talaromyces* | b | a | |
| *Torula* | a | | |
| *Toxicocladosporium* | | b | |
| *Trametes* | a, b | | |

(continued)

**Table 4.1** (continued)

| Isolated from book materials identified by cultural dependent[a] and culture independent[b] methods | | | |
|---|---|---|---|
| Genera of fungi | Paper | Parchment | References |
| *Trichocladium* | a | | |
| *Trichoderma* | a, b | a, b | |
| *Trichosporum* | b | a, b | |
| *Trichothecium* | a | a | |
| *Ulocladium* | a, b | | |
| *Verticillium* | a | a | |
| *Wallemia* | b | | |

is therefore necessary to verify the mechanisms behind paper and parchment biodeterioration in which both bacteria and fungi are at play, because as yet knowledge of the role played by different groups of microorganisms in the material's spoilage is limited. It is widely believed that fungi are the main actors in paper biodeterioration, while in the case of parchment the most effective taxa in biodegradation processes are bacteria. However, recent evidence indicates that mechanisms are more complex and that often the co-occurrence of bacteria and fungi is the norm, with clear implications for the choice of treatments to be applied during the conservation of materials. In the case of fungi, culture-dependent methods are still widely used in the laboratory, and research results indicate that 58% of fungal species can be identified using culturing methods, 26% using molecular methods and 15% using both methods (Gutarowska 2016).

So far though, only a few genera of microorganisms inhabiting book materials have been identified using both methods (culture-dependent and molecular), with just 18 and 12 genera of fungi, on paper and parchment, respectively, and 8 taxa of bacteria on paper that have been identified using both approaches. Microbial community analysis using the NGS method makes it possible to detect higher biodiversity than the culture-dependent approach yields (Kraková et al. 2018) and some early culture-independent approaches based on cloning (Michaelsen et al. 2006, 2009, 2010; Pangallo et al. 2009). Being in possession of exhaustive lists of the species of fungi and bacteria that attack various library and archival materials, and different parts of the objects in various types of environment, will certainly aid us in better understanding colonization and deterioration mechanisms and will facilitate statistical analysis of the frequency with which certain species occur. So far, this type of study has been restricted by the limited number of culturable organisms. However, in order to know which organisms are responsible for causing damage, it is necessary to have not only a list of candidates but also to identify their impact on the materials and the metabolites they produce. Tables 4.3 and 4.4 report the most common microorganisms inhabiting paper and parchment, for which biodeterioration mechanisms have been described.

**Table 4.2** Book materials (paper and parchment) contaminated with bacteria, identified by means of culture-dependent and independent methods

| Isolated from book materials identified by cultural dependent[a] and culture independent[b] methods | | | |
|---|---|---|---|
| Genera of bacteria | Paper | Parchment | References |
| *Actinomycetospora* | | a | |
| *Actinomycetospora* | b | | Michaelsen et al. (2010), Kraková et al. (2012), Piñar |
| *Actinoalloteichus* | b | | et al. (2015a, b), Lech (2016), Karakasidou et al. |
| *Amycolatopsis* | b | | (2018), Gutarowska (2016), Kraková et al. (2018), |
| *Aerococcus* | a,b | | Szulc et al. (2018), Migliore et al. (2019) |
| *Aeromonas* | b | | |
| *Acinetobacter* | b | a,b | |
| *Alkanindiges* | b | | |
| *Alloiococcus* | b | | |
| *Arthrobacter* | b | b | |
| *Bacillus* | a,b | a,b | |
| *Brevibacillus* | b | | |
| *Brevibacterium* | b | | |
| *Brevundimonas* | b | | |
| *Burkholderia* | b | | |
| *Caedibacter* | | b | |
| *Cellvibrio* | b | | |
| *Carnobacterium* | b | | |
| *Clostridium* | b | b | |
| *Corynebacterium* | b | | |
| *Curtobacterium* | b | | |
| *Dehalobacter* | b | | |
| *Delfia* | b | | |
| *Desulfotomaculum* | b | | |
| *Enterococcus* | b | | |
| *Erwinia* | b | | |
| *Facklamia* | b | | |
| *Fructobacillus* | b | b | |
| *Geomicrobium* | | b | |
| *Gluconobacter* | b | | |
| *Halobacillus* | | b | |
| *Halobacterium* | | b | |
| *Halomonas* | | | |
| *Ignavigranum* | b | | |
| *Jeotgalicoccus* | b | b | |
| *Kocuria* | a,b | | |
| *Lactobacillus* | b | b | |
| *Legionella* | | b | |
| *Leuconostoc* | b | | |

(continued)

**Table 4.2** (continued)

| Isolated from book materials identified by cultural dependent[a] and culture independent[b] methods | | | |
|---|---|---|---|
| Genera of bacteria | Paper | Parchment | References |
| *Lysinibacillus* | a,b | | |
| *Lysobacter* | b | b | |
| *Masillia* | b | | |
| *Methylococcus* | | | |
| *Microbacterium* | a,b | | |
| *Micrococcus* | b | | |
| *Micromonospora* | b | | |
| *Mycobacterium* | b | | |
| *Methylobacterium* | | b | |
| *Natronocella* | | b | |
| *Nocardiopsis* | | b | |
| *Pediococcus* | b | | |
| *Paenisporosarcina* | b | b | |
| *Pseudomonas* | b | b | |
| *Pseudonocardia* | b | | |
| *Psychrobacillus* | a | | |
| *Psychrobacter* | b | b | |
| *Propionibacterium* | b | b | |
| *Ralstonia* | b | | |
| *Rhodococcus* | b | | |
| *Rhizobium* | | b | |
| *Saccharopolyspora* | b | | |
| *Serratia* | b | | |
| *Sporosarcina* | a,b | | |
| *Staphylococcus* | a,b | b | |
| *Stenotrophomonas* | b | | |
| *Streptococcus* | a | | |
| *Streptomyces* | b | | |
| *Sulfobacillus* | b | b | |
| *Vibrio* | b | | |
| *Virgibacillus* | b | | |

# 3   Arrival, Colonization and Affirmation

The colonization of libraries, archives and other repositories is accomplished by organisms with particular characteristics which allow them to adopt suitable dispersion mechanisms. Water and nutrients are usually only sporadically available in artificial environments, which makes them rather fragmented and variable (Lazaridis et al. 2018). Fungal hyphae are better adapted than bacteria to cross nutrient-and water-deprived areas when searching for resources (Ritz 1995), thanks to their

**Table 4.3** Microbial colonizers of paper and biodeterioration mechanisms

| Microorganisms colonizing paper | Mechanisms of biodeterioration | References |
|---|---|---|
| **Fungi** | | |
| Acremonium sp. Alternaria sp. (A. alternata, A. infectoria) Aspergillus sp. (A. awamori, A. clavatus, A. flavus, A. fumigatus, A. jensenii, A. niger, A. penicillioides, A. pulvericola, A. ruber, A. versicolor) Chalastospora sp. (- Ch. gossypii) Cladonia sp. Cladosporium sp. Epicoccum sp. (E. nigrum) Eurotium sp. (E. herbariorum) Penicillium sp. (P. citrinum, P. chrysogenum, P. oxalicum, P. rubens) Phoma sp. Pseudocercospora sp. (P. chiangmaiensis) Trametes sp. (T. ochracea) Trichoderma sp. (T. longibrachiatum) | Production of organic acids and other metabolites, which can react with elements on paper leading to oxidative reaction and formation of brown foxing stains. Oxidation of cellulose. Reaction of cellulose degradation products with amino acids, auto-oxidation of lipids present in the fungal mycelium (Maillard reaction). | Florian (1996), Arai (2000), Corte et al. (2003), Choi (2007), Zotti et al. (2008), Karbowska-Berent et al. (2014), Piñar et al. (2015a), Modica et al. (2016), Szulc et al. (2018), Saada et al. (2018), Oetari et al. (2016), Karakasidou et al. (2018) |
| Alternaria sp. (A. solani) Aspergillus sp. (A. carneus, A. flavus, A. fumigatus, A. nidulans, A. niger, A. oryzae, A. melleus, A. penicillioides, A. sclerotium, A. tamarii, A. terreus, A. ustus, A. versicolor), Cladosporium sp. (C. cladosporioides, C. sphaerospermum), Chaetomium sp. (- Ch. globosum) Eurotium sp. (E. repens, E. rubrum, E. amstelodami, E. halophilicum, E. herbariorum) Penicillium sp. (P. chrysogenum, P. citrinum, P. commune, P. funiculosum, P. islandicum, P. notatum, P. purpurogenum, P. spinulosum) | Production of stains—polyketide quinones, carotenoids (main brown 54%, black 23%, green 6%, yellow 7%, purple 5% pink 5%) | Melo et al. (2019) |

(continued)

**Table 4.3** (continued)

| Microorganisms colonizing paper | Mechanisms of biodeterioration | References |
|---|---|---|
| *Trichoderma* spp. (*T. koningii, T. pseudokoningii, T. citrinoviride*) Others | | |
| **Bacteria** | | |
| *Ralstonia* sp. *Delfia* sp. | Enzymatic degradation of chemicals used in papermaking, production of coloured slimes, ability to produce diffusible brown pigments. | Väisänen et al. (1998), Szulc et al. (2018) |
| *Pseudomonas* sp. (*P. stutzeri*) *Burkholderia* sp. (*B. cepacia*) | Enzymatic degradation of starch, casein, carboxymethyl cellulose, ability to produce diffusible yellow pigments. | Szulc et al. (2018) |
| *Bacillus* sp. *Brevibacillus* sp. *Lysinibacillus* sp. | Enzymatic degradation of starch, casein, cellulose. | De Paolis and Lippi (2008), Szulc et al. (2018) |

ability to mechanically penetrate substrates and translocate nutrients. Active and passive dispersal mechanisms can produce very different results in indoor environments. Fungi and bacteria populate the airborne dust of indoor environments as resting spores or living propagules. These organisms can be introduced to buildings by different sources or be generated from "amplification sites". Damp walls, ventilation systems, spoiled food or rotting materials can host bacterial or fungal biofilms and act as rich sources of airborne contamination (Wang et al. 2016; Liu et al. 2018). The number of spores deriving from a contaminated material can vary considerably with the biology and ecology of individual species, and can also be influenced by chance events. However, not all microbes have evolved for air dispersion; there are several other mechanisms which favour the microbial colonization of new "niches". Some of the pests that attack organic materials in libraries and archives, such as wood, paper, leather and parchment, can feed on moulds or cohabit with them, and act as active or passive means of dispersal. Insect-mediated fungal dispersion has been widely documented in natural environments (Jacobsen et al. 2018), but it also occurs in indoor environments. Some indoor species such as silverfish (family *Lepismatidae*), booklice (family *Liposcelididae*) and mites (i.e. family *Pyroglyphidae*, genus *Dermatophagoides*) are often directly correlated to the moulds infecting materials (Green and Farman 2015) in mutually beneficial circumstances. In fact, the metabolic water produced by insects, their droppings and debris constitute an ideal substrate for fungal development (Pinzari and Montanari 2008, 2011). Humans constitute another factor in the contamination of materials and the dispersion of microorganisms.

Traces of the human microbiome are typically present on materials and thanks to molecular analysis this is becoming increasingly apparent, as the list of

**Table 4.4** Microbial colonizers of parchment and biodeterioration mechanisms

| Microorganisms colonizing paper | Mechanisms of biodeterioration | References |
|---|---|---|
| **Fungi** | | |
| *Alternaria* sp. (*A. alternata*) *Aspergillus* sp. (*A. versicolor, A. niger, A. fumigatus*) *Aureobasidium* sp. (*A. pullulans*) *Cladosporium* sp. (*C.cladosporioides*) *Chaetomium* sp. (*Ch. globosum*) *Epicoccum* sp. (*E. nigrum*) *Eurotium* sp. (*E. halophilicum*) *Mucor* sp. *Penicillium* sp. (*P. chrysogenum, P. citrinum, P. glabrum, P. spinulosum*) *Rhizopus* sp. (*R. oryzae*) *Trichoderma* sp. (*T. pseudokoningii, T. longibrachiatum*) *Talaromyces* sp. (*T. spectabilis*) | Enzymatic degradation of collagen, structural damage. Production of stains (different colours). Production of organic acids (fumaric, lactic, malic). | Piñar et al. (2015b, c), Lech (2016), Paiva de Carvalho et al. (2016), Saada et al. (2018) |
| **Bacteria** | | |
| *Streptomyces* sp. *Pseudonocardia* sp. *Bacillus cereus* *Acinetobacter lwoffii* | Enzymatic degradation of collagen, structural damage. | Karbowska–Berent and Strzelczyk (2000), Piñar et al. (2015c), Lech (2016) |
| *Nocardiopsis salina* *Saccharopolyspora* sp. | Enzymatic degradation of collagen, structural damage Production of stains (red, purple). | Piñar et al. (2015b, c) |
| *Halomonas* sp. *Halobacillus* sp. *Halobacterium* sp. | Production of stains (red, purple), in presence of hygroscopic salts (sulphates, chlorides) used in manufacturing of parchment, initial attack (halophilic species). | Piñar et al. (2015c), Migliore et al. (2019) |
| *Bacillus* sp. (*B. cereus*) *Acinetobacter* sp. (*A. lwoffii*) *Stenotrophomonas* sp. (*S. maltophilia*) | Enzymatic degradation of collagen, structural damage. | Piñar et al. (2015c) |
| *Arthrobacter* sp. *Methylobacter* sp. | Production of stains red-pigmented (carotenoids production). | Piñar et al. (2015c), Lech (2016) |

contaminants, even those that do not grow in culture, is growing longer and more accurate with the advent of new investigative techniques based on the extraction and sequencing of DNA.

The microorganisms that are identified in the air and on materials in libraries, archives and book repositories using massive sequencing methods include those typically found growing in indoor environments as well as those introduced from outdoors (Liao et al. 2004; Borrego et al. 2012; Nunes et al. 2013). Fungi that are obviously "outdoor" additions to the indoor environment often include ecto- and arbuscular-mycorrhizal genera that are obligate associates of host plants. The presence of mammalian mycobionts such as *Candida albicans*, *Malassezia* spp. and bacterial species belonging to the group *Propionibacterium/Cutibacterium* shows that humans contribute directly to the microbiome of indoor dust. Other bacteria of cutaneous origin, belonging to the genera *Corynebacterium* and *Staphylococcus* (Ramsey et al. 2016), have often been detected on archival documents in the past (Puškárová et al. 2016; Kraková et al. 2012, 2018).

The main limiting factor that permits fungal development on paper and parchment is water. Spore germination only occurs when some water is available. From an ecological standpoint, it is possible to imagine at least two different scenarios in which a microbiological attack on books and documents is possible (Figs. 4.2, 4.3, 4.4): (1) a sudden abundance of water, such as in the case of flooding or pipe rupture and (2) a situation in which a microclimate is established that encourages the development of particular organisms possessing the essential characteristics which make them eminently suited to occupying a particular ecological niche (Gu et al. 2013; Sato et al. 2014). These are very different situations that can lead to conflicting biodeterioration mechanisms as well as non-comparable effects on materials.

The microbial community that inhabits the indoor environment depends on the level of abiotic stress (e.g. low water availability, poor resources). The fungal and bacterial species that take advantage of a sudden availability of free water in a library are usually r-selected organisms capable of arriving, establishing themselves and persisting until reproduction and dissemination have been accomplished (Boddy and Hiscox 2016). Community change occurs when the initial colonizers are substituted by organisms that are either better able to tolerate conditions within the resource, or capable of using the resource modified by the initial colonizers and their residue. Competition for space and food may involve specialized species-specific interactions such as mycoparasitism, or be more generalized, but in any case will involve metabolic changes, reactive oxygen species production, acidic compounds and antibiotics release into the immediate environment, with the result that community development on materials is not a predictable and ordered sequence, but instead a complex, ever-changing and rather haphazard process (Boddy and Hiscox 2016).

The fast-growing fungal species that develop on paper and other organic materials stored in libraries and archives can generate strong odours (Pinzari and Montanari 2011; Micheluz et al. 2016), coloured stains (Melo et al. 2019) and toxic compounds (Micheluz et al. 2016). Substrate composition and the availability of degradable carbon sources such as sugars and starch or mineral nutrients like nitrogen, phosphorus and potassium can be determining factors in the growth of the

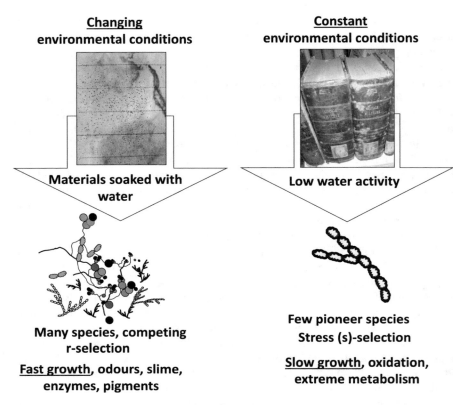

**Fig. 4.4** Diagram summarizing the two broad categories in which it is possible to separate the principal library and archival documents biodeterioration mechanisms. The main distinction is based on the availability of free water. A sudden availability of water is advantageous for organisms that are able to grow rapidly, and which have evolved to resist environments that change continuously, such as those that survive as airborne spores. They are species that produce many spores, compete with each other, and exploit the substrate in an efficient way (i.e. by secreting powerful enzymes). The opposite category consists of species of microorganisms that have evolved to colonize limiting environments, in the absence of competitors but under constant conditions. These organisms are often pioneering species that grow rather slowly and exhibit a metabolism adapted to extreme conditions, that is with little water availability or high osmotic pressure (i.e. when high salts concentrations are present in the substrate)

mycelia and the production of fruiting bodies and spores by the fungi. The competition between microbial species for water, nutrients and space in indoor environments also accounts for the time needed by the fungus to germinate or grow when a favourable situation arises. Library micro-environments characterized by transient high humidity (such as air conditioning units, or walls that absorb humidity from heavy downpours) typically support the growth of fungi capable of re-starting their growth from the dry hyphal tips within a few hours following a re-wetting event; this is the case with *Alternaria*, *Aureobasidium*, *Cladosporium*, *Phoma*, and *Ulocladium* genera (Nielsen 2002, 2003). The development of the microorganisms that require high water activity typically takes longer, subsequent to a sudden wetting event. The growth of mycelium and production of spores directly in the materials after they have been soaked with water usually takes a couple of days and depends on the ambient temperature. In fact, after flooding, librarians and conservators know that water-impregnated materials must be refrigerated or dried hastily so as to arrest the development of fungi and bacteria.

By contrast with what has been described so far for organisms capable of taking advantage of randomly and suddenly favourable situations, it is also possible to encounter very different species living on book materials and in libraries. The microorganisms that establish themselves on materials when environmental conditions are extreme, but constant, are typically slow-growing species that are particularly resistant to stress (Fig. 4.4). These species are defined as "stress-tolerant" or "s-selected"; they are specialized in occupying specific niches and their particular advantage is the ability to survive and disperse in a very hostile environment. The "s-selected" fungal species that inhabit indoor environments can tolerate desiccation and grow in restrictive conditions (Boddy 2000).

# 4 Random Events, Common Denominators, Model Mechanisms

## 4.1 Actors and Mechanisms

The wide diffusion and frequency of some bacterial and fungal species as confirmed spoilers of archival and library materials have only recently begun to emerge thanks to an increase in the number of diagnostic investigations based on massive DNA sequencing techniques that extend beyond the limits of culture-dependent investigations. Over the last ten years or so, various articles have been published which have clarified some biodeterioration mechanisms and the ecology of particular species responsible for phenomena that, although widespread, were previously unknown (Montanari et al. 2012; Piñar et al. 2015a, b, c; Pinzari et al. 2012a, 2018, Melo et al. 2019; Migliore et al. 2019).

Even if it is impossible to imagine perfectly repeatable patterns of material biodeterioration caused by the different groups of microorganisms, we can describe recurrent situations where phenomena can be parameterized, so as to model and forecast the risks posed by one or another biodeterioration mechanism, based on the presence of specific environmental conditions or the assemblage of microbial species (Fig. 4.4).

The initial creation of two broad categories, as mentioned above, is based on the availability of free water. If there is plenty of water available, such as in a sudden and unexpected event (a flood, for example), the microorganisms that attack the paper or the parchment grow rapidly, produce pigmented stains and strong odours, abundant exopolymeric material, and exhibit significant enzymatic activity that results in the

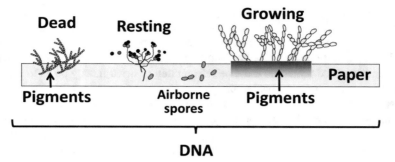

**Fig. 4.5** Diagram showing the different conditions in which microorganisms can be found in a material such as paper. Some can be active and grow, and produce pigments, whereas others can be viable but dormant; others still can be dead remains. Finally, it is possible to have many microorganisms represented by individual spores, deposited from the air, viable or otherwise. All the organisms present in the various physiological states are detectable using massive DNA sequencing analysis. In order to be able to distinguish among a long list of candidates that are responsible for damage, it is necessary to add to the genetic analysis further diagnostic techniques, capable of distinguishing the viable species from the dead ones or to describe a causative mechanism behind the damage

destruction of cellulose fibres in the case of paper and of collagen in the case of parchment (Figs. 4.2, 4.3, 4.4). In this biodeterioration scenario the species, both bacterial and fungal, can be numerous and multifarious. They are often the same as those found in dust, although sometimes unusual saprophytes are identified that are transported by chance on the materials and suddenly placed in a situation where they can germinate and grow at the expense of large quantities of organic matter (Fig. 4.5) (Pinzari and Montanari 2008). During the slow drying of the sodden materials a succession of organisms occurs that gradually take over from each other by taking advantage of the partially degraded matter and the remains of the first settlers. This represents a natural succession mechanism that can produce very similar results, albeit depending (in each instance) on a different assortment of species, all of which are capable of rapid growth, in addition to being endowed with enzymes and metabolites diffusible in abundant water and different competition mechanisms (e.g. the production of antimicrobial compounds or volatile products with allopathic functions) (Pinzari et al. 2004; Canhoto et al. 2004; Sawoszczuk et al. 2015; Micheluz et al. 2015, 2016).

Cellulolytic and proteolytic enzymes cause structural damage by means of a precisely targeted attack of the covalent chemical bonds of macromolecules (i.e. cellulose and collagen), whilst organic acids promote oxidation mechanisms and chelate microelements, such as calcium and iron (Fig. 4.1). Metals and salts can precipitate as secondary compounds between the fibres, thereby increasing spatial heterogeneity, which in turn facilitates the coexistence of multiple species of spoiling microorganisms, each linked to a particular microenvironment and capable of causing a different type of damage to the material in question (Pinzari 2018).

Table 4.2 lists some of the mechanisms involved in paper and parchment biodeterioration caused by bacteria and fungi. *Aspergillus* and *Penicillium* fungal species are among those most frequently associated with the production of organic acids and other metabolites, which can react with various components of paper (Fig. 4.2). In fact, not only cellulose can be affected by oxidative reactions and enzymatic degradation. The presence of starch, gelatin, rosin and other glues, or that of salts and mineral compounds can exert an influence on fungal metabolism and determine, for example, the kind of stains produced (Fig. 4.2). The fungal pigments that affect paper have been associated to polyketide quinones, carotenoids and other compounds whose synthesis or colour can depend on the availability of nitrogen, the prevailing pH and the presence of other limiting nutrients and enzyme cofactors such as metals and cations (i.e. Fe, K, Ca, Mg, Mn) (Melo et al. 2019). However, in order for fungi to be able to produce abundant enzymes and synthesize complex metabolites, there must be no limiting factors, and therefore water must be available together with the organic matter.

When free water is scarce, the microbiological attack on the materials takes place differently (Figs. 4.3, 4.4). It is slower, and there are only a few or single extremophile species that can germinate and grow. In such situations, generally linked to specific microenvironmental conditions, the secretion of enzymes, organic acids and pigments also occurs, albeit to a lesser extent and with limited diffusion, and often producing different effects depending on the species concerned. The appearance of damage on the material, alterations, and also identifying the presence of a biodeterioration phenomenon in progress in such cases may not be immediate, but instead occur over a long period of time. While in the case of a flood the development of microorganisms is a virtual inevitability, the outcome of other crises is far less predictable, owing to local phenomena, micro-environments or extremophile species that involve mechanisms that are more complex and difficult to predict.

An example of slow-forming damage that gradually makes itself apparent is foxing (Choi 2007; Nunes et al. 2015; Modica et al. 2019). The formation of foxing stains on paper and other materials can be caused by fungi as a result of limited water availability. Absolute tonophilic fungi germinate on paper in the presence of very low water availability ($a_w < 0.80$) and release cello-oligosaccharides, aminobutyric acid and amino acids. These compounds condensate in a spontaneous chemical reaction (Maillard condensation) to form brown-coloured compounds known as "melanoidins" (Arai 1987, 2000; Arai et al. 1990), which in many cases are responsible for foxing stains.

Pinzari (2018) described three types of biodeterioration mechanisms with increasing levels of complexity: (1) the damage is caused by the enzymatic activity of microorganisms and is focused on the main chemical component of the material, as in the case of cellulolytic fungi that attack paper soaked with water; (2) the damage is caused by the presence of peculiar conditions that trigger species with very particular requirements, e.g. extremophilic species that grow at the presence of salts; (3) the damage is the result of several related events such as the succession of species that feed on each other remains, with a primary colonizer which is often able to drastically modify the initial conditions of the material (as in mechanisms 1 and

2). Understanding what causes damage, which species are involved, the resulting metabolites and the chemical changes undergone by the materials are not a mere speculative exercise. The microorganisms' ecology, their growth requirements and the overall biodeterioration mechanism can all have a very significant effect on the outcomes of the conservation treatments. Here we describe three examples where the biodeterioration mechanisms have been examined in detail (Figs. 4.3, 4.6): (1) a rare fungal species, linked to the droppings of mites and the presence of salts, which attacked the preparatory layer of an illuminated parchment and caused its detachment; (2) a fungus, a solitary colonizer with extremophile characteristics, which found its ideal niche in libraries; (3) a complex turnover of species that recurs in ancient parchments all over the world, consistently causing the same type of damage (Figs 4.3 a, b).

## 4.2   A Rare Fungal Species

The halophilic fungus *Diploöspora rosea* (Tanney et al. 2015) was observed growing copiously on a twentieth-century illuminated parchment (Fig. 4.6). The framed parchment leaf was on display in a museum, hung on a wall affected by rising damp. The fungus was documented on the thick preparative layer of gypsum and calcium carbonate, and was seen to be capable of developing on the green pigments used for the illumination. The pigments were prepared using chromium, iron and arsenic salts. *D. rosea* grew not only in the presence of toxic metals but also showed the ability to produce fruiting structures on very low water activity media, including malt extract agar amended with 60% sucrose. Despite the extreme xerophilic nature of this fungus, its structures caused the detachment of large parts of the artwork's preparative layer and the overlying illumination. *D. rosea* is an onygenalean fungus, of uncertain taxonomic position, basal or sister to the *Gymnoascaceae* that seems to occur in archival and domestic environments subject to periodic wetting. This fungus was previously identified in 1913, on damp cardboard covered with several fungi. The dominant fungus was unknown and was described as the type species of the new hyphomycete genus *Diploöspora*. In the 100 years following its first collection, it was reported exclusively in association with illuminated parchments and in indoor settled dust in Micronesia. In the biodeteriorated parchment samples, *D. rosea* grew with other unidentified fungi and house mite exuvia and faeces pellets. Mite faeces were also observed on the holotype (i.e. the 1913 cardboard). This hints at an insect/mite-mediated dispersal mechanism for this fungus. The high salt and nitrogen concentrations that characterize mite droppings may represent a selective niche for osmophilic fungal species. The vast geographical distance between specimens suggested that *D. rosea* may have a broad distribution. Its apparent rarity may reflect its xerophilic nature and slow growth rate rather than its actual prevalence. However, its rarity in indoor settled dust samples was confirmed by the detection of only two similar sequences using pyrosequencing. This case study shows how some fungal and bacterial species, rarely found in nature, can

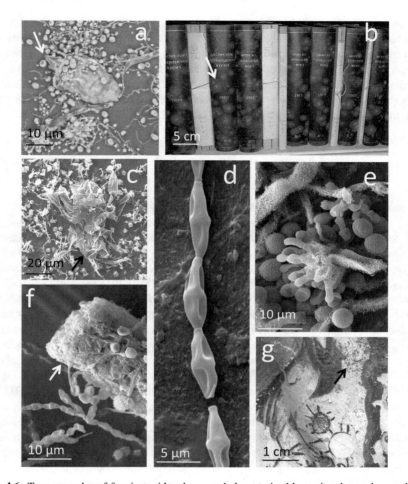

**Fig. 4.6** Two examples of fungi considered rare and characterized by a singular ecology, which have been shown to be capable of shifting with ease from extreme natural environments to man-made niches: *Eurotium halophilicum* and *Diploöspora rosea*. Both are osmophilic species, growing only on low water activity media. They are both associated with mites for their dispersion; (**a**) SEM image (variable pressure, non-coated sample), mycelium and spores of *E. halophilicum* infiltrating the remains of a mite (arrow = head of the mite); (**b**) *E. halophilicum* as a monospecific infection, forming white colonies on the spines of volumes in a library; (**c**) *D.rosea* and other fungal species close to the exuvial remains of a mite (arrow = head of the mite); (**d**) *D. rosea* chains of conidia (high vacuum SEM image on a gold-coated sample); (**e**) mycelium, head of conidiophore and spores of *E. halophilicum* (high vacuum SEM image on a gold-coated sample); (**f**) the dung of a mite containing spores of an *Aspergillus* species and infiltrated by the mycelium of *D. rosea* (high vacuum SEM image on a gold-coated sample); (**g**) the illuminated parchment upon which *D. rosea* developed. The gypsum-based layer and the pigments deposited on the parchment proved to be a perfect niche for this fungus

find in the niches created by man their ideal environment where they can develop and proliferate, sometimes without competition. In the case of the *D. rosea*, however, several factors seem to have played a role, such as the possible repeated wetting events suffered by the material, the presence of mites and their droppings, the mineral nature of the substrate that allowed the fungus to thrive, despite being an osmophilic species, probably at the expense of the glues used to consolidate the preparatory layer, subsequently causing the detachment of the illuminations.

## 4.3  A Solitary Colonizer

Occasionally, some fungi considered rare and of singular ecology, not only occupy a niche of the anthropic environment, but also find effective means of diffusion, and therefore become particularly invasive. They are often intrusive species, capable of shifting with alarming ease from extreme natural environments to man-made niches. A good example of this behaviour is provided by the spread of the fungus *Eurotium halophilicum* in Italian libraries and archives (Pinzari and Montanari 2011; Montanari et al. 2012) (Fig. 4.6). *Eurotium halophilicum* is a xerophilic fungus, with high tolerance to water stress; it was first isolated from dry food and indoor dust in association with other xerophilic fungal species and dust mites (Montanari et al. 2012; Micheluz et al. 2015) (Fig. 4.6). Since the first reports of it as a book contaminant, it has been identified in dozens of libraries and archives in Italy as the main, if not the sole fungus flourishing on books and archival materials (Pinzari and Montanari 2011; Montanari et al. 2012; Micheluz et al. 2015; Polo et al. 2017). It has also been linked to the appearance of foxing-like stains on materials (Piñar et al. 2015c; Sclocchi et al. 2016). The minimum water activity value observed for its successful germination and growth is 0.675 (Christensen et al. 1959). It has repeatedly been found in association with the covers of books and other surfaces within museums, libraries and archives, even when the overall environmental conditions are in line with those recommended for the conservation of the materials (i.e. relative humidity values ranging from 50 to 60%, and a temperature ranging from 20 to 22°C). The niches preferred by this fungus are characterized by infrequent ventilation and water condensation events after a drop in temperature or night/day thermic cycles. The inadequacy of sampling procedures and the very slow growth of the fungus on most media probably account for previous underestimations. Another xerophilic fungus, *Aspergillus penicillioides*, has often been isolated from books in association with *E. halophilicum*, probably due to its similar ecology. Other fungi have frequently been isolated from *E. halophilicum* mycelium: *Aspergillus creber*, *A. protuberus*, *Penicillium chrysogenum* and *P. brevicompactum* (Micheluz et al. 2015). These species grow on the dead mycelium of *E. halophilicum*. It has been observed that their presence is associated with historic infections of the fungus. As long as the mycelium is alive, the fungus remains the only organism on materials, and therefore represents the primary colonizer. Once established and mature, its mycelium serves as a substrate for other xerophilic organisms. Dead structures of the fungus *E. halophilicum* have also been found in foxing spots on paper and other materials. Metal oxalates have often been documented in association with fungal hyphae (i.e. calcium oxalates) (Pinzari et al. 2010). These findings are consistent with the hypothesis that absolute tonophilic fungi germinate on paper, releasing organic acids, oligosaccharides and amino acids. The damage that this fungal species causes to materials consists of pale brown to dull grey stains and other forms of decolouration (Figs. 4.3c, 4.6). When decomposition occurs, it is limited to small areas and consists of a sort of superficial erosion. This fungus also produces volatile organic compounds (Micheluz et al. 2016) and represents a potential hazard for workers who manage infected materials. GC-MS analysis showed that *E. halophilicum* produces at least 20 different volatile compounds, with acetone and 2-butanone being the main products. A total of eight secondary metabolites were detected through LC/MS-MS, e.g. deoxybrevianamid E, neoechinulin A and tryprostatin B (Micheluz et al. 2015).

## *4.4  A Complex Turnover*

In the two examples of the microbiological attack described above, the agents responsible for the colonization of the material and the main spoiling effects it results in have been identified. The other organisms associated with them have been attributed to secondary phases that only played a marginal role in the mechanism. The situation in which all or most of the organisms present in the alterations play a role and are part of a complex mechanism whose phases are difficult to reconstruct is quite different. An example of a complex mechanism is described in the works of Gallo and Strzelczyk (1971), Petushkova and Koestler (1996), Karbowska–Berent and Strzelczyk (2000), Piñar et al. (2015a, b, c), Migliore et al. (2017, 2019) (to cite only the main ones), where more than once and by means of different and progressively more sophisticated methods the cause of the formation of purple stains on parchment leaves was sought (Fig. 4.3a, b).

The parchment alteration that has long intrigued many experts consists of red or purple nucleated maculae with peripheral halos, isolate or coalescent, often associated with perforations and a loss of material which is more severe on the "flesh side" of a leaf. Karbowska–Berent and Strzelczyk (2000) isolated species of *Streptomyces* from the purple stains marring several documents of different ages and origins. They found that many of the strains had proteolytic enzymes that are capable of destroying collagen through their hydrolytic activity and were alkaliphiles, hence able to develop on skins prepared with lime and chalk (Figs. 4.1b, 4.3a, b). The isolated bacteria, however, did not reproduce the purple stains when inoculated on modern parchment.

Piñar et al. (2015a) analysed five documents with different histories and origins, all marred by purple spots. Both scanning electron microscopy and molecular analysis, based on culture-independent techniques (total DNA extraction, cloning and sequencing of bacterial 16S and fungal ITS rRNA genes), detected the presence of bacteria and fungi in the damaged areas. Halophilic, proteolytic bacterial species were found in all the documents. Moreover, as common microbial denominators, members of the *Actinobacteria*, mainly *Saccharopolyspora* spp. and species of the fungal genus *Aspergillus*, were detected in all investigated cases. Piñar et al. (2015a) proposed that a relationship exists between the phenomenon of purple spots appearing on ancient parchments and that of so-called red heat, known to affect some leather and animal skin products manufactured with marine salt and associated with the presence in the salt of extremophilic living bacteria.

Migliore et al. (2017) used 16S rRNA gene analysis in conjunction with 454-pyrosequencing to study the purple spots on a parchment roll dated to 1244 A.D. They hypothesized a two-phase model mechanism of parchment colonization, where halophilic *Archaea* colonize parchment as pioneers, followed by *Gammaproteobacteria*, and fungi as the last colonizers. In a second study (Migliore et al. 2019), the microbiomes of three parchment manuscripts affected by purple spots were analysed by means of next-generation sequencing (Illumina platform). In this instance, the authors attributed to *Halobacterium salinarum* the role of common denominator in the purple stains, along with that of the primary, triggering cause of the phenomenon. They identified as haloarchaeal bacterioruberin and bacteriorho-dopsin the pigment within the purple spots using RAMAN spectroscopy. Migliore et al. (2019) proposed a further multi-phase microbial succession model leading to progressive degradation of the parchment's collagen. In the suggested mechanism, the first phase "pioneer" colonizers are the halophilic and halotolerant microorgan-isms inhabiting the brines used to cure the skins. The second phase "late" colonizers consist of other bacteria and fungi that can vary among different environments and according to the life history of each document. The halophilic and halotolerant microbes from the marine salt enter the hides during the manufacturing process, forming the core of the purple spot damage (Perini et al. 2019). They develop inside the parchment and release proteolytic and lipolytic enzymes which attack and degrade the collagen's fibres. The formation of the nucleated purple stains was explained as the result of the lysis of halobacterial cells, with the subsequent release of bacteriorhodopsin and nutrients that can trigger colonization by other halotolerant organisms of the proteobacterial taxa (mainly *Gammaproteobacteria*) and *Firmicutes*. In the model formulated by Migliore et al. (2019), both actinobacteria and fungi are involved in the damage caused to skins because of their ability to attack not only the collagen but other compounds too, although they only participate in the last phases of the succession. Confirmation of a late role played by fungi in the colonization of parchment in areas already affected by purple spots came from different studies. Pinzari et al. (2012a) quantified both the adenosine triphosphate (ATP) and the fungal $\beta(1\text{-}4)$-N-acetyl-D-glucosaminidase activity in damaged and undamaged samples, concluding that the viable fungal mycelium was present, albeit not strictly associated with the purple stains. Additionally, Piñar et al. (2015b) quantified the $\beta$-actin gene through real-time polymerase chain reaction analyses (qPCR) in parchment samples affected by purple spots (taken from the Archimedes Palimpsest). They identified a greater abundance of fungi on degraded areas in comparison to healthy ones. A broader statistic on the communities of microorganisms associated with purple spots in the future will make it possible to draw clearer conclusions and fill the current gaps in the "puzzle", such as a potential specific role of the actinobacterium *Saccharopolyspora,* found on practically all the documents affected by purple spots (Piñar et al. 2015a, b; Teasdale et al. 2017, Migliore et al. 2017).

## 5   Perspectives

Systems biology is the study of the interactions and behaviour between the components of biological systems, and how these interactions give rise to the function of that system (for example, the enzymes and metabolites in a metabolic pathway). Systems biology also offers a useful way of thinking and of conducting research, with a more in-depth focus on complex interactions within biological systems, using a holistic approach, instead of reductionism. The future of research in the field of biodeterioration of cultural heritage, and in particular of paper and parchment documents lies, in fact, in a systemic vision of processes, and in a merging of current knowledge with all the multiple and multidisciplinary clues that will be obtained through the investigations conducted during new case studies. One of the aims of systems biology, namely to model and discover emergent properties, when applied to the study of paper and parchment biodeterioration, will facilitate reaching a better understanding of the mechanisms involved.

In the not too distant future, system biology tools will be developed to uncover the functions of the metabolites present in materials and to trace them back to the organisms that produced them. It should also be possible to predict the behaviour of communities when many environmental variables change in concert, as well as to understand the role played by individual taxa in complex communities.

The application of innovative molecular biology techniques, correlative microscopy and chemical mass spectroscopy methods has already permitted some in-depth analyses of paper and parchment biodeterioration mechanisms, as reported in some recent studies (Kraková et al. 2012, 2018; Piñar et al. 2015a, b, c; Teasdale et al. 2017; Sawoszczuk et al. 2017; Sanmartín et al. 2018; Cicero et al. 2018, Szulc et al. 2018, Mazzoli et al. 2018; Melo et al. 2019, Migliore et al. 2019; Marvasi et al. 2019). In recent times, all the interest focused on the study of biodeterioration phenomena has been directed towards a group of technologies known as "omics", which have revolutionized the approach to environmental studies. These include metagenomics, transcriptomics, metabolomics and proteomics; they are used to study total genomes and transcripts, to identify enzymes and proteins in very small samples, and to recognize the metabolites and the metabolic pathways active in materials. It is undeniable that omics tools have ushered in a new era in biological and chemical studies and have rapidly become the basic tools for all the environmental microbiology studies. They have proved to be particularly useful where there are interactions between microbiological consortia or between individual microorganisms or microorganisms and the environment. Although developed over the last ten years or so, the application of omics to the study of cultural heritage and, in particular, to archival documents and materials is only now beginning to gain ground, as are the concepts of systemic biology. Indeed, very recently Illumina MiSeq was used for massive DNA sequencing to analyse the microbiological biodiversity in historical samples of paper and parchment. The technique was applied in order to better understand phenomena which up till now have remained controversial. In fact, it was applied for the purpose of comparing brown foxing spots on paper and purple spots on parchment (Szulc et al. 2018, Migliore et al. 2019). The study identified the contribution of new microorganisms (*Phoma, Cladonia* moulds and *Gluconobacter, Ralstonia* bacteria) to the foxing phenomena observed on 19th-century paper (Szulc et al. 2018); it also helped to establish the mechanism underlying the microbial succession that occurs in purple stains on parchment, with haloarchaea establishing initially, followed by halotolerant bacteria, then actinobacteria, and finally fungi arriving at the end of the deterioration process (Migliore et al. 2019). In addition to massive DNA sequencing techniques, metabolomic techniques have been shown to provide useful elements for understanding the mechanisms underlying the biodeterioration of materials (Fig. 4.5). Metabolomics encompasses a comprehensive qualitative and quantitative analysis of small molecule substances with various properties that contribute to the metabolic pathways in the system under study. Metabolomics and proteomics provide information on biological mechanisms and potential biomarkers in samples. Recently, metabolomic analysis based on the AuNPET SALDI-ToF-MS method was applied in the study of foxing stains marring some 19th-century papers. This technique enabled the authors (Szulc et al. 2018) to demonstrate the occurrence in the stained areas of several metabolic pathways, including sugar degradation, amino acid and protein metabolism, ubiquinone and other terpenoid-quinone biosynthesis, 2-methyl-6-phytylquinol and delta-, gamma-, beta-tocopherols (responsible for the yellowish-brown colour of foxing spots) and 3-hydroxy-L-kynurenine (a fluorescent, yellow compound). These pigments can all contribute to the mechanism underlying the appearance of foxing caused by microorganisms (Szulc et al. 2018).

Notwithstanding the considerable body of knowledge with respect to the metabolic processes occurring in organisms, the number of identified metabolites in databases, particularly for microorganisms involved in cultural heritage deterioration, remains limited. Having a complete database that also includes the metabolism of taxa of non-biomedical or food interest is a goal yearned for by all microbiologists working in the field of environmental metabolomics (Marvasi et al. 2019). Similarly, public databases that facilitate the alignment of diagnostic sequences for the identification of fungi and bacteria are often lacking for taxa whose species are poorly studied, of little interest, and troublesome to cultivate in vitro, and hence poorly represented in living culture collections. This is a common situation for microorganisms that attack cultural heritage. Molecular ecology studies are very effective in capturing biodiversity, but the fungal and bacterial isolates conserved as living collections allow for studies of growth, morphology, secondary metabolism, genomics and other traits (Paiva de Carvalho et al. 2016). Without the existence of a living culture that makes it possible to couple morphological observations and DNA extraction, many molecular sequences are of little use since they often result in poor matches with sequences stored in public databases. This state of affairs underscores the importance of actively collecting and accessioning sequences of described species that are currently unrepresented in sequence databases (new or unsequenced species). Populating databases, such as GenBank, with verified sequences obtained from organisms isolated in cultural heritage biodeterioration studies will improve future amplicon-based metagenomics studies. In addition to meagre databases, the environmental sector and especially the cultural heritage sector suffer from a paucity of computational biology tools that would make it possible to combine genetic and chemical data, or informatics tools that would be of help in modelling biodeterioration mechanisms, so as to be able to predict the processes before they happen (Sterflinger et al. 2018).

If we had adequate and specifically earmarked funds for research in the field of biodeterioration of cultural heritage, we could envisage the development, in the not too distant future, of bioinformatics tools capable of utilizing the masses of data already available and those that are rapidly being generated, so as to enable us to identify unique chemical or genetic markers associated with the deterioration of materials (e.g. early diagnostic systems that predict harmful mechanisms based on a few metabolomic or genomic clues).

The quantity of data obtained during omics analyses is enormous: a single, relatively simple experiment aimed at analysing a microbial metabolome generates tens of thousands of spectra. These data, as well as those obtained from studies of metagenomics and transcriptomics, which can be directly associated with the biodeterioration mechanisms of materials and chemical data on the materials themselves, should by rights already be collected within a single public database. This would permit the development of appropriate tools for bioinformatics and networks analysis, and therefore promote and stimulate further studies on the mechanisms and preventative methods, as well as early diagnosis of cultural heritage biodeterioration.

**Acknowledgements** We would like to thank Mark Livesey for his helpful suggestions during the preparation of the English text, and the anonymous reviewer for the meaningful comments.

# References

Arai H (1987) Microbiological studies on the conservation of paper and related cultural properties. Part 5. Physiological and morphological characteristics of fungi isolated from foxing, formation mechanisms and countermeasures. Sci Conserv 26:43–52

Arai H (2000) Foxing caused by fungi: twenty–five years of study. Int Biodeter Biodegr 46:181–188. https://doi.org/10.1016/S0964-8305(00)00063-9

Arai H, Matsumura N, Murakita H (1990) Microbiological studies on the conservation of paper and related cultural properties, part 9, induction of artificial foxing. Hozon Kagaku 29:25–34. (in Japanese)

Ayu DC, Teja TP (2016) Occurrence of fungi on deteriorated old dluwang manuscripts from Indonesia. Int Biodeter Biodegrad 114:94–103. https://doi.org/10.1016/j.ibiod.2016.05.025

Blyskal B, Lenart–Borod A, Borod P (2017) Approaches to taxonomic studies of actinomycetes isolated from historic and contemporary materials. J Pure Appl Microbiol 11:637–648

Boddy L (2000) Interspecific combative interactions between wood decaying basidiomycetes. FEMS Microbiol Ecol 31:185–194

Boddy L, Hiscox J (2016) Fungal ecology: principles and mechanisms of colonization and competition by saprotrophic fungi. Microbiol Spectrum 4: FUNK-0019-2016

Borrego S, Lavin P, Perdomo I, Gómez de Saravia S, Guiamet P (2012) Determination of indoor air quality in archives and biodeterioration of the documentary heritage. ISRN Microbiol 2012:680598

Brunner I, Fischer M, Rüthi J, Stierli B, Frey B (2018) Ability of fungi isolated from plastic debris floating in the shoreline of a lake to degrade plastics. PLoS ONE 13:e0202047. https://doi.org/10.1371/journal.pone.0202047

Canhoto O, Pinzari F, Fanelli C, Magan N (2004) Application of electronic nose technology for the detection of fungal contamination in library paper. Int Biodeter Biodegr 54:303–309. https://doi.org/10.1016/j.ibiod.2004.04.001

Choi S (2007) Foxing on paper: a literature review. J Am Inst Conserv 46:137–152. https://doi.org/10.1179/019713607806112378

Christensen C, Papavizas GC, Benjamin CR (1959) A new halophilic species of *Eurotium*. Mycologia 51:636–640

Cicero C, Pinzari F, Mercuri F (2018) 18th Century knowledge on microbial attacks on parchment: analytical and historical evidence. Int Biodeter Biodegr 134:76–82. https://doi.org/10.1016/j.ibiod.2018.08.007

Cordero RJB, Casadevall A (2017) Functions of fungal melanin beyond virulence. Fungal Biol Rev 31:99–112

Coronado-Ruiz C, Avendaño R, Escudero–Leyva E, Conejo-Barboza G, Chaverri P, Chavarría M (2018) Two new cellulolytic fungal species isolated from a 19th–century art collection. Sci Rep 8:1–9. https://doi.org/10.1038/s41598-018-24934-7

Corte AM, Ferroni A, Salvo VS (2003) Isolation of fungal species from test samples and maps damaged by foxing, and correlation between these species and the environment. Int Biodeterior Biodegradation 51:167–173. https://doi.org/10.1016/S0964-8305(02)00137-3

De Paolis MR, Lippi D (2008) Use of metabolic and molecular methods for the identification of a Bacillus strain isolated from paper affected by foxing. Microbiol Res 163:121–131. https://doi.org/10.1016/j.micres.2007.06.002

El Bergadi F, Laachari F, Elabed S, Mohammed IH, Ibnsouda SK (2014) Cellulolytic potential and filter paper activity of fungi isolated from ancients manuscripts from the Medina of Fez. Ann Microbiol 64:815–822. https://doi.org/10.1007/s13213-013-0718-6

Florian ML (1996) The role of the conidia of fungi in fox spots + rusty irregularly shaped areas on rag paper in 16th-century to 19th-century books. Stud Conserv 41:65–67. https://doi.org/10.2307/1506518

Florian M–LE, Manning L (2000) SEM analysis of irregular fungal spot in an 1854 book: population dynamics and species identification. Int Biodeterior Biodegrad 46:205–220

Florian M–LE (2002) Fungal facts: solving fungal problems in heritage collections. Archetype publications. London

Florian M–LE (2007) Protein facts. Fibrous proteins in cultural and natural history artifacts. Archetype Publications, London

Gallo F, Strzelczyk A (1971) Indagine preliminare sulle alterazioni microbiche della pergamena. Boll Ist Patol Libro 30:71–87

Green PWC, Farman DI (2015) Can paper and glue alone sustain damaging populations of booklice, *Liposcelis bostrychophila*? J Conserv Museum Stud 13:3

Gu JD, Kigawa R, Sato Y, Katayama Y (2013) Addressing the microbiological problems of cultural property and archive documents after earthquake and tsunami. Int Biodeter Biodegr 85:345–346. https://doi.org/10.1016/j.ibiod.2013.08.018

Gutarowska B (2016) A modern approach to biodeterioration assessment and the disinfection of historical book collections. Lodz University of Technology

Hyde KD, Xu J, Rapior S et al (2019) The amazing potential of fungi: 50 ways we can exploit fungi industrially. Fungal Divers 97:1. https://doi.org/10.1007/s13225-019-00430-9

Jacobsen RM, Sverdrup–Thygeson A, Kauserud H, Birkemoe T (2018) Revealing hidden insect–fungus interactions; moderately specialized, modular and anti–nested detritivore networks. Proc R Soc B 285:20172833. https://doi.org/10.1098/rspb.2017.2833

Jayasekara S, Ratnayake R (2019) Microbial cellulases: an overview and applications. In: Rodríguez Pascual A, Eugenio Martín ME (eds) Cellulose. IntechOpen, London. https://doi.org/10.5772/intechopen.84531.

Karakasidou K, Nikolouli K, Amoutzias GD, Pournou A, Manassis C, Tsiamis G, Mossialos D (2018) Microbial diversity in biodeteriorated Greek historical documents dating back to the 19th and 20th century: a case study. Microbiol Open 7:e00596. https://doi.org/10.1002/mbo3.596

Karbowska–Berent J, Strzelczyk A (2000) The role of streptomycetes in the biodeterioration of historic parchment. Nicolaus Copernicus University, Torun

Karbowska-Berent J, Jarmiłko J, Czuczko J (2014) Fungi in fox spots of a drawing by Leon Wyczółkowski. Restaurator 35:159–179. https://doi.org/10.1515/res-2014-1000

Koul B, Upadhyay H (2018) Fungi–mediated biodeterioration of household materials, libraries, cultural heritage and its control. In: Gehlot P, Singh J (eds) Fungi and their role in sustainable development: current perspective. Springer, Singapore, pp 597–615

Kraková L, Chovanová K, Selim SA, Šimonovičová A, Puškarová A, Maková A, Pangallo D (2012) A multiphasic approach for investigation of the microbial diversity and its biodegradative abilities in historical paper and parchment documents. Int Biodeter Biodegr 70:117–125. https://doi.org/10.1016/j.ibiod.2012.01.011

Kraková L, Šoltys K, Otlewska A, Pietrzak K, Purkrtová S, Savická D, Puškárová A, Bučková M, Szemes T, Budiš J, Demnerová K, Gutarowska B, Pangallo D (2018) Comparison of methods for identification of microbial communities in book collections: culture–dependent (sequencing and MALDI–TOF-MS) and culture–independent (Illumina MiSeq). Int Biodeter Biodegr 131:51–59. https://doi.org/10.1016/j.ibiod.2017.02.015

Larsen R (2002) Microanalysis of parchment. Archetype Publications, London

Lax S, Cardona C, Zhao D, Winton VJ, Goodney G, Gao P, Gottel N, Hartmann EM, Henry C, Thomas PM, Kelley ST, Stephens B, Gilbert JA (2019) Microbial and metabolic succession on common building materials under high humidity conditions. Nat Commun 10:1767. https://doi.org/10.1038/s41467-019-09764-z

Lazaridis M, Katsivela E, Kopanakis I, Raisi L, Mihalopoulos N, Panagiaris G (2018) Characterization of airborne particulate matter and microbes inside cultural heritage collections. J Cult Herit 30:136–146. https://doi.org/10.1016/j.culher.2017.09.018

Lech T (2016) Evaluation of a parchment document, the 13th century incorporation charter for the city of Krakow, Poland, for microbial hazards. Appl Environ Microbiol 82:2620–2631. https://doi.org/10.1128/AEM.03851-15

Liao C-M, Luo W-C, Chen S-C, Chen J-W, Liang H-M (2004) Temporal/seasonal variations of size–dependent airborne fungi indoor/outdoor relationships for a wind–induced naturally ventilated airspace. Atmos Environ 38:4415–4419. https://doi.org/10.1016/j.atmosenv.2004.04.029

Liu Z, Zhang Y, Zhang F, Hu C, Liu G, Pan J (2018) Microbial community analyses of the deteriorated storeroom objects in the Tianjin Museum using culture–independent and culture–dependent approaches. Front Microbiol 9:802. https://doi.org/10.3389/fmicb.2018.00802

Mallo AC, Nitiu DS, Elíades LA, Saparrat MCN (2017) Fungal degradation of cellulosic materials used as support for cultural heritage. Int J Conserv Sci 8:619–632

Marvasi M, Cavalieri D, Mastromei G, Casaccia A, Perito B (2019) Omics technologies for an in–depth investigation of biodeterioration of cultural heritage. Int Biodeter Biodegr 144:104736. https://doi.org/10.1016/j.ibiod.2019.104736

Mazzoli R, Giuffrida MG, Pessione E (2018) Back to the past: "find the guilty bug—microorganisms involved in the biodeterioration of archeological and historical artifacts". Appl Microbiol Biotechnol 102:6393–6407. https://doi.org/10.1007/s00253-018-9113-3

Melo D, Sequeira SO, Lopes JA, Macedo MF (2019) Stains versus colourants produced by fungi colonising paper cultural heritage: a review. J Cult Herit 35:161–182. https://doi.org/10.1016/j.culher.2018.05.013

Michaelsen A, Pinzari F, Ripka K, Lubitz W, Piñar G (2006) Application of molecular techniques for identification of fungal communities colonising paper material. Int Biodeter Biodegr 58:133–141. https://doi.org/10.1016/j.ibiod.2006.06.019

Michaelsen A, Piñar G, Montanari M, Pinzari F (2009) Biodeterioration and restoration of a 16th–century book using a combination of conventional and molecular techniques: a case–study. Int Biodeter Biodegr 63:161–168. https://doi.org/10.1016/j.ibiod.2008.08.007

Michaelsen A, Piñar G, Pinzari F (2010) Molecular and microscopical investigation of the microflora inhabiting a deteriorated Italian manuscript dated from the 13th–century. Microb Ecol 60:69–80. https://doi.org/10.1007/s00248-010-9667-9

Micheluz A, Manente S, Tigini V, Prigione V, Pinzari F, Ravagnan G, Varese GC (2015) The extreme environment of a library: xerophilic fungi inhabiting indoor niches. Int Biodeter Biodegr 99:1–7. https://doi.org/10.1016/j.ibiod.2014.12.012

Micheluz A, Manente S, Rovea M, Slanzi D, Varese GC, Ravagnan G, Formenton G (2016) Detection of volatile metabolites of moulds isolated from a contaminated library. J Microbiol Methods 128:34–41. https://doi.org/10.1016/j.mimet.2016.07.004

Migliore L, Thaller MC, Vendittozzi G, Mejia AY, Mercuri F, Orlanducci S, Rubechini A (2017) Purple spot damage dynamics investigated by an integrated approach on a 1244 A.D. parchment roll from the secret Vatican archive. Sci Rep 7:9521. https://doi.org/10.1038/s41598-017-05398-7

Migliore L, Perini N, Mercuri F, Orlanducci S, Rubechini A, Thaller MC (2019) Three ancient documents solve the jigsaw of the parchment purple spot deterioration and validate the microbial succession model. Sci Rep 9:1623. https://doi.org/10.1038/s41598-018-37651-y

Modica AMB, Di Bella M, Alberghina MF, Brai MDF, Tranchina L (2016) Characterization of foxing stains in early twentieth century photographic and paper materials. Nat Prod Res 33:987–996. https://doi.org/10.1080/14786419.2016.1180600

Modica A, Bruno M, Di Bella M, Alberghina MF, Brai M, Fontana D, Tranchina L (2019) Characterization of *foxing* stains in early twentieth century photographic and paper materials. Nat Prod Res 33:987–996. https://doi.org/10.1080/14786419.2016.1180600

Montanari M, Melloni V, Pinzari F, Innocenti G (2012) Fungal biodeterioration of historical library materials stored in compactus movable shelves. Int Biodeter Biodegr 75:83–88. https://doi.org/10.1016/j.ibiod.2012.03.011

Nielsen KF (2002) Mould growth on building materials. Secondary metabolites, mycotoxins and biomarkers. PhD Thesis. BioCentrum–DTU, Technical University of Denmark.

Nielsen KF (2003) Review: Mycotoxin production by indoor molds. Fungal Genet Biol 39:103–117

Nol L, Henis Y, Kenneth RG (2001) Biological factors of foxing in postage stamp paper (Reprinted). Int Biodeterior Biodegradation 48:98–104. https://doi.org/10.1016/S0964-8305(01)00072-5

Nosanchuk JD, Stark RE, Casadevall A (2015) Fungal Melanin: what do we know about structure? Front Microbiol 6:1463. https://doi.org/10.3389/fmicb.2015.01463

Nunes I, Mesquita N, Cabo Verde S, Leitao Bandeira AM, Carolino MM, Portugal A, Botelho ML (2013) Characterization of an airborne microbial community: a case study in the archive of the University of Coimbra, Portugal. Int Biodeter Biodegr 79:36–41. https://doi.org/10.1016/j.ibiod.2013.01.013

Nunes M, Relvas C, Figueira F, Campelo J, Candeias A, Caldeira AT, Ferreira T (2015) Analytical and microbiological characterization of paper samples exhibiting foxing stains. Microsc Microanal 21:63–77. https://doi.org/10.1017/S143192761500001X

Oetari A, Susetyo–Salim T, Sjamsuridzal W, Suherman EA, Monica M, Wongso R, Fitri R, Nurlaili DG, Ayu DC, Teja TP (2016) Occurrence of fungi on deteriorated old *dluwang* manuscripts from Indonesia. Int Biodeter Biodegr 114:94–103. https://doi.org/10.1016/j.ibiod.2016.05.025

Paiva de Carvalho H, Mesquita N, Trovão J, Peixoto da Silva J, Rosa B, Martins R, Bandeira AML, Portugal A (2016) Diversity of fungal species in ancient parchments collections of the archive of the University of Coimbra. Int Biodeter Biodegr 108:57–66. https://doi.org/10.1016/j.ibiod.2015.12.001

Pangallo D, Chovanova K, Simonovicova A, Ferianc P (2009) Investigation of microbica community isolated from indoor artworks and their environment: identification, biodegradative abilities, and DNA typing. Can J Microbiol 55:277–287. https://doi.org/10.1139/W08-136

Perini N, Mercuri F, Thaller MC, Orlanducci S, Castiello D, Talarico V, Migliore L (2019) The stain of the original salt: red heats on chrome tanned leathers and purple spots on ancient parchments are two sides of the same ecological coin. Front Microbiol 10:2459. https://doi.org/10.3389/fmicb.2019.02459

Petushkova JP, Koestler R (1996) Biodeterioration studies on parchment and leather attacked by bacteria in the Commonwealth of Socialist States. In: Federici C, Munafò PF (eds) International conference on conservation and restoration of archival and library materials. Erice, 22nd-29th April 1996. G.P. Palumbo, Palermo, pp 195–211

Piñar G, Sterflinger K, Pinzari F (2015a) Unmasking the measles–like parchment discoloration: molecular and microanalytical approach. Environ Microbiol 17:427–443. https://doi.org/10.1111/1462-2920.12471

Piñar G, Sterflinger K, Ettenauer J, Quandt A, Pinzari F (2015b) A combined approach to assess the microbial contamination of the Archimedes Palimpsest. Microb Ecol 69:118–134. https://doi.org/10.1007/s00248-014-0481-7

Piñar G, Tafer H, Sterflinger K, Pinzari F (2015c) Amid the possible causes of a very famous foxing: molecular and microscopic insight into Leonardo da Vinci's self–portrait. Environ Microbiol Rep 7:849–859. https://doi.org/10.1111/1758-2229.12313

Pinzari F (2011) Microbial ecology of indoor environments. The ecological and applied aspects of microbial contamination in archives, libraries and conservation environments. In: Abdul-Wahab SA (ed) Sick building syndrome in public buildings and workplaces. Springer, Berlin, pp 153–178

Pinzari F (2018) Microbial processes involved in the deterioration of paper and parchment. In: Mitchell R, Clifford J (eds) Biodeterioration and preservation in art, archaeology and architecture. Archetype Publications, London, pp 33–56

Pinzari F, Montanari M (2008) A substrate utilisation pattern (SUP) method for evaluating the biodeterioration potential of micro -flora affecting libraries and archival materials. In: Joice H, Townsend L, Toniolo F, Cappitelli F (eds) Conservation science. Archetype Publications, London, pp 236–241

Pinzari F, Montanari M (2011) Mould growth on library materials stored in compactus–type shelving units. In: Abdul-Wahab SA (ed) Sick building syndrome in public buildings and workplaces. Springer, Berlin, pp 193–206

Pinzari F, Canhoto O, Fanelli C, Magan N (2004) Electronic nose for the early detection of moulds in libraries and archives. Indoor Built Environ 13:387–395. https://doi.org/10.1177/1420326X04046948

Pinzari F, Pasquariello G, De Mico A (2006) Biodeterioration of paper: a SEM study of fungal spoilage reproduced under controlled conditions. Macromol Symp 238:57–66. https://doi.org/10.1002/masy.200650609

Pinzari F, Zotti M, De Mico A, Calvini P (2010) Biodegradation of inorganic components in paper documents: formation of calcium oxalate crystals as a consequence of *Aspergillus terreus* Thom growth. Int Biodeter Biodegr 64:499–505. https://doi.org/10.1016/j.ibiod.2010.06.001

Pinzari F, Troiano F, Piñar G, Sterflinger K, Montanari M (2011) The contribution of microbiological research in the field of book, paper and parchment conservation. In: Engel P, Schirò J, Larsen R, Moussakova E, Kecskeméti I (eds) New approaches to book and paper conservation–restoration. Verlag Berger, Horn/Wien, pp 575–594

Pinzari F, Cialei V, Piñar G (2012a) A case study of ancient parchment biodeterioration using variable pressure and high vacuum scanning electron microscopy. In: Meeks N, Cartwright C, Meek A, Mongiatti A (eds) Historical technology, materials and conservation: SEM and microanalysis. Archetype Publications, London, pp 93–99

Pinzari F, Colaizzi P, Maggi O, Persiani AM, Schütz R, Rabin I (2012b) Fungal bioleaching of mineral components in a twentieth–century illuminated parchment. Anal Bioanal Chem 402:1541–1550. https://doi.org/10.1007/s00216-011-5263-1

Polo A, Cappitelli F, Villa F, Pinzari F (2017) Biological invasion in the indoor environment: the spread of *Eurotium halophilicum* on library materials. Int Biodeter Biodegr 118:34–44. https://doi.org/10.1016/j.ibiod.2016.12.010

Principi P, Villa F, Sorlini C, Cappitelli F (2011) Molecular studies of microbial community structure on stained pages of Leonardo da Vinci's Atlantic Codex. Microb Ecol 61:214–222. https://doi.org/10.1007/s00248-010-9741-3

Puškárová A, Bučková M, Habalová B, Kraková L, Maková A, Pangallo D (2016) Microbial communities affecting albumen photography heritage: a methodological survey. Sci Rep 6:20810. https://doi.org/10.1038/srep20810

Ramsey MM, Freire MO, Gabrilska RA, Rumbaugh KP, Lemon KP (2016) *Staphylococcus aureus* shifts toward commensalism in response to *Corynebacterium* species. Front Microbiol 7:1230. https://doi.org/10.3389/fmicb.2016.01230

Rakotonirainy MS, Heude E, Lavedrine B (2007) Isolation and attempts of biomolecular characterization of fungal strains associated to foxing on a 19th century book. J Cult Herit 8:126–133. https://doi.org/10.1016/j.culher.2007.01.001

Ritz K (1995) Growth responses of some soil fungi to spatially heterogeneous nutrients. FEMS Microbiol Ecol 16:269–280. https://doi.org/10.1016/0168-6496(94)00090-J

Saada NS, Abdel-Maksound G, Youssef AM, Abdel-Aziz MS (2018) The hydrolytic activities of two fungal species isolated from historical quranic parchment manuscript. J Soc Leath Tech Ch 102:141–148

Sanmartín P, DeAraujo A, Vasanthakumar A (2018) Melding the old with the new: trends in methods used to identify, monitor, and control microorganisms on cultural heritage materials. Microb Ecol 76:64–80. https://doi.org/10.1007/s00248-016-0770-4

Sato YY, Aoki M, Kigawa R (2014) Microbial deterioration of tsunami–affected paper–based objects: a case study. Int Biodeter Biodegr 88:142–149. https://doi.org/10.1016/j.ibiod.2013.12.007

Sawoszczuk T, Syguła–Cholewińska J, del Hoyo-Meléndez JM (2015) Optimization of headspace solid phase microextraction for the analysis of microbial volatile organic compounds emitted by fungi: application to historical objects. J Chromatogr A 1409: 30-45. doi:https://doi.org/10.1016/j.chroma.2015.07.059

Sawoszczuk T, Syguła-Cholewińska J, del Hoyo-Meléndez JM (2017) Application of HS–SPME–GC–MS method for the detection of active moulds on historical parchment. Anal Bioanal Chem 409:2297–2307. https://doi.org/10.1007/s00216-016-0173-x

Sclocchi MC, Kraková L, Pinzari F, Colaizzi P, Bicchieri M, Šaková N (2016) Microbial life and death in a foxing stain: a suggested mechanism of photographic prints defacement. Microb Ecol 73:1–12. https://doi.org/10.1007/s00248-016-0913-7

Sterflinger K, Pinzari F (2012) The revenge of time: Fungal deterioration of cultural heritage with particular reference to books, paper and parchment. Environ Microbiol 14:559–566. https://doi.org/10.1111/j.1462-2920.2011.02584.x

Sterflinger K, Little B, Piñar G, Pinzari F, de los Rios A, Gu JD (2018) Future directions and challenges in biodeterioration research on historical materials and cultural properties. Int Biodeter Biodegr 129:10–12. https://doi.org/10.1016/j.ibiod.2017.12.007

Szczepanowska H, Mathia TG, Belin P (2014) Morphology of fungal stains on paper characterized with multi-scale and multi-sensory surface metrology. Scanning 36:76–85. https://doi.org/10.1002/sca.21095

Szczepanowska HM, Jha D, Mathia TG (2015) Morphology and characterization of Dematiaceous fungi on a cellulose paper substrate using synchrotron X–ray microtomography, scanning electron microscopy and confocal laser scanning microscopy in the context of cultural heritage. J Anal Atom Spectrom 30:651–657. https://doi.org/10.1039/c4ja00337c

Szulc J, Otlewska A, Ruman T, Kubiak K, Karbowska-Berent J, Kozielec T, Gutarowska B (2018) Analysis of paper foxing by newly available omics techniques. Int Biodeter Biodegr 132:157–165. https://doi.org/10.1016/j.ibiod.2018.03.005

Tanney JB, Nguyen HDT, Pinzari F, Seifert KA (2015) A century later: rediscovery, culturing and phylogenetic analysis of *Diploöspora rosea*, a rare onygenalean hyphomycete. Anton Leeuw Int J G 108:1023–1035. https://doi.org/10.1007/s10482-015-0555-7

Teasdale MD, Fiddyment S, Vnouček J, Mattiangeli V, Speller C, Binois A, Carver M, Dand C, Newfield TP, Webb CC, Bradley DG, Collins MJ (2017) The York Gospels: a 1000–year biological palimpsest. R Soc Open Sci 4:170988. https://doi.org/10.1098/rsos.170988

Treseder KK, Lennon JT (2015) Fungal traits that drive ecosystem dynamics on land. Microbiol Mol Biol Rev 79:243–262. https://doi.org/10.1128/MMBR.00001-15

Vaisanen OM, Weber A, Bennasar A, Rainey FA, Busse HJ, Salkinoja-Salonen MS (1998) Microbial communities of printing paper machines. J Appl Microbiol 84:1069–1084

Wang XW, Houbraken J, Groenewald JZ, Meijer M, Andersen B, Nielsen KF, Crous PW, Samson RA (2016) Diversity and taxonomy of *Chaetomium* and *Chaetomium*–like fungi from indoor environments. Stud Mycol 84:145–224. https://doi.org/10.1016/j.simyco.2016.11.005

Zotti M, Ferroni A, Calvini P (2008) Microfungal biodeterioration of historic paper: preliminary FTIR and microfungal analyses. Int Biodeterior Biodegradation 62:186–194. https://doi.org/10.1016/j.ibiod.2008.01.005

Zotti M, Ferroni A, Calvini P (2011) Mycological and FTIR analysis of biotic foxing on paper substrates. Int Biodeterior Biodegradation 65:569–578. https://doi.org/10.1016/j.ibiod.2010.01.011

Zyska B (1997) Fungi isolated from library materials: a review of the literature. Int Biodeter Biodegr 40:43–51. https://doi.org/10.1016/S0964-8305(97)00061-9

# Part II
# Green Methods Again Biodeterioration

# Chapter 5
# Novel Antibiofilm Non-Biocidal Strategies

**Francesca Cappitelli and Federica Villa**

**Abstract** Subaerial biofilm (SAB) formation on cultural heritage objects is often considered an undesirable process in which microorganisms and their by-products, e.g., enzymes and pigments, cause damage or alteration to a surface. Since biofilms are widespread phenomena, there has been a high demand for preventive and control strategies that resist their formation or reduce their negative effects once formed. Up to date, the main strategy to control biofilms has been the use of biocides. Because of their intrinsic properties, biocidal products can pose risks to humans, animals, and the environment. In this chapter, the authors call "green" only those alternative strategies to biocides able to prevent/control biofilms but that do not kill microorganisms, i.e., irrespective of the use of natural compounds. Here, we describe some of the methods that are most commonly used to test the effectiveness of antibiofilm compounds with multiple-species biofilm model systems. A unified terminology and well described protocols and guidelines are still required to compare and test the effectiveness of traditional or novel compounds against biofilms retrieved on heritage surfaces.

**Keywords** Antibiofilm · Sublethal · Green alternatives · Prevention · Control · Lab models

## 1 Biocides

Biofilms are highly structured communities of microbial cells that adhere to a surface and are embedded in extracellular polymeric substances (EPS) consisting of polysaccharides, nucleic acids, proteins, and lipids. Biofilm is a natural and ubiquitous form of bacterial and fungal growth. The fact that EPS are commonly observed close to their cells also suggests that lichens are biofilms (Banfield et al. 1999; De Los Rios

F. Cappitelli (✉) · F. Villa
Department of Food, Environmental and Nutritional Sciences, Università degli Studi di Milano, Milano, Italy
e-mail: francesca.cappitelli@unimi.it; federica.villa@unimi.it

© The Author(s) 2021

E. Joseph (ed.), *Microorganisms in the Deterioration and Preservation of Cultural Heritage*, https://doi.org/10.1007/978-3-030-69411-1_5

117

et al. 2005; Grube and Berg 2009; Casano et al. 2015). The concept of the biofilm implies a need to take into consideration the interactions established among microorganisms, and of the microorganisms with the heritage substratum and the immediate environment, a view that overcomes the value-laden identification of microorganisms and their activities as single taxon.

In the scientific literature, biofilms have been mainly investigated when growing on a solid surface and exposed to a liquid phase. Indeed, most studies have focused on human infecting biofilms, typically at the solid–liquid interface of the inner body and in marine biofouling environments. Medical literature addressing the solid–air interface is considerably less substantial and concerns (partially) only skin wound biofilms. Despite many similarities between biofilms grown at a liquid–solid interface and at an air–solid interface, there are marked differences that have been noted only relatively recently. In contrast to the solid–air interface, *Bacillus subtilis* forms sessile pellicles at liquid–air and solid–liquid interfaces (Vlamakis et al. 2013). In this scenario, the work of cultural heritage microbiologists that mainly addresses sessile life at solid–air interfaces is pivotal to study a biofilm aspect that is less investigated.

Subaerial biofilm (SAB) formation on cultural heritage objects is often considered an undesirable process in which microorganisms and their by-products, e.g., enzymes and pigments, cause damage or alteration to a surface. Since biofilms are widespread phenomena, there has been a high demand for preventive and control strategies that resist their formation or reduce their negative effects once formed. Up to date, the main strategy to control biofilms has been the use of biocides. According to the European Regulation 528/2012, "biocidal product" means "*any substance or mixture, in the form in which it is supplied to the user, consisting of, containing or generating one or more active substances, with the intention of destroying, deterring, rendering harmless, preventing the action of, or otherwise exerting a controlling effect on, any harmful organism by any means other than mere physical or mechanical action*" (Regulation (EU) No 528/2012 of the European Parliament and of the Council of 22 May 2012). In the past, biocides have been freely used to control biofilms on cultural heritage surfaces. However, the use of biocides has been proven to have major drawbacks: biocides harm non-target populations of the surrounding environment and they carry a risk of the development of resistance to themselves as well as cross-resistance to antibiotics (Villa et al. 2020). Thus, because of their intrinsic properties, biocidal products can pose risks to humans, animals, and the environment. As a result, the EU has set up strict rules and procedures to minimize these risks and all biocidal products have to be authorized by a competent authority before they are placed on the market (https://ec.europa.eu/health/biocides/overview_en).

Moreover, specifically in relation to conservation, before deciding whether removal of microbial colonization with biocides is actually beneficial, we should consider the effects of this so-called biodeterioration in comparison with the physico-chemical deterioration induced by the use of the biocidal agent. In outdoor conditions, once structures and objects have been cleaned, recolonization is inevitable and may begin shortly after treatment. In many cases no well-defined differences in the microbial community between the untreated and the treated surfaces are seen as only a biofilm component is affected (Urzì et al. 2016). In other cases, subsequent colonization might be more aggressive or disfiguring than the one eradicated (Martin-Sanchez et al. 2012).

In conservation as well as in all other fields, the new approach is to find more sustainable ways to use biocides for the removal of deleterious biofilms from relevant abiotic surfaces and, at the same time, to search for feasible sound alternatives.

## 2   What Do We Mean by Green Alternatives?

In this chapter we define green alternatives those able to prevent/control biofilms without affecting cell growth. One of the many advantages is that, if the antibiofilm effect is not due to a biocidal effect, it is possible to prevent the generation of microbial drug-resistance.

At first glance, physical techniques seem a good green alternative for the removal of biodeteriogens. These methods dislodge or cause the lysis of some microorganisms with the help of tools, such as brushes (Caneva et al. 1991; Borderie et al. 2015). In contrast to biocides, the killing does not lead to resistance but the material can be discolored or damaged (López et al. 2010; Vujcic et al. 2019). Additionally, the effect is mainly on the surface, e.g., endolithic growth is generally not affected. Therefore, physical cleaning is not the focus of this chapter even if its use in combination with other methods discussed here can be considered.

Antibiofilm agents target various stages of biofilm growth, namely adhesion, maturation, and dispersion. Any action that inhibits adhesion is considered a preventive approach, whereas any action that interferes with maturation and dispersion is included in a control strategy. Prevention can be operated with two strategies: either through the employment of specific substances in the repository environment or by modifying the surface of the heritage object. For instance, essential oils in the vapor phase or blended with some solvent have the potential to be used as antibiofilm substances and can be considered an interesting alternative (Borrego et al. 2012).

From an ecological point of view, the selection of plants as source materials of antimicrobial compounds is a good approach, since plants produce a wide range of secondary metabolites that naturally defend them against microorganisms. Silva et al. (2016) reported a comprehensive list of plant metabolites that have been proven effective against microorganisms and their products, including pigments. We can argue about whether this is the green alternative we need. "Natural," including plant-derived compounds, is not a synonym of non-toxic or non-biocidal. In the past, thymol vapor was used extensively by a number of conservators in "thymol cabinets." However, in respect to heritage materials thymol softens varnishes and resins, renders parchment brittle, degrades some paper supports, watercolor binders, and iron gall ink, and, with respect to human health, librarians and archivists suspected it to be a carcinogenic substance (Holben Ellis 1995; Isbell 1997). Indeed, the conclusion of the peer review of the pesticide risk assessment of the active substance thymol by the European Food Safety Authority, reached on the basis of the evaluation of the representative uses of thymol as a fungicide on table and wine grapes, reported that "*a high risk was identified for aquatic organisms, leading to a critical area of concern*" and "*the formulated product contains the impurity methyleugenol which is a genotoxic carcinogen*" (EFSA 2012).

The use of alternative substances can be considered after making all the preliminary tests to prove the worth of them. The first step is to evaluate the threshold above which the compounds inhibit microbial growth. The sixteenth century scientist Paracelsus was the first to state that all things are poisons, and that the degree of toxicity is only caused by the dose. In microbiology, the minimum inhibitory concentration (MIC) identifies the minimum amount of the compound that is required to inhibit microbial growth, under defined laboratory conditions. MIC has been used for decades and measures the concentration required to inhibit growth or to kill planktonic microorganisms. Importantly, it is now known that for some substances, the resistance of biofilm bacteria may be a thousand times greater than that of planktonic bacteria of the same strain (Olson et al. 2002). Studying the photocatalytic titanium dioxide ($TiO_2$) nanopowder and $TiO_2$ thin film, Polo et al. (2011) showed one order of magnitude reduction of *Pseudomonas aeruginosa* planktonic cells in 2 h and an almost complete eradication of *P. aeruginosa* planktonic cells, respectively. In contrast, neither the photocatalytic treatment with $TiO_2$ film nor that with $TiO_2$ nanopowder had any effect on *P. aeruginosa* biofilms. Nevertheless, when studying non-biocidal antibiofilm substances it is possible to use MIC. In fact, if the selected compound does not kill planktonic cells at the concentration adopted, it is even less likely that it will kill sessile cells.

Before use, the environmentally friendly organic compounds have to be proven not to act as carbon and energy sources for the target microorganisms. The rationale of the experiment is that, if a substance does not kill a microorganism, it can function as a nutrient. To investigate this matter the substance can be supplied as the only carbon and energy source for the target microorganisms in various amounts.

Another concern of using phytochemicals, such as oils and plant extracts, is that their composition varies over time. For instance, Nezhadali et al. (2014) claimed that the composition and quantity of essential oil from a specific thyme species can be considerably influenced by harvesting season, geographical location, and other agronomic factors. Additionally, based upon soil type variations, distinct differences among chemotypes can be found over a few meters. Moreover, extrinsic factors related to the extraction method affect their chemical composition (Dhifi et al. 2016). The apparently different results of the compounds once extracted from the same plant sources often found in literature are easily explained by the fact that the tested materials have a different composition. Consequently, all these plant-derived mixtures must be chemically characterized or otherwise other researchers cannot repeat the proposed experiments. Otherwise chemically synthetized compounds can be purchased from the market. In this respect, it is also worth noting that some phytochemicals, such as essential oils, are very expensive. Therefore, a valuable alternative to essential oils is to use the main pure synthetic counterpart that is also generally more stable (Rakotonirainy and Lavédrine 2005). Indeed the use of essential oils has been proposed in conservation literature as a "green alternative" (Macro et al. 2018). However, in this chapter, although they can be effective, we do not define as a green approach any alternative to common biocides that uses products of natural origin but we define green a strategy that does not kill microorganisms.

Disrupting the biofilm not killing the cells is not yet a reality in conservation. In this line of thought, no data are currently provided on the effectiveness of non-biocidal strategies as alternatives to control biofilms. However, at present, it is possible to suggest multiple approaches in addition to biocides in order to reach an effective clearance of biofilms without a massive use of toxic substances. This has been proven with nitric oxide (NO). NO is a signaling molecule involved in the modulation of quorum sensing (QS), a method of cell-to-cell communication, able to elicit bacterial dispersal (Kyi et al. 2014). Microorganisms isolated from the biodeteriorated wooden sculpture *So It's Come To This* (1986) by Bruce Armstrong, at the University of Melbourne headquarters, have been treated with a nitroxide, a compound of which the antibiofilm mechanism is similar to that of NO, while it is less expensive and with a longer life than NO-donors (Alexander et al. 2015; Alexander and Schiesser 2017). A 24 h treatment with 50 μM nitroxide followed by 2 h treatment with 0.001% w/v benzalkonium chloride effectively eradicated biofilms. Importantly, in this study, the biocide was used at a concentration much lower than those usually employed (2% w/v).

In regard to microbiological issues, enzymes have been employed in conservation for removing microbial staining (Konkol et al. 2009), for monitoring biodeterioration (Rosado et al. 2013), and for killing microbial cells (Valentini et al. 2010). In the Chinese literature (Wu and Lou 2016), chitinases were also used to inhibit the development of filamentous fungi growing on word walls and canoes at the Cross Lake Bridge ruins in Xiaoshan. However, the authors of this chapter could not understand in depth this research (in particular, whether toxicity was evaluated and fungal biofilm was formed) as only the abstract of the Chinese manuscript has been translated into English. Another application with the potential for green antibiofilm technology is the use of enzymes to degrade the extracellular matrix for biofilm dispersal, including glycosidases, proteases, and deoxyribonucleases (Kaplan 2010; Shadia and Aeron 2014), also immobilized to a surface (Spadoni Andreani et al. 2017). To the best of our knowledge, matrix-degrading enzymes have never been tested to disperse biofilms in the conservation field but EPS inhibitors, such as a mixture of bismuth nitrate and dimercaprol, have been successfully used during the EU project BIODAM (Robertson et al. 2004).

Metal cations, such as calcium, magnesium, and iron have been implicated in maintaining matrix integrity. Antibiofilm formulations incorporating ethylenediaminetetraacetic acid (EDTA) and other permeabilizers have shown efficacy on in vitro biofilms in synergism with antimicrobial agents (Robertson et al. 2004). Nuclear fast red and methylene blue, two photodynamic agents investigated in combination with hydrogen peroxide in the EU project BIODAM (Young et al.

2008), showed the potential to destroy cyanobacteria on stone samples and, since photodynamic agents are themselves subsequently degraded by visible light, the substratum is not discolored. Neither light alone nor the presence of $H_2O_2$ led any change in the fluorescence of *Synechococcus leopoliensis*. Combining methylene blue with hydrogen peroxide resulted in a decrease of the fluorescence of 40%.

Interestingly to mention in this chapter, even if some natural products were selected with the aim to kill the target microorganisms they showed much lower toxicity in respect to non-target microorganisms in comparison with traditional biocides. This is the case of metabolites produced by *Bacillus* spp. that showed no lethality against brine shrimp and Swiss mice through administration of 5000 mg/kg acute dose (Silva et al. 2016b). In contrast, Preventol® caused acute toxicity with 10 times minor concentration dose administrated in the same conditions.

Unfortunately, at present, only few molecular mechanisms of action related to some antibiofilm agents are known. Proteomic analysis of biofilm exposed to the antibiofilm zosteric acid sodium salt and salicylic acid revealed that a number of proteins were up- and downregulated, and these proteins were associated with stress, motility, cell-to-cell communication, reactive oxygen species accumulation and metabolism (Villa et al. 2012; Cattò et al. 2017). Recently, the zosteric acid sodium salt and the usnic acid have been encapsulated in silica nanosystems commonly used to protect stone surfaces (Ruggiero et al. 2020). The antifouling activity was successfully assessed against planktonic microorganisms from biopatinas colonizing the Aurelian Walls in Rome. Hopefully, in the near future, the above coatings will be tested *in situ* and their antibiofilm properties will be investigated.

# 3 Lab Biofilm Systems to Test the Efficacy of an Antibiofilm Compound/Mixture

On the one hand, the use of typical laboratory planktonic microorganisms for selection of biocides is inappropriate. On the other hand, biofilms on cultural heritage surfaces are inherently heterogeneous. Consequently, researchers have made use of laboratory biofilm models that capture the salient features of target sessile cells while providing numerous and consistent biofilm samples. Therefore, first of all, a biofilm at the solid–air interface (SAB) has to be formed.

In fact, while field experiments are clearly instrumental to evaluate the efficacy of an antibiofilm compound, their design and execution face several challenges: small sample size, unrepeatable samples, and complex sample structure. Furthermore, many field-based studies are time-limited as many biofilm processes relevant to the performance evaluation of antibiofilm agents occur over a very long time, for example, processes such as biofilm regrowth and succession as well as response to various environmental conditions.

The model systems are in contrast with field systems in terms of simplicity and accessibility. The purpose of a laboratory model system is not to represent a mini

**Fig. 5.1** Schematic representation of the colony biofilm system

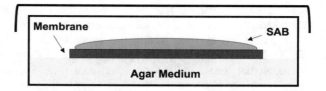

version of field systems, but rather to simplify nature so its processes can be easily understood (Jessup et al. 2004).

In the biofilm community it has long been debated whether a bacterial colony on a solid growth medium is to be considered a biofilm. Definitely, few colony morphologies can effectively reflect some of the important attributes of biofilms (Haussler and Fuqua 2013). However, as this question is not convincingly answered yet and, in conservation, a solid surface is better for mimicking a cultural heritage substratum, other systems have been considered better performing to achieve this aim.

A colony biofilm is grown on a semipermeable membrane that sits on an agarized medium (Fig. 5.1). The membrane is a solid interface and it is semipermeable to allow nutrients to reach the cells. This method has been especially devised in the study of the antibiotic-resistant properties of cells (Merritt et al. 2005). Sessile cells can be given a new supply of nutrients by relocating the membrane with biofilm on a fresh agar plate. One can therefore easily assess the effect of the antibiofilm compound at the time of adhesion (inoculum) or when the biofilm is already mature. The membrane can be also modified for binding antibiofilm agents. The colony biofilms can grow quickly, are easy to handle, and require inexpensive laboratory materials. Recently, protocols for the growth of cyanobacterial and fungal colony biofilms have been set up by Sanmartín et al. (2015) and Gambino et al. (2017), respectively.

Biofilms that are composed of a mono-species are relatively rare in nature; rather, microorganisms are generally found in complex multispecies communities. In line with this observation, Miller et al. (2008, 2009) cultivated on lithic surfaces a natural green biofilm from an enriched microbial consortium residing on a limestone monument. However, the main question is how the original community adapts and evolves under the artificial conditions created in the laboratory systems. Recently, Vázquez-Nion et al. (2016) developed a multispecies liquid culture from natural SABs inhabiting granitic historic buildings, with the aim to use it as the inoculum for lab-scale biofilms. The researchers observed that after one year, the microbial community of the culture was stable and dominated by members of the *Chlorophyta* and *Cyanobacteria*, which were not part of the core microbiome of the original SABs.

Furthermore, natural biofilm communities are difficult to analyze at molecular level. This may limit the comprehension of the mechanism of antibiofilm compounds, since omics-based technologies ideally require microorganisms with available genetic and physiological information (Noack-Schönmann et al. 2014).

To overcome the limitations arising from the complex phototrophic SABs, elegant dual-species biofilm models based on phototroph–heterotroph associations were developed, starting from two species that do not originate from an environmental microbial consortium and have never been introduced to each other.

A model system comprising the cyanobacterium *Nostoc punctiforme* strain ATCC 29133 (PCC 73102) as phototroph, and the well-studied marble-derived microcolonial fungus A95 *Knufia petricola* (syn. *Sarcinomyces petricola*) as the heterotrophic component was developed by Gorbushina and Broughton (2009). This model allowed Seiffert et al. (2014, 2016) to successfully study the biological impact of the consortium on weathering granite and related minerals in a geomicrobiologically modified percolation column. This system is promising for testing antibiofilm agents, because the compound can percolate into the column and interact with the biofilm. At the end of the treatment, the biofilm biomass can be removed from the system for a further analysis.

Villa et al. (2015) used a laboratory model of SABs composed of the unicellular cyanobacterium *Synechocystis* sp. strain PCC 6803 and the chemoheterotroph *Escherichia coli* K12 (Fig. 5.2b). A modified drip flow reactor (DFR) was used to mimic a monument surface (Fig. 5.2a). SABs on outdoor stone monuments follow the water flow downward, experiencing low fluid velocity over the surface, while the biomass is continuously exposed to the air. Similarly, biofilms in the DFR are under low-shear/laminar flow as the medium drips on a 10-degree surface, and a high gas transfer environment as the biofilm is continuously exposed to the air in the head space. This setup was effective in reproducing SABs as it was able to capture typical features of biofilms on outdoor stone monuments such as: (i) microcolonies of aggregated bacteria; (ii) a network-like structure following surface topography; (iii) cooperation between phototrophs, heterotrophs, and cross-feeding processes; (iv) ability to change the chemical parameters that characterize the microhabitats, and (v) survival under desiccation stress. By using the SAB model system, Villa et al. (2015) compared the susceptibility of biofilm and planktonic cells to a disinfection treatment based on a quaternary ammonium salt. The results revealed that the biofilm retained more viable cells than the planktonic cells after 45 min of antimicrobial exposure. These results suggested that the dual-species SAB exhibits an antimicrobial tolerance, which is the hallmark of the biofilm mode of growth.

While SAB models on stone are well developed, few lab-scale systems are available for other heritage substrata. Rakotonirainy and Lavédrine (2005) designed a system to reproduce fungal SABs on paper (Fig. 5.3). At the beginning a piece of paper is inoculated with a single strain or a mixture of many strains, letting the mycelium cover the entire surface and dry before inserting it in a book. Then the closed books are placed some centimeters high on a grid in a transparent test-chamber (Rakotonirainy and Lavédrine 2005). The tested antibiofilm agent impregnates a paper disc in an open plate on the bottom of the chamber. Humidity and temperature are set at the desired level. After exposure of the biofilms on paper to the vapor of the antibiofilm for a specific time cells can be investigated. Rakotonirainy and Lavédrine (2005) demonstrated that, at the concentrations tested, linalool, the main component in lavender oil, is fungistatic and could be employed in preventing fungal growth. Interestingly, this compound cannot be used as a fungicide as higher concentrations lead to a decrease in paper pH.

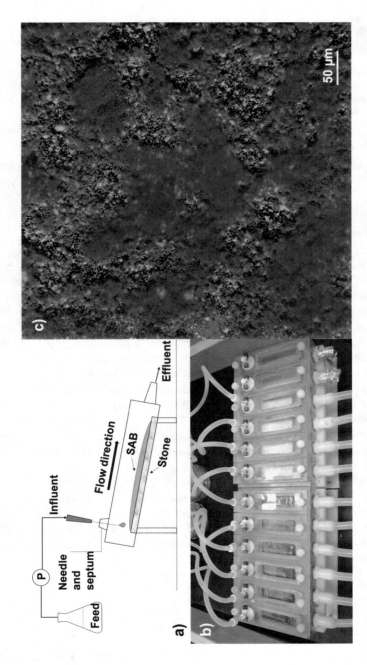

**Fig. 5.2** Panel (**a**) is the schematic representation of the modified drip flow reactor (DFR) used by Villa et al. (2015). Panel (**b**) displays a picture of DFRs with a glass lid (panel **b**). Panel (**c**) shows the 3D reconstruction of dual-species SABs grown in the DFR following the method reported by Villa et al. (2015). Color legend: *E. coli* cells are green (GFP); *Synechocystis* cells are red (autofluorescence); a reflection from inorganic materials is gray

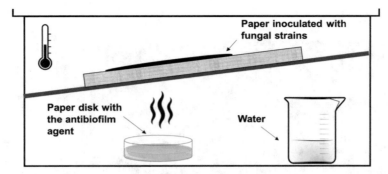

**Fig. 5.3** Schematic representation of the system designed by Rakotonirainy and Lavédrine (2005) to reproduce biofilms on paper

## 4 Methods for Testing the Effectiveness of Antibiofilm Substances

Despite the vast use of traditional and novel antibiofilm agents in conservation of cultural heritage, these practices are being increasingly questioned because of the lack of correlation between results obtained with conventional susceptibility tests at laboratory-scale and those obtained from in-field applications. The reason for this discrepancy is based on the use of planktonic cells for testing antibiofilm at laboratory-scale rather than sessile cells. As previously mentioned, it is now well known that biofilms are ∼1000-fold more resistant than their planktonic counterparts (Hall and Mah 2017). Furthermore, the lack of simple reliable and standard assays has remained a major limitation in selecting adequate antibiofilm treatments.

In view of the previous considerations, biofilm model systems are instrumental for testing the effectiveness of antibiofilm substances before an application in-field. Once the biofilm is grown, it can be exposed to different treatments and environmental conditions. Furthermore, biofilm model systems can be used to study biofilm regrowth when the treatment is concluded.

The antibiofilm agents can affect biofilm biomass, impair the biofilm matrix, or affect biofilm thickness and morphology (e.g., inability of fungi to form filamentous structures), while cells remain metabolically active. Thus, several methods can be applied to evaluate the impacts of a compound against the biofilm. Here, we briefly describe some of the methods that are most commonly used to test the effectiveness of antibiofilm compounds.

### 4.1 Plate Count Assay

To be able to assess the effectiveness of antibiofilm compounds, it is essential to determine the amount of viable cells in the biofilm, or the relative reduction in viable cells after the treatment. The most widely used technique to determining bacterial viability is counting the number of colony forming units (CFU) after plating cultures on agar media. This method to assess biofilm viability requires the dislodgment of the biomass from the surface and the disaggregation of cell clusters due to the presence of the biofilm matrix. Procedures such as vortexing, sonication, or matrix-dissolving enzymes can be applied to separate bacterial cells from the matrix. However, these procedures may affect bacterial viability, leading to significant errors in the quantification of the biomass (Azeredo et al. 2017). Furthermore, many strains cannot be cultured, as well as metabolically inactive cells (such as dormant cells and spores) commonly present within the biofilm (Sebestyen et al. 2016). Finally, while CFU counting can be a valuable tool for quantifying a bacterial biofilm, it is not suitable for fungal biofilms that develop filamentous structures and spores.

## 4.2 XTT Cell Viability Assay

The tetrazolium salt assay (XTT assay) is another useful technique for assessing the viability of biofilm cells in vitro, as well as their susceptibility to antimicrobials. The number of viable biofilm cells can be deduced by measuring the absorbance of the supernatant after the metabolic reduction of XTT into the water-soluble formazan (Pantanella et al. 2013). Since the formazan product is water soluble and easily measured in cellular supernatants, the XTT assay allows the investigation of biofilm agent susceptibility without disruption of the biofilm structure (Kuhn et al. 2003). The main limitation of this method is related to the biofilm heterogeneity with gradients of substrata and products, which lead to cells with different metabolic states. Furthermore, the biofilm matrix might slow down the reduction of XTT or partially retain and release the formazan.

## 4.3 ATP-Bioluminescence Assay

The bioluminescence assay, based on the use of the North American firefly luciferase, sees the conversion of intracellular ATP, the main energy carrier in cells of all living organisms, into light. The light generated is recorded by a luminometer and quantified in relative light units (Nante et al. 2017). Intracellular ATP content is a good indicator of cell viability. In fact, upon cell death, ATP synthesis is immediately arrested, while its hydrolysis can continue for some time. The main drawback of this method is related to the variation in ATP production among diverse microbial taxa (Ivanova et al. 2006).

## 4.4  Spectrofluorometric Assay

Fluorescein diacetate (3',6'-diacetyl-fluorescein, FDA) is a pre-fluorophore that can be hydrolyzed by a wide spectrum of nonspecific extracellular and membrane-bound enzymes. Once hydrolyzed, FDA is converted to fluorescein that shows a yellow-green color absorbing at 490 nm. The method generally assumes the determination of FDA-hydrolyzing enzyme activity of cells that are detached from the carrier. Konkol et al. (2010) reported the use of beta-N-acetylhexosaminidase activity as a rapid and reliable means of fungal detection on a variety of cultural heritage materials. Adapted for use on cultural heritage materials, fluorogenic 4-methylumbelliferyl (MUF) labeled substrate N-acetyl-beta-d-glucosaminide (NAG) was used to detect beta-N-acetylhexosaminidase activity in fungi that were actively growing on library materials (Konkol et al. 2010). Being a metabolic assay, a significant limitation is the fact that the microorganisms in biofilms do not all display the same metabolic activity. Moreover, metabolic fluorimetric-based assays are often calibrated against planktonic cultures, introducing significant error as the metabolic rates differ greatly between the planktonic and the biofilm states (Welch et al. 2012).

## 4.5  Real-Time Reverse Transcription PCR (qRT-PCR)

Real-time quantitative-reverse transcription PCR (qRT-PCR) is one of the most powerful and sensitive gene analysis techniques available so far, in which the fluorescent signal is measured in real-time at each amplification cycle and is directly proportional to the number of amplicons generated. Moreover, since qRT-PCR detects mRNA, with its short half-life, it is a promising indicator of cell viability (Guilbaud et al. 2005; Xie et al. 2011). Thus, qRT-PCR can be applied not only to detect but also to quantify viable microorganisms in the biofilm (Ettenauer et al. 2015; Piñar et al. 2015). The main limitations are related to sample preparation, including the extraction of RNA.

## 4.6  Chlorophyll

Chlorophyll-a is a good estimator for quantifying microalgal and cyanobacterial biofilm biomass (Jesus et al. 2006; Sanmartín et al. 2010; Sendersky et al. 2017; Chamizo et al. 2018). The method requires the dislodgment of the biomass from the surface and the extraction of chlorophyll-a in DMSO followed by spectrophotometric measurements (Fernandez-Silva et al. 2011). Alternatively, biofilm biomass measurement can be achieved by measuring chlorophyll fluorescence. Recently, pulse amplitude modulated (PAM) fluorometry was used by Vázquez-Nion et al.

(2018) to measure in vivo the fluorescence signal in a non-destructive way. The authors developed an indicator of fitness for photosynthetic organisms by quantifying the maximum photochemical efficiency of photosystem II. To this end, they measured the minimal fluorescence signal of dark-adapted cells and the maximal fluorescence signal after a saturating light pulse in dark-adapted cells and they developed a standard curve that allowed the correlation of the minimal fluorescence signal of dark-adapted cells with the amount of chl-a content as a biofilm biomass estimator.

## 4.7 Proteins Quantification

Protein cellular content can be used to estimate the total biofilm biomass. After cellular lysis, the released proteins can be quantified by colorimetric assays such as quantification, including the Bradford, Lowry, and bicinchoninic acid (BCA) methods. The most used method is the Bradford assay, which consists of adding an acidic Bradford reagent containing Coomassie Brilliant Blue G-250 dye to the lysed sample. Once the protein binds to the dye, changing the color from brown to blue occurs. The change in absorbance at 595 nm is recorded and converted to a total protein concentration through a standard curve (Bradford 1976).

## 4.8 Biofilm Staining and Microscopy

One of the most important advances in the study of biofilm has been the ability to visualize the effects of antibiofilm compounds on hydrated living biofilms in three dimensions, over time, using confocal laser scanning microscopy (CLSM).

The use of fluorescent probes allows imaging by multiple fluorochromes for evaluating the impacts of a compound on individual biofilm components simultaneously. Thus, specific fluorescent stains are also used in combination with CLSM for imaging biofilm cellular and extracellular matrix material. Furthermore, the intense development of specialized image processing software with user-friendly interfaces and the implementation of advanced CLSM techniques such as fluorescence lifetime imaging (FLIM), fluorescence correlation spectroscopy (FCS), and fluorescence recovery after photobleaching (FRAP) in commercially available confocal microscopes have improved the development of quantitative methods for the analyses of biofilm images.

The fluorescent stains are generally designed to bind a specific cellular component, such as DNA (e.g., propidium iodide, SYBR-green, and ToTo-1) or protein (e.g., Sypro-Ruby, 3-(4-carboxybenzoyl) quinoline-2-carboxaldehyde (CBQCA) or the NanoOrange) (Cattò and Cappitelli 2019). Furthermore, the intrinsic natural autofluorescence of some phototrophic microorganisms can be exploited for imaging differentiation. A comprehensive list of fluorescent stains commonly used in biofilm research has been provided by Neu and Lawrence (2014).

The LIVE/DEAD BacLight assay is a dual staining kit composed of two fluorophores, namely SYTO9 and propidium iodide (PI). The green-fluorescent nucleic acid stain SYTO9 enters live and dead bacterial cells and binds to DNA of both Gram-positive and Gram-negative bacteria. The red-fluorescent nucleic acid stain PI intercalates to DNA and it is commonly used for identifying dead cells in a population and as a counterstain in multicolor fluorescent techniques because it is supposed to penetrate only cells with disrupted membranes and it is generally excluded from viable cells. Thus, the SYTO9/PI combination allows one to detect bacterial viability based on the detection of membrane integrity. The main advantages of using the LIVE/DEAD BacLight assay are related to the rapid procedure, the possibility to perform quantitative analyses, as well as to measure the fluorescent signal using various instruments such as flow cytometers and microplate readers. The principal drawback of this method is the need to observe a statistically relevant portion of the sample to obtain representative information on the total population.

The coupling of CLSM with fluorescence in situ hybridization (FISH) probes leads to more information about the identification and localization of specific microbial taxa before and after a biocide treatment. In fact, FISH-CLSM allows microorganisms to be specifically labeled within the 3D extracellular matrix by using non-invasive imaging of fully hydrated biofilms. In short, biofilm cells are fixed, permeabilized to facilitate access of the probe to the target site and then hybridized with nucleic acid probes labeled with a fluorochrome. The classic FISH technique relied on ribosomal RNA as probe target, and thus it is traditionally applied for the phylogenetic identification of microorganisms in mixed assemblages without prior cultivation. Since the FISH signal is related to the cellular rRNA content, which reflects the cell activity, this technique can be used to evaluate the biocidal effect of a compound or a treatment towards target microorganisms.

Despite these promising features, the classic FISH protocol might generate weak fluorescent signals in metabolically active cells. This drawback can be attributed to the biofilm matrix that hinders the access of the probes into the cell, as well as to the low ribosome content in slowly growing or metabolically inactive biofilm cells. To overcome these limitations, different ways for increasing the FISH signal have been developed, including the use of multiple probes for one target microorganism, the use of peptide nucleic acid probes (PNA-FISH), and the use of catalyzed reporter deposition (CARD-FISH), just to name a few. Recently, a method based on combinatorial labeling and spectral imaging FISH (CLASI-FISH) has been developed to detect hundreds of different microbial taxa in single microscopy imaging (Valm et al. 2011; Behnam et al. 2012; Valm et al. 2012). Another disadvantage of FISH approaches is that the samples have to be processed with several treatments prior to the probe hybridization, which may disturb the structure of the biofilm and makes time-course studies difficult.

Most fluorescent dyes used in biofilm studies stain cellular components. Visualizing the extracellular matrix by CSLM has been more challenging. A few fluorescent stains are becoming available for CSLM studies of biofilm matrix components.

In particular, fluorescent lectins bind to specific sugars and can be used to stain certain extracellular polysaccharides. Since lectins are large molecules, they do not penetrate biofilms well. Ideally, small molecule stains will become available for staining the polysaccharide component of biofilms, such as Calcofluor, which is used to stain biofilms of strains that produce cellulose (Spiers et al. 2003). Another component of the biofilm extracellular matrix material is extracellular DNA. Various stains are available that bind DNA, with the TOTO-1 iodide stain providing excellent contrast between the biofilm eDNA component and the biofilm cells (Gloag et al. 2013). By comparing the biovolumes of the matrix components before and after the treatments, it is possible to evaluate the effectiveness of an antibiofilm compound.

In addition to staining biofilm components, fluorescent stains are available for studying metabolic activities within biofilms. For example, tetrazolium salts precipitate when reduced by the biofilm cells, forming a zone of fluorescence around the active cells (Franklin et al. 2015). Calcein AM is a fluorogenic, cell-permeant fluorescent probe that indicates cellular viability in biofilms. The probe is nonfluorescent until acted upon by nonspecific esterases present in live cells. Thus, cleaving the AM ester allows the probe to emit a fluorescent signal that is proportional to cell vitality.

Calcein AM green stain has also been used to observe the permeabilization of biofilm cells by a biocide with time-lapse CLSM (Davison et al. 2010). The technique allowed for the simultaneous imaging of changes in biofilm structure and disruption of cellular membrane integrity through the loss of the intracellular green signal, generated by the cleavage of the AM ester prior to the antimicrobial exposure.

Daddi Oubekka et al. (2012) used a set of advanced fluorescence microscopic tools such as fluorescence recovery after photobleaching, fluorescence correlation spectroscopy, and fluorescence lifetime imaging, to characterize the dynamics of fluorescently labeled vancomycin in biofilms.

Thus, imaging techniques can be applied to elucidate the effectiveness of a biocide treatment by watching, through a microscope, the antimicrobial attack.

# 5  Conclusion

Biofilms are the dominant lifestyle of microorganisms in all environments, either natural or manmade, including heritage. The development of effective strategies to combat biofilms is a challenging task.

These emerging novel antibiofilm strategies are still in the nascent phase of development, and more research is urgently needed to validate these approaches, which may eventually lead to effective prevention and control of biofilms. Until now, the research and application of antibiofilm compounds have often been questioned owing to the diversity of the testing methods available and the variations of the results reported in the literature vs those obtained in-field. Thus, numerous innovative antibiofilm approaches have been published, but it is difficult to reliably compare all these strategies.

Some factors still hamper the testing and screening of antibiofilm compound such as, among others, the scarcity of homogenized testing protocols, the lack of normalized vocabulary, the difficulty of testing repeatability and reproducibly. Thus, a key aspect of future antibiofilm research is the need for standards: a unified terminology and well described protocols and guidelines are required to test the effectiveness of traditional or novel compounds against biofilms retrieved on heritage surfaces. These protocols and guidelines should be a preliminary step in the direction of a potential code of green practices.

# References

Alexander S-A, Schiesser CH (2017) Heteroorganic molecules and bacterial biofilms: controlling biodeterioration of cultural heritage. Arkivoc part ii: 180–222

Alexander S-A, Rouse EM, White JM, Tse N, Kyi C, Schiesser CH (2015) Controlling biofilms on cultural materials: the role of 3-(dodecane-1-thiyl)-4-(hydroxymethyl)-2, 2, 5, 5-tetramethyl-1-pyrrolinoxyl. Chem Commun 51:3355–3358

Azeredo J, Azevedo NF, Briandet R, Cerca N, Coenye T, Costa AR, Desvaux M, Di Bonaventura G, Hébraud M, Jaglic Z, Kačániová M, Knøchel S, Lourenço A, Mergulhão F, Meyer RL, Nychas G, Simões M, Tresse O, Sternberg C (2017) Critical review on biofilm methods. Crit Rev Microbiol 43:313–351

Banfield JF, Barker WW, Welch SA, Taunton A (1999) Biological impact on mineral dissolution: application of the lichen model to understanding mineral weathering in the rhizosphere. Proc Natl Acad Sci USA 96:3404–3411

Behnam F, Vilcinskas A, Wagner M, Stoecker K (2012) A straightforward DOPE (double labeling of oligonucleotide probes)-FISH (fluorescence in situ hybridization) method for simultaneous multicolor detection of six microbial populations. Appl Environ Microbiol 78:5138–5142

Borderie F, Alaoui-Sossé B, Aleya L (2015) Heritage materials and biofouling mitigation through UV-C irradiation in show caves: state-of-the-art practices and future challenges. Environ Sci Pollut Res 22:4144–4172

Borrego S, Valdes O, Vivar I et al (2012) Essential oils of plants as biocides against microorganisms isolated from Cuban and Argentine documentary heritage. ISRN Microbiol 2012:826786

Bradford MM (1976) A rapid and sensitive method for the quantitation of microgram quantities of protein utilizing the principle of protein-dye binding. Anal Biochem 72:248–254

Caneva G, Nugari MP, Salvadori O (1991) Biology in the conservation of works of art. ICCROM, Rome

Casano LM, Braga MR, Álvarez R, del Campo EM, Barreno E (2015) Differences in the cell walls and extracellular polymers of the two *Trebouxia* microalgae coexisting in the lichen *Ramalina farinacea* are consistent with their distinct capacity to immobilize extracellular Pb. Plant Sci 236:195–204

Cattò C, Cappitelli F (2019) Testing anti-biofilm polymeric surfaces: where to start? Int J Mol Sci 20:3794

Cattò C, Grazioso G, Dell'Orto SC, Gelain A, Villa S, Marzano V, Vitali A, Villa F, Cappitelli F, Forlani F (2017) The response of *Escherichia coli* biofilm to salicylic acid. Biofouling 33:235–251

Chamizo S, Adessi A, Mugnai G, Simiani A, De Philippis R (2018) Soil type and cyanobacteria species influence the macromolecular and chemical characteristics of the polysaccharidic matrix in induced biocrusts. Microb Ecol 78:482–493

Daddi Oubekka S, Briandet R, Fontaine-Aupart MP, Steenkeste K (2012) Correlative time-resolved fluorescence microscopy to assess antibiotic diffusion-reaction in biofilms. Antimicrob Agents Chemother 56:3349–3358

Davison WM, Pitts B, Stewart PS (2010) Spatial and temporal patterns of biocide action against *Staphylococcus epidermidis* biofilms. Antimicrob Agents Chemother 54:2920–2927

De Los Ríos A, Wierzchos J, Sancho LG, Green TGA, Ascaso C (2005) Ecology of endolithic lichens colonizing granite in continental Antarctica. Lichenologist 37:383–395

Dhifi W, Bellili S, Jazi S, Bahloul N, Mnif W (2016) Essential oils' chemical characterization and investigation of some biological activities: a critical review. Medicines (Basel) 3:25

EFSA (2012) Conclusion on the peer review of the pesticide risk assessment of the active substance thymol. EFSA J 10:2916

Ettenauer J, Piñar G, Tafer H, Sterflinger K (2015) Quantification of fungal abundance on cultural heritage using real time PCR targeting the β-actin gene. Front Microbiol 5:262

Fernandez-Silva I, Sanmartín P, Silva B, Moldes A, Prieto B (2011) Quantification of phototrophic biomass on rocks: optimization of chlorophyll-a extraction by response surface methodology. J Ind Microbiol Biotechnol 38:179–188

Franklin MJ, Chang C, Akiyama T, Bothner B (2015) New technologies for studying biofilms. Microbiol Spectrum 3. https://doi.org/10.1128/microbiolspec

Gambino M, Ahmed MAA, Villa F, Cappitelli F (2017) Zinc oxide nanoparticles hinder fungal biofilm development in an ancient Egyptian tomb. Int Biodeter Biodegr 122:92–99

Gloag ES, Turnbull L, Huang A, Vallotton P, Wang H, Nolan LM, Mililli L, Hunt C, Lu J, Osvath SR, Monahan LG, Cavaliere R, Charles IG, Wand MP, Gee ML, Prabhakar R, Whitchurch CB (2013) Self-organization of bacterial biofilms is facilitated by extracellular DNA. Proc Natl Acad Sci USA 110:11541–11546

Gorbushina AA, Broughton WJ (2009) Microbiology of the atmosphere–rock interface: how biological interactions and physical stresses modulate a sophisticated microbial ecosystem. Ann Rev Microbiol 63:431–450

Grube M, Berg G (2009) Microbial consortia of bacteria and fungi with focus on the lichen symbiosis. Fungal Biol Rev 23:72–85

Guilbaud M, de Coppet P, Bourion F, Rachman C, Prévost H, Dousset X (2005) Quantitative detection of *Listeria monocytogenes* in biofilms by realtime PCR. Appl Environ Microbiol 71:2190–2194

Hall CW, Mah TF (2017) Molecular mechanisms of biofilm-based antibiotic resistance and tolerance in pathogenic bacteria. FEMS Microbiol Rev 41:276–301

Haussler S, Fuqua C (2013) Biofilms 2012: new discoveries and significant wrinkles in a dynamic field. J Bacteriol 195:2947–2958

Holben Ellis M (1995) The care of prints and drawings. Altamira Press, Walnut Creek

Isbell LH (1997) The effects of thymol on paper, pigments, and media. Abbey Newsletter 21:39–43

Ivanova EP, Alexeeva YV, Pham DK, Wright JP, Nicolau DV (2006) ATP level variations in heterotrophic bacteria during attachment on hydrophilic and hydrophobic surfaces. Int Microbiol 9:37–46

Jessup CM, Kassen R, Forde SE, Kerr B, Buckling A, Rainey PB, Bohannan BJM (2004) Big questions, small worlds: microbial model systems in ecology. Trends Ecol Evol 19:189–197

Jesus B, Perkins RG, Mendes CR, Brotas V, Paterson DM (2006) Chlorophyll fluorescence as a proxy for microphytobenthic biomass: Alternatives to the current methodology. Mar Biol 150:17–28

Kaplan JB (2010) Biofilm dispersal: mechanisms, clinical implications, and potential therapeutic uses. J Dent Res 89:205–218

Konkol N, McNamara C, Sembrat J, Rabinowitz M, Mitchell R (2009) Enzymatic decolorization of bacterial pigments from culturally significant marble. J Cult Herit 10:362–366

Konkol N, McNamara CJ, Mitchell R (2010) Fluorometric detection and estimation of fungal biomass on cultural heritage materials. J Microbiol Methods 80:178–182

Kuhn DM, Balkis M, Chandra J, Mukherjee PK, Ghannoum MA (2003) Uses and limitations of the XTT assay in studies of *Candida* growth and metabolism. J Clin Microbiol 41:506–508

Kyi C, Sloggett R, Schiesser CH (2014) Preliminary investigations into the action of nitric oxide in the control of biodeterioration. AICCM Bull 35:60–68

López AJ, Rivas T, Lamas J, Ramil A, Yáñez A (2010) Optimisation of laser removal of biological crusts in granites. Appl Phys A-Mater Sci Process 100:733–739

Macro N, Sbrana C, Legnaioli S., Galli E (2018) Antimicrobial activity of essential oils: a green alternative to treat cultural heritage. In: Mosquera MJ, Almoraima Gil ML (eds) Conserving cultural heritage. Proceedings of the 3rd International Congress on Science and Technology for the (TechnoHeritage 2017), May 21–24, 2017, Cadiz, Spain, pp 291–293

Martin-Sanchez PM, Nováková A, Bastian F, Alabouvette C, Saiz-Jimenez C (2012) Use of biocides for the control of fungal outbreaks in subterranean environments: the case of the Lascaux Cave in France. Environ Sci Technol 46:3762–3770

Merritt JH, Kadouri DE, O'Toole GA (2005) Growing and analyzing static biofilms. Curr Protoc Microbiol Chapter 1, Unit–1B.1, 1B.1.1–1B.1.17

Miller AZ, Laiz L, Gonzalez JM, Dionísio A, Macedo MF, Saiz-Jimenez C (2008) Reproducing stone monument photosynthetic–based colonization under laboratory conditions. Sci Total Environ 405:278–285

Miller AZ, Liaz L, Dionisio A, Macedo MF, Saiz-Jimenez C (2009) Growth of phototrophic biofilms from limestone monuments under laboratory conditions. Int Biodeter Biodegr 63:860–867

Nante N, Ceriale E, Messina G, Lenzi D, Manzi P (2017) Effectiveness of ATP bioluminescence to assess hospital cleaning: a review. J Prev Med Hyg 58:E177–E183

Neu TR, Lawrence JR (2014) Investigation of microbial biofilm structure by laser scanning microscopy. Adv Biochem Eng Biotechnol 146:1–51

Nezhadali A, Nabavi M, Rajabian M, Akbarpour M, Pourali P, Amini F (2014) Chemical variation of leaf essential oil at different stages of plant growth and in vitro antibacterial activity of *Thymus vulgaris* Lamiaceae, from Iran. Beni-Suef Univ J Basic Appl Sci 3:87–92

Noack-Schönmann S, Bus T, Banasiak R, Knabe N, Broughton WJ, Dulk-Ras HD, Hooykaas PJJ, Gorbushina AA (2014) Genetic transformation of *Knufia petricola* A95 – a model organism for biofilm–material interactions. AMB Express 4:80

Olson ME, Ceri H, Morck DW, Buret AG, Read RR (2002) Read biofilm bacteria: formation and comparative susceptibility to antibiotics. Can J Vet Res 66:86–92

Pantanella F, Valenti P, Natalizi T, Passeri D, Berlutti F (2013) Analytical techniques to study microbial biofilm on abiotic surfaces: pros and cons of the main techniques currently in use. Ann Ig 25:31–42

Piñar G, Sterflinger K, Ettenaueret J, Quandt A, Pinzari F (2015) A combined approach to assess the microbial contamination of the archimedes palimpsest. Microb Ecol 69:118–134

Polo A, Diamanti MV, Bjarnsholt T, Høiby N, Villa F, Pedeferri MP, Cappitelli F (2011) Effects of photoactivated titanium dioxide nanopowders and coating on planktonic and biofilm growth of *Pseudomonas aeruginosa*. Photochem Photobiol 87:1387–1394

Rakotonirainy MS, Lavédrine B (2005) Screening for antifungal activity of essential oils and related compounds to control the biocontamination in libraries and archives storage areas. Int Biodeter Biodegr 55:141–147

Robertson PKJ, Alakomi HL, Arrien N, Gorbushina AA, Krumbein WE, Maxwell I, McCullagh C. Ross N, Saarela M, Valero J, Vendrell M, Young ME (2004) Inhibitors of biofilm damage on mineral materials (Biodam). In: Kwiatkowski D, Löfvendahl R (eds) 10th International Congress on Deterioration and Conservation of Stone, Stockholm, Sweden, 27 June-2 July 2004. ICOMOS Sweden, pp 399–406

Rosado T, Martins MR, Pires M, Mirão J, Candeias A, Caldeira AT (2013) Enzymatic monitorization of mural paintings biodegradation and biodeterioration. Int J Conserv Sci 4:603–612

Ruggiero L, Bartoli F, Fidanza MR, Zurlo F, Marconi E, Gasperi T, Tuti S, Crociani L, Di
    Bartolomeo E, Caneva G, Ricci MA, Sodo A (2020) Encapsulation of environmentally-friendly
    biocides in silica nanosystems for multifunctional coatings. Appl Surf Sci 514:145908
Sanmartín P, Aira N, Devesa-Rey R, Silva B, Prieto B (2010) Relationship between color and
    pigment production in two stone biofilm-forming cyanobacteria (*Nostoc* sp PCC 9104 and
    *Nostoc* sp PCC 9025). Biofouling 26:499–509
Sanmartín P, Villa F, Polo A, Silva B, Prieto B, Cappitelli F (2015) Rapid evaluation of three
    biocide treatments against the cyanobacterium *Nostoc* sp. PCC 9104 by color changes. Ann
    Microbiol 65:1153–1158
Sebestyen P, Blanken W, Bacsa I, Tóth G, Martin A, Bhaiji T, Dergez Á, Kesseru P, Koós Á, Kiss I
    (2016) Upscale of a laboratory rotating disk biofilm reactor and evaluation of its performance
    over a half-year operation period in outdoor conditions. Algal Res 18:266–272
Seiffert F, Bandow N, Bouchez J, von Blanckenburg F (2014) Microbial colonization of bare rocks:
    laboratory biofilm enhances mineral weathering. Procedia Earth Planet Sci 10:123–129
Seiffert F, Bandow N, Kalbe U, Milke R, Gorbushina AA (2016) Laboratory tools to quantify
    biogenic dissolution of rocks and minerals: a model rock biofilm growing in percolation
    columns. Front Earth Sci 4:31
Sendersky E, Simkovsky R, Golden SS, Schwarz R (2017) Quantification of chlorophyll as a proxy
    for biofilm formation in the cyanobacterium *Synechococcus elongatus*. Bio-Protoc 7:14
Shadia MAA, Aeron A (2014) Bacterial biofilm: dispersal and inhibition strategies. SAJ Biotechnol
    1:105
Silva LN, Zimmer KR, Macedo AJ, Silva Trentin D (2016a) Plant natural products targeting
    bacterial virulence factors. Chem Rev 116:9162–9236
Silva M, Salvador C, Candeias MF, Teixeira D (2016b) Toxicological assessment of novel green
    biocides for cultural heritage. Int J Conserv Sci 7:265–272
Spadoni Andreani E, Villa F, Cappitelli F, Krasowska A, Biniarz P, Łukaszewicz M, Secundo F
    (2017) Coating polypropylene surfaces with protease weakens the adhesion and increases the
    dispersion of *Candida albicans* cells. Biotechnol Lett 39:423–428
Spiers AJ, Bohannon J, Gehrig SM, Rainey PB (2003) Biofilm formation at the air-liquid interface
    by the *Pseudomonas fluorescens* SBW25 wrinkly spreader requires an acetylated form of
    cellulose. Mol Microbiol 50:15–27
Urzì C, De Leo F, Krakova L, Pangallo D, Bruno L (2016) Effects of biocide treatments on the
    biofilm community in Domitilla's catacombs in Rome. Sci Total Environ 572:252–262
Valentini F, Diamanti A, Palleschi G (2010) New bio-cleaning strategies on porous building
    materials affected by biodeterioration event. Appl Surf Sci 256:6550–6563
Valm AM, Mark Welch JL, Rieken CW, Hasegawa Y, Sogin ML, Oldenbourg R, Dewhirst FE,
    Borisy GG (2011) Systems-level analysis of microbial community organization through com-
    binatorial labeling and spectral imaging. Proc Natl Acad Sci USA 108:4152–4157
Valm AM, Mark Welch JL, Borisy GG (2012) CLASI-FISH: principles of combinatorial labeling
    and spectral imaging. Syst Appl Microbiol 35:496–502
Vázquez-Nion D, Rodríguez-Castro J, López-Rodríguez MC, Fernández-Silva I, Prieto B (2016)
    Subaerial biofilms on granitic historic buildings: microbial diversity and development of
    phototrophic multi–species cultures. Biofouling 32:657–669
Vazquez-Nion D, Silva B, Prieto B (2018) Bioreceptivity index for granitic rocks used as construc-
    tion material. Sci Total Environ 633:112–121
Villa F, Remelli W, Forlani F, Vitali A, Cappitelli F (2012) Altered expression level of *Escherichia
    coli* proteins in response to treatment with the antifouling agent zosteric acid sodium salt.
    Environ Microbiol 14:1753–1761
Villa F, Pitts B, Lauchnor E, Cappitelli F, Stewart PS (2015) Development of a laboratory model of
    a phototroph–heterotroph mixed–species biofilm at the stone/air interface. Front Microbiol
    6:1251
Villa F, Gulotta D, Toniolo L, Borruso L, Cattò C, Cappitelli F (2020) Aesthetic alteration of marble
    surfaces caused by biofilm formation: effects of chemical cleaning. Coatings 10:122

Vlamakis H, Chai Y, Beauregard P, Losick R, Kolter R (2013) Sticking together: building a biofilm the *Bacillus subtilis* way. Nat Rev Microbiol 11:157–168

Vujcic I, Masic S, Medic M (2019) The influence of gamma irradiation on the color change of wool, linen, silk, and cotton fabrics used in cultural heritage artifacts. Radiat Phys Chem 156:307–313

Welch K, Cai Y, Strømme M (2012) A method for quantitative determination of biofilm viability. J Funct Biomater 3:418–431

Wu J, Lou W (2016) Study of biological enzymes' inhibition of filamentous fungi on the Crosslake Bridge ruins. Sci Archaeol Conserv:25–29

Xie Z, Thompson A, Kashleva H, Dongari-Bagtzoglou A (2011) A quantitative real-time RT-PCR assay for mature *C. albicans* biofilms. BMC Microbiol 11:93

Young ME, Alakomi H-L, Fortune I, Gorbushina AA, Krumbein WE, Maxwell I, McCullagh C, Robertson P, Saarela M, Valero J, Vendrell M (2008) Development of a biocidal treatment regime to inhibit biological growths on cultural heritage: BIODAM. Environ Geol 56:631–641

# Chapter 6
# Green Mitigation Strategy for Cultural Heritage Using Bacterial Biocides

Ana Teresa Caldeira

**Abstract** The microbiota present in cultural heritage objects, made by diverse inorganic and organic materials and inserted into particular environment, represents a complex and dynamic ecosystem composed by bacteria, cyanobacteria, fungi, algae and lichens, which can induce decay by biological mechanisms. To control the microbial growth several methods are being applied such as mechanical and physical processes and chemical biocides. However, these methods have several weaknesses like be dangerous to handle, material incompatibility or produce environmental and health hazards. Therefore, the identification of effectively biodeteriogenic agents and the design of mitigation strategies directed to these agents without prejudice to historical materials, to the environment and to operators, taking into account the microbial community's dynamics, is an important challenge to control biodeterioration of cultural heritage. Bacteria, in particular *Bacillus* spp. are worth for the creation of new green biocides solutions because they produce a great variety of secondary metabolites including ribosomally and non-ribosomally synthesized antimicrobial peptides, known to possess antagonistic activities against many biodeteriogenic fungi and bacteria. The discovery of new safe active compounds and green nanotechnology for direct application in cultural heritage safeguard can in a close future contribute to potentiate a new generation of biocides and safe sustainable methods for cultural heritage.

**Keywords** Biodegradation/biodeterioration · Bacterial biocides · Cultural heritage · Green biocides · Sustainability

A. T. Caldeira (✉)
HERCULES Laboratory, Évora University, Évora, Portugal

Chemistry Department, School of Sciences and Technology, Évora University, Évora, Portugal
e-mail: atc@uevora.pt

© The Author(s) 2021
E. Joseph (ed.), *Microorganisms in the Deterioration and Preservation of Cultural Heritage*, https://doi.org/10.1007/978-3-030-69411-1_6

# 1    Biodeterioration of Cultural Heritage Materials

Biodeterioration of Cultural Heritage materials has been neglected for a long time since it was previously believed that detriment was mainly due to chemical and physical processes. Over the last decades, it has been recognized that the action of microorganisms is a critical factor in the deterioration of Cultural Heritage that needs to be considered (Caldeira et al. 2015). To fully understand the role of biodeterioration/biodegradation processes and the deleterious effects on cultural assets it is fundamental to characterize the microbiota and to identify the microorganisms present. Furthermore, it is a key step to define efficient preventive conservation approaches and strategies to protect monuments and artworks from microbial re/colonization (Barresi et al. 2017; Rosado et al. 2013a, 2017; Piñar and Sterflinger 2009; Salvador et al. 2016, 2017).

The characterization of microbial communities, through detection and identification of microorganisms present on cultural assets can be carried out by means of specific complementary methods and new processes and strategies are constantly being developed. In the near past, research carried out in this field relied mainly on classical culture-based methodologies which are time consuming and omit slow growing and uncultivable microorganisms that may account for more than 90% of the microflora present (Amann et al. 2001; Moter and Göbel 2020). Recently, this major drawback has been tackled by the introduction of culture-independent methods based in molecular approaches to study microorganisms which are more sensitive and require smaller amounts of samples than the former (Rosado et al. 2013a, 2014a) including Random Amplified Polymorphic DNA (RAPD), Micro-Satellite Primer-Polymeric Chain Reaction (MSP-PCR), restriction fragment length polymorphism (RFLP), Denaturing Gradient Gel Electrophoresis (DGGE) (González and Saiz-Jiménez 2005), RNA-FISH (Gonzalez et al. 2014, 2017; Vieira et al. 2014, 2018) and Next Generation Sequencing (NGS) (Caldeira et al. 2015; Dias et al. 2018; Rosado et al. 2014a, 2015, 2020). Furthermore, one has to take into account that the colonization and proliferation of microorganisms (like bacteria, fungi, algae and lichens) on heritage materials are influenced by microclimatic conditions such as relative humidity, temperature and light and by the intrinsic chemical nature of the support material (Pangallo et al. 2009).

Figure 6.1 shows several examples of microbial presence in some heritage materials with obvious signs of biodegradation and aesthetic damage.

Although nowadays biodeterioration is well recognized in the overall deterioration process of cultural heritage, the specific role of the different microbial species that compose the most biodeteriogenic agents in communities is not yet well understood. Many microorganisms produce serious damage in historic materials, which are decomposed by the action of specific enzymes and organic acids (Urzì and De Leo 2007; Rosado et al. 2013a, 2014b, 2015). Fungi are especially dangerous due to the fact their hyphae can easily proliferate inside heritage materials and their spores, in a dormant state, are usually present and may germinate. Furthermore, fungi can produce carboxylic acids (e.g., oxalic, citric, malic, acetic, gluconic and

**Fig. 6.1** Microbial presence in several heritage materials like (**a**) mortars and mural paintings, (**b**) textiles, (**c**) wood, (**d**) cellulose acetate, (**e**) parchment, (**f**) canvas and easel paintings, (**g**) marble and granite, (**h**) ivory, with obvious signs of biodegradation and aesthetic damage. Adapted from Caldeira et al. (2015)

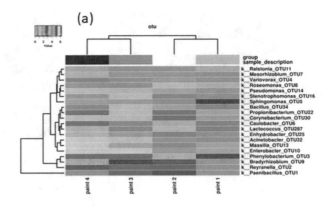

**Fig. 6.2** Bacterial genera present in four easel paintings, with visible signs of degradation (**a**) heatmap with quantitative visualization of bacterial community composition including the 20 most abundant genus; (**b**) core microbiome, with taxa abundance across sample groups, including the most abundant bacterial genus

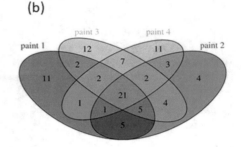

tartaric acids) which can enhance biochemical attack (Fomina et al. 2010; Hien et al. 2012; Rosado et al. 2013b, c). As reported in the literature a wide variety of fungi of the genera *Alternaria, Aureobasidium, Chaetomium, Cladosporium, Curvularia, Drechslera, Fusarium, Gliomastix, Penicillium, Trichoderma*, are abundant in deteriorated objects (Rosado et al. 2013a, 2014a; Sterflinger 2010). The development of fungi can induce discolouration and deterioration of surfaces, leading to the appearance of stains that alter the colour and hyphae penetration in materials may lead to detachment of fragments. Bacterial growth is frequently associated to the formation of biofilms, promoting discolouration of materials and pigments (Abdel-Haliem et al. 2013; Guiamet et al. 2011; Milanesi et al. 2006) but some metabolic compounds like oxalates and carotenes are also attributed to bacterial presence (Rosado et al. 2013a, b, 2014a, 2016) and the production of extracellular enzymes can also affect important materials, namely proteinaceous compounds in easel paintings (Salvador et al. 2019).

As an illustration, Figs. 6.1 and 6.2 show the major bacterial and fungal genera present in four easel paintings, with visible signs of degradation, all stored in the same reserve room (Fig. 6.3).

Bacterial core microbiome for the most abundant genera is composed by *Bacillus, Brevundimonas, Caulobacter, Corynebacterium, Enhydrobacter, Lactococcus, Mesorhizobium, Methylobacterium, Micrococcus Paenibacillus, Phenylobacterium*,

**Fig. 6.3** Fungi genera
present in four easel
paintings, with visible signs
of degradation (**a**) heatmap
with quantitative
visualization of fungal
community composition
including the 20 most
abundant genus and (**b**) core
microbiome, with taxa
abundance across sample
groups, including the most
abundant fungal genus

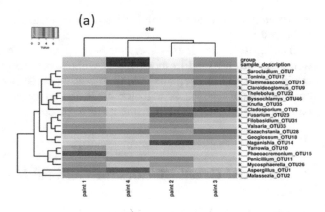

*Phyllobacterium, Propionibacterium, Pseudomonas, Ralstonia, Reyranella, Roseomonas, Sphingomonas, Staphylococcus, Stenotrophomonas* and *Variovorax.* Fungal core shows the presence of the genus *Aspergillus, Cladosporium, Fusarium, Kazachstania, Malassezia, Penicillium* and *Zygosaccharomyces.*

In fact, biodeteriogenic microorganisms cause serious aesthetical and structural damages in building materials, paintings, books or objects of inestimable value (see Fig. 6.1) which bring high expenses for museums and city councils (Allsopp 2011; Pangallo et al. 2009; Sterflinger 2010; Caldeira et al. 2018). To control the microbial growth several approaches are being applied such as mechanical and physical methods and chemical biocides (Allsopp et al. 2004; Barresi et al. 2017; Favero-Longo et al. 2017; Kakakhel et al. 2019; Pena-Poza et al. 2018; Quagliarini et al. 2018; Sanza et al. 2015).

However, these methods have several weaknesses: (i) physical methods can be dangerous to handle and promote deterioration of the object to preserve (Abdel-Haliem et al. 2013; Bosch-Roig et al. 2013; Scheerer et al. 2009; Tiano 2002); (ii) mechanical methods do not completely eradicate the microorganisms and it is not possible to apply to all materials (Sanza et al. 2015); and (iii) most chemical biocides while effective produce environmental and health hazards (Allsopp et al. 2004; Young et al. 2008; Cámara et al. 2011). So, the importance of carrying out proper remediation action for microbiologically contaminated historic materials is of vital importance.

The identification of effectively biodeteriogenic agents and the design of mitigation strategies directed to these agents without bias to historical materials, to the environment and to operators, taking into account the perspective of the microbial communities dynamics is a major challenge to control the biodeterioration of cultural heritage and where much remains to be done.

## 2 Bacterial Green Biocides for Cultural Heritage

According to the European legislation, a biocide can be defined as "a chemical substance or microorganism intended to destroy, deter, render harmless or exert a controlling effect on any harmful organism by chemical or biological means". Unfortunately, many traditional biocides used in cultural heritage may be dangerous for human health and the environment, while they differ in their toxicological profile, as EU databases indicate (https://echa.europa.eu). The protective solutions based on the use of these toxic chemicals face increasing restrictions due to the possibility of accumulation in animal tissues (Leifert et al. 1995).

Many commercial biocides are synthetic, mainly composed by mixtures with a range of formulation additives, generally formed by quaternary ammonium compounds or other nitrogen-containing compounds, urea, benzalkonium chloride and phenol derivatives (Barresi et al. 2017; Coors et al. 2018; Favero-Longo et al. 2017; Fernandes 2006). Copper, gold, platinum, silver, titanium, zinc metal nanoparticles are also applied in the last few years (Barresi et al. 2017; Ruggiero et al. 2020).

There are also naturally occurring biocides classified as natural biocides, derived from, e.g., bacteria (Caldeira et al. 2011a; Chena et al. 2020; Colombo et al. 2019; Junier and Joseph 2017; Mariana et al. 2020; Silva et al. 2014), fungi (Hooker et al. 1994; Massart and Jijakli 2007; Suárez et al. 2007) and plants (Arantes et al. 2019a, b; Chien et al. 2019; Jeong et al. 2018; Silva et al. 2019 Jeong et al. 2018; Palla et al. 2020). A review by Fidanza and Caneva (2019) gives a description of important natural biocides for the conservation of stone in cultural heritage. Controlled release systems of biocides by encapsulation in silica were also developed for several biocides, both commercial and natural, like Biotin T and New Des 50 (Dresler et al. 2017), zosteric acid sodium salt, usnic acid (Ruggiero et al. 2020), among others.

In heritage context, biocides are intended to limit the microbial growth and be useful in extending the life of heritage objects but material compatibility is a critical factor to be considered. Moreover, green and non-toxic biocides are eco-friendly and promising alternatives to chemical industrial biocides that allow to avoid unnecessary costs and risks. Their production and application do not imply the implementation of a risk-management protocol for toxic compounds which is an advantage in the field of conservation and restoration.

Therefore, it is crucial to develop green biocides compatible with historic materials that allow the elimination, inhibition and prevention of microbiological contamination.

Over the last years, some bioactive metabolites with potential applications in agriculture and pharmaceutical industries (Lowes et al. 2000; Walker et al. 1995), with a large spectrum of antimicrobial action have been reported. In agriculture the biological control of some pests has been carried out for decades through the introduction of bacteria in the soil or directly in the cultures. In recent years, interest in the biological control of phytopathogenic fungi has increased considerably because of the public concern to decrease the use of chemical pesticides and because there are some diseases that are hardly controllable by other strategies (Caldeira et al. 2011a). Besides the biocide effect, surfactins and fengycins of *Bacillus subtilis* prevented also the adhesion of competitive organisms on the surface of plant roots (Falardeau et al. 2013).

Mechanism and metabolite studies involved in this biological control often choose the use of bacteria (Beltran-Gracia et al. 2017), namely *Bacillus* spp. These microorganisms produce a variety of secondary metabolites, specifically antibiotics of peptide nature with antimicrobial properties (Caldeira et al. 2006, 2007, 2008, 2011a, b). The antimicrobial, fungicidal, insecticidal, anticarcinogenic and immunomodulating properties of lipopeptides can help to solve problems in medicine and agriculture, and the creation of an ecologically balanced environment (Maksimova et al. 2020).

A successful methodology to produce new green biocide solutions was recently developed (Silva et al. 2015, 2016a, b, 2017, 2019; Salvador et al. 2016; Rosado et al. 2019) exploring the potentialities of bioactive metabolites produced by *Bacillus* spp. in the cultural heritage safeguard field.

The *Bacillus* species are rich in biodiversity, constituting important repository bioresources, being worth for the creation of green biocides because they produce a great variety of secondary metabolites including ribosomally and non-ribosomally synthesized antimicrobial peptides, known to possess antagonistic activities against many biodeteriogenic fungi and bacteria (Berezhnayaa et al. 2019; Leifert et al. 1995; Jin et al. 2020; Medeot et al. 2020; Mikkola et al. 2004; Yazgan et al. 2001). *Bacillus subtilis; B. amyloliquefaciens, B. pumilus, B. licheniformis, B. methylotrophicus, B. atrophaeus, B. laterosporus, B. paralicheniformis, B. lehensis, Paenibacillus terrae, P. peoriae, P. polymyxa, P. larvae, P. mucilaginosus,* and *P. bovis* are probably the main producers of lipopeptides, which are known for their antifungal properties (Berezhnayaa et al. 2019; Caldeira et al. 2006, 2007, 2011a; Cawoy et al. 2015; Meena and Kanwar 2015; Moyne et al. 2001; Ongena and Jacques 2008; Tsuge et al. 2001; Zhao and Kuipers 2016).

Usually, three different classes of bioactive peptides can be distinguished: antimicrobial lanthipeptides ribosomally synthesized, such as subtilin, entianin, ericins, clausin, subtilomycin, thuricins, mersacidin, amylolysin, haloduracin, lichenicidin and cerecidins (Barbosa et al. 2015); and non-ribosomally synthesized antifungal peptides, such as bacilysin and rhizocticin or antifungal lipopeptides, such as surfactins, iturins and fengycins, bacillomycin, bacilysin, ericin, mersacidin, mycosubtilin, zwittermicins and kurstakins (Caldeira et al. 2008; Pabel et al. 2003; Rai et al. 2020).

Surfactin, iturin, bacillomycin D, fengycin and lichenysin from *Bacillus* species, rhamnolipids from *Pseudomonas aeruginosa* and daptomycin from *Streptomyces roseosporus* are among the most studied lipopeptides (Beltran-Gracia et al. 2017) and the gene clusters of the *Bacillus* lipopeptides encoding the surfactin, iturin and fengycin, families have been described and summarized in recent papers (Aleti et al. 2015; Abderrahmani et al. 2011; Deng et al. 2020; Othoum et al. 2018). These non-ribosomally synthesized compounds are amphiphilic cyclic biosurfactants that have many advantages over other biocides: low toxicity, biodegradability and environmentally friendly characteristics (Caldeira et al. 2011a). It is thought that lipopeptides act in the cell membrane of the target organism forming pores and producing an imbalance in the movement of ions both into and out, damaging the cell and producing a lethal effect. This mode of action makes it difficult for any target microorganism to gain resistance to the lipopeptide, which is a great advantage compared to other toxic compounds (Inès and Dhouha 2015; Płaza and Achal 2020; Silva et al. 2016a). These interesting properties are bringing lipopeptides to the spotlight of heritage research and testing (Caldeira et al. 2015).

The genus *Bacillus* includes a wide variety of species used industrially. *Bacillus* spp. present a great metabolic versatility, developing in almost all natural environments, the high area/volume ratio of these microorganisms allows a quick transfer of nutrients between the external environment and the interior of the cells. The resistance of spores to temperature, dehydration, radiation and other adverse environmental conditions confer a selective advantage on their survival. Some of these microorganisms can be grown in extreme temperature and pH conditions and produce stable products in these environments. The great physiological diversity of *Bacillus* spp., associated with the fact that most are not pathogenic, are easy to manipulate, good secretors of proteins and other metabolites and easily cultivable, makes them quite attractive for industrial applications (Arbige et al. 1993). Cultures of these microorganisms can be carried out in bioreactors, promoting a production of the active compounds. Some studies limiting their physiological growth by nutritional conditions, namely the carbon and nitrogen source, suggest an association between the lipopeptide production and *Bacillus* sporulation (Besson et al. 1987; Caldeira et al. 2008; Chevanet et al. 1986, Gonzalez-Pastor et al. 2003) evidencing higher antagonistic compounds production in cultures. Caldeira et al. (2011b, c), using an artificial intelligence approach to *Bacillus amyloliquefaciens* cultures, performed an Artificial Neural Network model addressing the antifungal activity of compounds produced, the *Bacillus* sporulation and biomass concentration obtained with different aspartic acid concentration like nitrogen source and different incubation time of cultures. This model was used to establish the operating conditions that maximize production of iturinic antimicrobial compounds. The combination of bioinformatics technology with proteomics based on mass spectrometry has shown that the synthesis of secondary metabolites from *Bacillus* can be regulated by quorum-sensing systems (Sandrin et al. 2013), allowing to increase the production yields.

The control of biodegradation processes in heritage context is of utmost importance, so the high scale production of these green compounds is a promising alternative. However, only few studies (Borrego et al. 2012; Favero-Longo et al. 2017; Silva et al. 2016a; Sterflinger 2010; Urzì and De Leo 2007) have been conducted to assess the efficacy of biocides in heritage context and these are mostly on the application of industrial chemical biocides which present toxicity. To meet the public concern to reduce the use of general chemical pesticides as biocides and the professional need for more effective and safer biocides, the discovery of new bioactive molecules and the development of environmentally friendly and safe new biocides tailored by biotechnological approaches to be used in conservation prevention and treatment and as additive for new products (e.g. paints, binders and consolidants) is fundamental.

A review by Rai et al. (2020) presents a comprehensive overview of significant lipopeptides produced by plant microbiome and their role in disease control and antagonism against phytopathogens in order to evaluate their potential for future explorations as antimicrobial agents.

In the field of heritage preservation, research undertaken at HERCULES Lab has enabled the production of new biocides from liquid cultures of *Bacillus amyloliquefaciens* (GENBank AY785775) previously isolated from healthy *Quercus suber* and displaying high levels of antagonistic properties against filamentous fungi that attack forest industrial products (Caldeira et al. 2011a). These amphiphilic biosurfactants based on cyclic lipopeptides (Fig. 6.4), non-ribosomal produced by multi-enzymatic complexes, have shown a large spectrum of fungal and bacterial inhibition displaying inhibition halos greater than commercial products and lower minimal inhibitory concentrations.

Furthermore, the new biocides are harmless and handling safe, acting through a membrane breakdown mechanism of the biodeteriogen films rather than an inhibitory/toxicity mechanism. These kind of compounds, made of amino acids and a fatty acid, are easily biodegradable (Cho et al. 2003) and non-toxic (Silva et al. 2016a).

Another interesting characteristic of antagonistic lipopeptides produced by bacteria is that the same bacterial strain can synthesize several lipopeptides (see Fig. 6.4). For instance, a *Bacillus* sp. strain reported by Abdellaziz et al. (2018) produces surfactin, pumalicidin, lichenisin, kurstakin, and various isoforms of fengycin and many other examples can be found in the literature (Zhao and Kuipers 2016).

These types of metabolites of peptide origin synthesized non-ribosomally by multi-enzymatic complexes form a set of microbial natural products that can contain residues of unusual amino acids in proteins. These peptides can be modified, namely by glycosylations and ring formation and generate lactones or esters. The great diversity in the structure of these compounds makes them sometimes quite active, with antimicrobial properties but also as bio-tensioactive, anti-inflammatory, antiviral, cytostatic and antitumor agents (Cameotra and Makkar 2004; Carrillo et al. 2003; Davis et al. 2001; Hsieh et al. 2004; Kim et al. 1998; Kluge et al. 1988; Marahiel 1997; Ohno et al. 1992) and the culture broth often holds a synergetic effect of these mixed components.

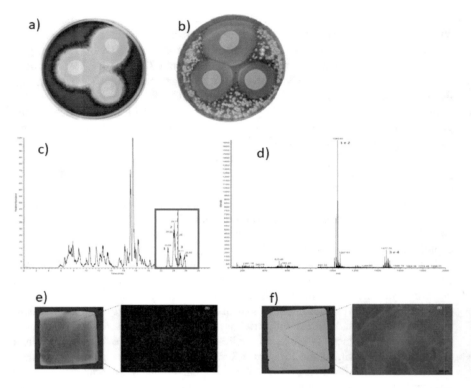

**Fig. 6.4** Lipopeptide biocide inhibition against *Penicillium* sp. (**a**), and *Cladosporium* sp. (**b**). Total ion current chromatogram (TIC) of *Bacillus* sp. culture broth sample (**c**) with mass spectra of lipopeptide peaks 1–4 (**d**) and bioactive compounds applied in marble samples (**e** and **f**). After sterilization, marble slabs were inoculated with a mixture composed of 1 mL of *Penicillium* $10^5$ CFU/mL spore suspension, 500 μL of malt extract, and 500 μL of sterilized water (**e**) or 500 μL of *Bacillus* sp. CCLBH 1053 liquid culture supernatant, which contained the bioactive compounds produced (**f**)

In the last few years, data generated by the different high-throughput technologies has expanded exponentially allowing genome sequencing of different species. Misra et al. (2019) discuss the possible development of standardized analytical pipelines that could be embraced by the omics research community highlighting recent methodologies, existing tools and potential limitations in the integration of omics datasets. The review of Mohana et al. (2018) discusses the recent progress in omics-based applications for biodiscovery of microbial natural products in antibiotic resistance era. Recently, Kalkreuter et al. (2020) presented different examples of successful strategies based on the combination of genome sequencing and bioinformatics that have been used for natural product discovery from cultivatable bacteria and discussed the opportunities and challenges that come from the association of genome conversion into natural products.

In fact, multi-omics methods enable us to study the potential of *Bacillus* antimicrobial compounds at both their compositional and functional levels.

Further insight into the potential metabolic processes, biosynthetic capabilities and stress adaptations can be inferred from genomic-scale comparison of strains. With this in mind, it is possible to screen and select strains with higher industrial potential capable to produce new desired active compounds or improve their production. Genomic approaches capable to identify the gene clusters that are localized in the genome can shed new light in the evaluation of the latent production capabilities of specific selected strains. The study of Yang Lu et al. (2019) provides a genomic and growth proteomic analysis of bacillomycin Lb biosynthesis during the *B. amyloliquefaciens* X030 growth cycle, making available a database about biosynthesis gene clusters, key regulatory enzymes, and proteins. The full-genome sequencing of a few genomes of bacteria, in particular, of the genus *Bacillus,* nowadays available in international databases like NCBI, allows determining the location of genetic clusters responsible for peptide synthesis. These biosynthetic gene clusters include non-ribosomal peptide synthetases (NRPSs), polyketide synthases (PKSs), and ribosomally synthesized and post-translationally modified peptides (RiPPs).

The complete genome sequence of *Bacillus* strains and a comparative analysis between the genome sequences revealed differences in the lipopeptide synthesis genes paving the way for the design of new solutions combining the potential of various strains of *Bacillus.*

Actually, we explore hypogenic environments, a robust microbial cell factory, with unexplored biosynthetic potential of strains, by omics-based approaches, to improve the antagonistic compounds (data not shown). New active compounds and green nanotechnology for direct application in cultural heritage safeguard can in a close future contribute to potentiate a future generation of biocides and safe sustainable methods.

Microorganisms are responsible for the biodeterioration of cultural heritage but can also be an important source of solutions. It's up to figure out the way. . . .

**Acknowledgements** Project SCREAM: Touchstone for Heritage Endangered by Salt Crystallization – a Research Enterprise on the Art of Munch (ALT20-03-0145-FEDER-031577); project 0483_PROBIOMA_5_E, Prospecion en ambientes subterraneos de compuestos bioactivos microbianos con uso potencial para la medicina, agricultura y medio ambiente. Co-financed by the European Regional Development; and the City University of Macau endowment to the Sustainable Heritage Chair.

**Ethical Statement   Conflict of Interest:** Authors declare that they have no conflict of interest in this study.

**Ethical Approval:** This article does not contain any studies with human participants or animals.

# References

Abdel-Haliem ME, Sakr AA, Ali MF, Ghaly MF, Sohlenkamp C (2013) Characterization of *Streptomyces* isolates causing colour changes of mural paintings in ancient Egyptian tombs. Microbiol Res 168:428–437

Abdellaziz L, Chollet M, Abderrahmani A, Bechet M, Yaici L, Chataigne G, Arias A, Leclere V, Jacques P (2018) Lipopeptide biodiversity in antifungal *Bacillus* strains isolated from Algeria. Arch Microbiol 200:1205–1216

Abderrahmani A, Tapi A, Nateche F, Chollet M, Leclere V, Wathelet B, Hacene H, Jacques P (2011) Bioinformatics and molecular approaches to detect NRPS genes involved in the biosynthesis of kurstakin from *Bacillus thuringiensis*. Appl Microbiol Biotechnol 92:571–581

Aleti G, Sessitsch A, Brader G (2015) Genome mining: prediction of lipopeptides and polyketides from *Bacillus* and related Firmicutes. Comput Struct Biotechnol J 13:192–203

Allsopp D (2011) Worldwide wastage: the economics of biodeterioration. Microbiol Today 38:150–153

Allsopp D, Swal K, Gaylande C (2004) The control of biodeterioration. In: Introduction to biodeterioration. Cambridge University Press, Cambridge, pp 203–223

Amann R, Fuchs B, Behrens S (2001) The identification of microorganisms by fluorescence in situ hybridisation. Curr Opin Biotechnol 12:231–236

Arantes S, Caldeira AT, Candeias A, Martins MR (2019a) Antimicrobial potential of *Lamiaceae* essential oils against heritage biodeteriogenic strains. Congress of Microbiology and Biotechnology, Book of abstracts Microbiotec' 19:97, Coimbra, Portugal (http://microbiotec19.net/test/wpcontent/uploads/2019/07/Book_abstract_online.pdf)

Arantes S, Guerreiro M, Piçarra A, Caldeira AT, Martins RM (2019b). Antimicrobial potential of essential oils of Portuguese autochthonous *Lavandula* against resistant bacteria. Congress of Microbiology and Biotechnology, Book of abstracts Microbiotec' 19, pp 313, Coimbra, Portugal (http://microbiotec19.net/test/wp-content/uploads/2019/07/Book_abstract_online.pdf)

Arbige MV, Bulthuis BA, Schultz J, Crabb D (1993) Fermentation of Bacillus. In: Sonenshein AL, Hoch JA, Losick R (eds) Bacillussubtilis and other gram-positive bacteria: biochemistry, physiology, and molecular genetics. American Society for Microbiology, Washington, DC, pp 871–895

Barbosa J, Caetano T, Mendo S (2015) Class I and class II lanthipeptides produced by *Bacillus* spp. J Nat Prod 78:2850–2866

Barresi G, Cammarata M, Palla F (2017) Biocide. In: Palla F, Barresi G (eds) Biotechnology and conservation of cultural heritage. Springer, Switzerland, pp 49–66

Beltran-Gracia E, Macedo-Raygoza G, Villafaña-Rojas J, Martinez-Rodriguez A, Chavez-Castrillon Y, Espinosa-Escalante F, Di Mascio P, Ogura T, Beltran-Garcia M (2017) Production of lipopeptides by fermentation processes: Endophytic bacteria, fermentation strategies and easy methods for bacterial Selection. In: Angela Faustino J (ed) Fermentation Processes, IntechOpen, London. https://doi.org/10.5772/64236

Berezhnayaa A, Evdokimovaa O, Valentovicha L, Sverchkovaa N, Titoka M, Kolomiyets E (2019) Molecular genetic and functional analysis of the genome of bacteria *Bacillus velezensis* BIM B-439D. Appl Biochem Microbiol 55:386–396

Besson F, Chevanet C, Michel G (1987) Influence of the culture medium on the production of iturin A by *Bacillus subtilis*. J Gen Microbiol 3:767–772

Borrego S, Valdés O, Vivar I, Lavin P, Guiamet P, Battistoni P, Gómez de Saravia S, Borges P (2012) Essential oils of plants as biocides against microorganisms isolated from cuban and argentine documentary heritage. Int Sch Res Notices 2012:7., Article ID 826786. https://doi.org/10.5402/2012/826786

Bosch-Roig P, Regidor-Ros JL, Montes-Estelles R (2013) Biocleaning of nitrate alterations on wall paintings by *Pseudomonas stutzeri*. Int Biodeter Biodegr 84:266–274

Britta K, Joachim V, Johann S, Klaus E (1988) Studies on the biosynthesis of surfactin, a lipopeptide antibiotic from Bacillus subtilis ATCC 21332. FEBS Lett:231. https://doi.org/10. 1016/0014-5793(88)80712-9

Caldeira AT, Savluchinske-Feio S, Arteiro JM, Roseiro J (2006) Antimicrobial activity of steady-state cultures of *Bacillus* sp. CCMI 1051 against wood contaminant fungi. Biochem Eng J 30:231–236

Caldeira AT, Savluchinske-Feio S, Arteiro JM, Roseiro J (2007) *Bacillus amyloliquefaciens* CCMI 1051 in vitro activity against wood contaminant fungi. Ann Microbiol 57:29–34

Caldeira AT, Savluchinske-Feio S, Arteiro JM, Coelho A, Roseiro J (2008) Environmental dynamics of *Bacillus amyloliquefaciens* CCMI1051 antifungal activity under different nitrogen patterns. J Appl Microbiol 104:808–816

Caldeira AT, Arteiro JM, Coelho A, Roseiro J (2011a) Combined use of LC–ESI-MS and antifungal tests for rapid identification of bioactive lipopeptides produced by *Bacillus amyloliquefaciens* CCMI 1051. Process Biochem 46(9):1738–1746

Caldeira AT, Vicente H, Arteiro JM, Roseiro J, Neves J (2011b) An artificial intelligence approach to *Bacillus amyloliquefaciens* CCMI 1051 cultures: application to the production of antifungal compounds. Bioresour Technol 102:1496–1502

Caldeira AT, Vicente H, Arteiro JM, Roseiro J, Neves J (2011c) Production of bioactive compounds against wood contaminant fungi: an artificial intelligence approach. In: Caldeira F (ed) Minimizing the environmental impact of the forest products industries. UFP Editions, Porto, pp 131–137

Caldeira AT, Rosado T, Silva M, Gonzalez M, Candeias A (2015) Microrganismos e Património - Novas abordagens. Magazine da Sociedade Portuguesa de Microbiologia 4:1–7

Caldeira AT, Salvador C, Rosado T, Teixeira D (2018) Biodeterioration of easel paintings – an overview. In: Mitchell R, Clifford J (eds) Biodeterioration and preservation in art, archaeology and architecture. Archetype Publications, London, pp 71–78

Cámara B, De los Ríos A, Urizal M, Buergo M, Varas MJ, Fort R, Ascaso C (2011) Characterizing the microbial colonization of a dolostone quarry: implications for stone biodeterioration and response to biocide treatments. Microb Ecol 62:299–313

Cameotra SS, Makkar RS (2004) Recent applications of biosurfactants as biological and immunological molecules. Curr Opin Microbiol 7:262–266

Carrillo PM, Robinson HS, Anumba CJ, Al-Ghassani AM (2003) IMPaKT: a framework for linking knowledge management to business performance. Electron J Knowl Manag 1(1):1–12

Cawoy H, Debois D, Franzil L, De Pauw E, Thonart P, Ongena M (2015) Lipopeptides as main ingredients for inhibition of fungal phytopathogens by *Bacillus subtilis/amyloliquefaciens*. Microb Biotechnol 8:281–295

Chena J, Xue Q, Mac CL, Tan X (2020) *Streptomyces pactum* may control *Phelipanche aegyptiaca* in tomato. Appl Soil Ecol 146:103369

Chevanet C, Besson F, Michel G (1986) Effect of various growth conditions on spore formation and bacillomycin L production in *Bacillus subtilis*. Can J Microbiol 32:254–258

Chien H, Kuo C, Kao L, Lin G, Chen P (2019) Polysaccharidic spent coffee grounds for silver nanoparticle immobilization as a green and highly efficient biocide. Int J Biol Macromol 140:168–176

Cho SJ, Lee SK, Cha BJ, Kim YH, Shin KS (2003) Detection and characterization of the *Gloeosporium gloeosporioides* growth inhibitory compound iturin A from *Bacillus subtilis* strain KS03. FEMS Microbiol Lett 223:47–51

Colombo EM, Kunova A, Pizzatti C, Saracchi M, Cortesi P, Pasquali M (2019) Selection of an endophytic streptomyces sp. strain DEF09 from wheat roots as a biocontrol agent against *Fusarium graminearum*. Front Microbiol 10:1–12

Coors A, Vollmar P, Heim J, Sacher F, Kehrer A (2018) Environmental risk assessment of biocidal products: identification of relevant components and reliability of a component based mixture assessment. Environ Sci Eur 30(3):1–15

Davis DA, Lynch HC, Varley J (2001) The application of foaming for the recovery of Surfactin from B. subtilis ATCC 21332 cultures. Enzym Microb Technol 28:346–354. https://doi.org/10.1016/S0141-0229(00)00327-6

Deng Q, Wang R, Sun D, Sun L, Wang Y, Pu Y, Fang Z, Xu D, Liu Y, Ye R, Yin S, Xie S, Gooneratne R (2020) Complete genome of Bacillus velezensis CMT-6 and comparative genome analysis reveals lipopeptide diversity. Biochem Genet 58:1–15

Dias L, Rosado T, Coelho A, Barrulas P, Lopes L, Moita P, Candeias A, Mirão J, Caldeira AT (2018) Natural limestone discolouration triggered by microbial activity—a contribution. AIMS Microbiol 4:594–607

Dresler C, Saladino M, Demirbag C, Caponetti E, Martino D, Alduina R (2017) Development of controlled release systems of biocides for the conservation of cultural heritage. Int Biodeterior Biodegrad 125:150–156

Falardeau J, Wise C, Novitsky L, Avis TJ (2013) Ecological and mechanistic insights into the direct and indirect antimicrobial properties of Bacillus subtilis Lipopeptides on plant pathogens. J Chem Ecol 39(7):869–878

Favero-Longo S, Benesperi R, Bertuzzi S, Bianchi E, Buffa G, Giordani P, Loppi S, Malaspina P, Matteucci E, Paoli L, Ravera S, Roccardi A, Segimiro A, Vannini A (2017) Species- and site-specific efficacy of commercial biocides and application solvents against lichens. Int Biodeterior Biodegrad 123:127–137

Fernandes P (2006) Applied microbiology and biotechnology in the conservation of cultural heritage materials. Appl Microbiol Biotechnol 73:291–296

Fidanza M, Caneva G (2019) Natural biocides for the conservation of stone cultural heritage: a review. J Cult Herit 38:271–286

Fomina M, Burford P, Hillier S, Kierans M, Gadd M (2010) Rock-building fungi. Geomicrobiol J 27:624–629

González JM, Saiz-Jiménez C (2005) Application of molecular nucleic acid-based techniques for the study of microbial communities in monuments and artworks. Int Microbiol 8:189–194

Gonzalez M, Vieira R, Nunes P, Rosado T, Martins S, Candeias A, Pereira A, Caldeira AT (2014) Fluorescence In Situ hybridization: a potentially useful technique for detection of microorganisms on mortars. e-conser J 2:44–52

Gonzalez M, Brinco C, Vieira R, Rosado T, Mauran G, Pereira A, Candeias A, Caldeira AT (2017) Dual phylogenetic staining protocol for simultaneous analysis of yeast and bacteria in artworks. Appl Phys A:123–142

Gonzalez-Pastor J, Hobbs E, Losick R (2003) Cannibalism by sporulating bacteria. Science 301:510–513

Guiamet P, Borrego S, Lavin P, Perdomo I, Saravia SGD (2011) Biofouling and biodeterioration in materials stored at the historical archive of the Museum of La Plata, Argentine and at the National Archive of the Republic of Cuba. Colloids Surf B Biointerf 85:229–234

Hien TH, Govin A, Guyonnet R, Grosseau P, Lors C, Garcia-Diaz E, Damidot D, Devès O, Ruot B (2012) Influence of the intrinsic characteristics of mortars on biofouling by Klebsormidium flaccidum. Int Biodeterior Biodegrad 70:31–39

Hooker JE, Jaizme-Vega M, Atkinson D (1994) Biocontrol of plant pathogens using arbuscular mycorrhizal fungi. In: Gianinazzi S, Schüepp H (eds) Impact of Arbuscular Mycorrhizas on sustainable agriculture and natural ecosystems. ALS Advances in Life Sciences. Birkhäuser, Basel, pp 191–200

Hsieh FC, Li MC, Lin TC, Kao SS (2004) Rapid detection and characterization of surfactin-producing Bacillus subtilis and closely related species based on PCR. Curr Microbiol 49:186–191. https://doi.org/10.1007/s00284-004-4314-7

Inès M, Dhouha G (2015) Lipopeptide surfactants: production, recovery and pore forming capacity. Peptides 71:100–112

Jeong S, Lee H, Kim DW, Chung Y (2018) New biocide for eco-friendly biofilm removal on outdoor stone monuments. Int Biodeterior Biodegrad 131:19–28

Jin P, Wang H, Tan Z, Xuan Z, Dahar GY, Li QX, Miao W, Liu W (2020) Antifungal mechanism of bacillomycin D from *Bacillus velezensis* HN-2 against *Colletotrichum gloeosporioides* Penz. Pestic Biochem Phys 163:102–107

Junier P, Joseph E (2017) Microbial biotechnology approaches to mitigating the deterioration of construction and heritage materials. Microb Biotechnol 10(5):1145–1148

Kakakhel I, Wu F, Gu JD, Feng H, Shah K, Wang W (2019) Controlling biodeterioration of cultural heritage objects with biocides: a review. Int Biodeterior Biodegrad 143(104721):1–10

Kalkreuter E, Pan G, Cepeda A, Shen B (2020) Targeting bacterial genomes for natural product. Discovery Trends Pharmacol Sci 41:13–26

Kim K, Jung SY, Lee DK, Jung JK, Park JK, Kim DK, Lee CH (1998) Suppression of inflammatory responses by surfactin, a selective inhibitor of platelet cytosolic phospholipase A(2). Biochem Pharmacol 55(7):975–985. https://doi.org/10.1016/S0006-2952(97)00613-8

Leifert C, Li H, Chidburee S, Hampson S, Workman S, Sigee D, Epton H, Harbour A (1995) Antibiotic production and biocontrol activity by *Bacillus subtilis* CL27 and *Bacillus pumilus* CL45. J Appl Bacteriol 78:97–108

Lowes KF, Shearman CA, Payne J, MacKenzie D, Archer DB, Merry RJ, Gasson MJ (2000) Prevention of yeast spoilage in feed and food by the yeast mycocin HMK. Appl Environ Microbiol 66(3):1066–1076. https://doi.org/10.1128/aem.66.3.1066-1076.2000

Lu JY, Zhou K, Huang WT, Zhou P, Yang S, Zhao X, Xie J, Xia L, Ding X (2019) A comprehensive genomic and growth proteomic analysis of antitumor lipopeptide bacillomycin Lb biosynthesis in *Bacillus amyloliquefaciens* X030. Appl Microbiol Biotechnol 103:7647–7662

Maksimova I, Singhb B, Cherepanovaa E, Burkhanovaa G, Khairullin R (2020) Prospects and applications of lipopeptide-producing bacteria for plant protection. Appl Biochem Microbiol 56 (1):15–28

Marahiel MA, Stachelhaus T, Mootz HD (1997) Modular peptide synthetases involved in nonribosomal peptide synthesis. Chem Rev 97:2651–2673

Mariana M, Ohnob T, Suzukic H, Kitamurac H, Kurodac K, Shimizua M (2020) A novel strain of endophytic Streptomyces for the biocontrol of strawberry anthracnose caused by *Glomerella cingulate*. Microbiol Res 234(126428):1–9

Massart S, Jijakli H (2007) Use of molecular techniques to elucidate the mechanisms of action of fungal biocontrol agents: a review. J Microbiol Methods 69(2):229–241

Medeot D, Fernandez M, Morales G, Jofré E (2020) Fengycins from *Bacillus amyloliquefaciens* MEP218 exhibit antibacterial activity by producing alterations on the cell surface of the pathogens *Xanthomonas axonopodis* pv. vesicatoria and *Pseudomonas aeruginosa* PA01. Front Microbiol 10(3107):1–12

Meena KR, Kanwar SS (2015) Lipopeptides as the antifungal and antibacterial agents: applications in food safety and therapeutics. Biomed Res Int 2015:473050

Mikkola R, Andersson M, Grigoriev P, Teplova V, Saris N, Rainey F, Salonen M (2004) Bacillus amyloliquefaciens strains isolated from moisture-damaged buildings produced surfactin and a substance toxic to mammalian cells. Arch Microbiol 181:314–323

Milanesi C, Baldi F, Borin S, Vignani R, Ciampolini F, Faleri C, Cresti M (2006) Biodeterioration of a fresco by biofilm forming bacteria. Int Biodeterior Biodegrad 57:168–173

Misra B, Langefeld C, Olivier M, Cox L (2019) Integrated omics: tools, advances and future approaches. J Mol Endocrinol 62:21–45

Mohana N, Rao H, Rakshith D, Mithun PR, Nuthan BR, Satish S (2018) Omics based approach for biodiscovery of microbial natural products in antibiotic resistance era. J Genet Eng Biotechnol 16(1):1–8

Moter A, Göbel U (2020) Fluorescence in situ hybridization (FISH) for direct visualization of microorganisms. J Microbiol Methods 41(2):85–112

Moyne A, Shelby R, Cleveland T, Tuzun S (2001) Bacillomycin D: an iturin with antifungal activity against *Aspergillus flavus*. J Appl Microbiol 90:622–629

Ongena M, Jacques P (2008) Bacillus lipopeptides: versatile weapons for plant disease biocontrol. Trends Microbiol 16:115–125

Ohno A, Takashi A, Shoda M (1992) Production of a lipopeptideantibiotic surfactin with recombinant Bacillus subtilis. Biotechnol Lett 14:1165–1168

Othoum G, Bougouffa S, Razali R et al (2018) In silico exploration of Red Sea Bacillus genomes for natural product biosynthetic gene clusters. BMC Genomics 19:382. https://doi.org/10.1186/s12864-018-4796-5

Pabel C, Vater J, Wilde C, Franke P, Hofemeisrer J, Adler B, Bringmann G, Hacker J, Hentschel U (2003) Antimicrobial activities and matrix-assisted laser desorption/ionization mass spectrometry of Bacillus isolates from the marine sponge *Aplysina aerophoba*. Mar Biotechnol 5:424–434

Palla F, Bruno M, Mercurio F, Tantillo A, Rotolo V (2020) Essential oils as natural biocides in conservation of cultural heritage. Molecules 25(730):1–11

Pangallo D, Chovanová K, Simonovicova A, Ferianc P (2009) Investigation of microbial community isolated from indoor artworks and air environment: identification, biodegradative abilities, and DNA typing. Can J Microbiol 55:277–287

Pena-Poza J, Ascaso C, Sanz M, Pérez-Ortega S, Oujja M, Wierzchos J, Souza-Egipsy V, Cañamares MV, Urizal M, Castillejo M, García-Heras M (2018) Effect of biological colonization on ceramic roofing tiles by lichens and a combined laser and biocide procedure for its removal. Int Biodeterior Biodegrad 126:86–94

Piñar G, Sterflinger K (2009) Microbes and building materials. In: Cornejo DN, Haro JL (eds) Building materials: properties, performance and applications. Nova Science Publishers, New York, pp 163–188

Płaza G, Achal V (2020) Biosurfactants: eco-friendly and innovative biocidesagainst biocorrosion. Int J Mol Sci 21:2152

Quagliarini E, Graziani L, Diso D, Licciulli A, D'Orazio M (2018) Is nano-TiO2 alone an effective strategy for the maintenance of stones in cultural heritage? J Cult Heritage 30:81–91

Rai J, Singh H, Brahmaprakash G (2020) Lesson from ecotoxity: revisiting the microbial lipopeptides for the management of emerging diseases for crop protection. Int J Environ Res Public Health 17(1434):1–27

Rosado T, Martins MR, Pires M, Mirão J, Candeias A, Caldeira AT (2013a) Enzymatic monitorization of mural paintings biodeterioration. Int J Cons Sci 4:603–612

Rosado T, Gil M, Mirão J, Candeias A, Caldeira AT (2013b) Oxalate biofilm formation in mural paintings due to microorganisms - a comprehensive study. Int Biodeterior Biodegrad 85:1–7

Rosado T, Mirão J, Gil M, Candeias A, Caldeira AT (2013c) Evaluation of mural paintings biodeterioration by oxalate formation. In: Rogerio-Candelera MA, Lazzari M, Cano E (eds) Science and technology for the conservation of cultural heritage. CRC Press/Balkema, London, pp 147–150

Rosado T, Mirão J, Candeias A, Caldeira AT (2014a) Microbial communities analysis assessed by pyrosequencing - a new approach applied to conservation state studies of mural paintings. Anal Bioanal Chem 406:887–895

Rosado T, Reis A, Mirão J, Candeias A, Vandenabeeled P, Caldeira AT (2014b) Pink! Why not? On the unusual colour of Évora Cathedral. Int Biodeterior Biodegrad 94:121–127

Rosado T, Mirão J, Candeias A, Caldeira AT (2015) Characterizing microbial diversity and damage in mural paintings. Microsc Microanal 21(1):78–83

Rosado T, Gil M, Mirão J, Candeias A, Caldeira AT (2016) Darkening on lead based pigments: microbiological contribution. Color Res Appl 41(3):294–298

Rosado T, Silva M, Dias L, Candeias A, Gil M, Mirão J, Pestana J, Caldeira AT (2017) Microorganisms and the integrated conservation-intervention process of the renaissance mural paintings from Casas Pintadas in Évora - know to act, act to preserve. J King Sau Univ – Sci 29 (4):478–486

Rosado T, Santos R, Silva M, Galvão A, Mirão J, Candeias A, Caldeira AT (2019) Mitigation approach to avoid fungal colonisation of porous limestone. Int J conser sci 10(1):3–14

Rosado T, Dias L, Lança M, Nogueira C, Santos R, Martins MR, Candeias A, Mirão J, Caldeira AT (2020) Assessment of microbiota present on a Portuguese historical stone convent using high throughput sequencing approaches. Microbiol Open e1030:1–18. https://doi.org/10.1002/mbo3. 1030

Ruggiero L, Bartoli F, Fidanza MR, Zurlo F, Marconi E, Gasperi T, Tuti S, Crociani L, Di Bartolomeo E, Caneva G, Ricci MA, Sodo A (2020) Encapsulation of environmentally-friendly biocides in silica nanosystems for multifunctional coatings. Appl Surf Sci 514(145908):1–9

Salvador C, Silva M, Rosado T, Vaz Freire R, Bordalo R, Candeias A, Caldeira AT (2016) Biodeterioration of easel paintings: development of new mitigation strategies. Conservar Património 23:119–124

Salvador C, Borlado R, Silva M, Rosado T, Candeias A, Caldeira AT (2017) On the conservation of easel paintings - evaluation of microbial contamination and artist materials. Appl Phys A 123 (80):1–13

Salvador C, Medeiros R, Candeias A, Caldeira AT (2019) Biodeterioration of easel paintings: a case study of Munch's paintings. In: Congress of microbiology and biotechnology, book of abstracts Microbiotec'19. Portugal, Coimbra, p 90. http://microbiotec19.net/test/wp-content/uploads/ 2019/07/Book_abstract_online.pdf

Sandrin TR, Goldstein GE, Schumaker S (2013) MALDI TOF MS profiling of bacteria at the strain level: a review. Mass Spectrom Rev 32:188–217

Sanza M, Oujja M, Ascaso C, Ríos A, Pérez-Ortega S, Souza-Egipsy V, Wierzchos J, Speranza M, Canamares M, Castillejo M (2015) Infrared and ultraviolet laser removal of crustose lichens on dolomite heritage stone. Appl Surf Sci 346:248–255

Scheerer S, Ortega-Morales O, Gaylarde C (2009) Microbial deterioration of stone monuments – an updated overview. Adv Appl Microbiol 66:97–139

Silva M, Teixeira D, Silva S, Candeias A, Caldeira AT (2014) Production of novel biocides for cultural heritage from Bacillus sp. In: Rogerio-Candelera MA (ed) Science, technology and cultural heritage. CRC Press/Balkema, London, pp 223–228

Silva M, Rosado T, Teixeira D, Candeias A, Caldeira AT (2015) Production of Green biocides for cultural heritage- novel biotechnological solutions. Int J Conerv Sci 6:519–530

Silva M, Vador C, Candeias MF, Teixeira D, Candeias A, Caldeira AT (2016a) Toxicological assessment of novel green biocides for cultural heritage. Int J Conserv Sci 7:265–272

Silva M, Pereira A, Teixeira D, Candeias A, Caldeira AT (2016b) Combined use of NMR, LC-ESIMS and antifungal tests for rapid detection of bioactive lipopeptides produced by Bacillus. Adv Microbiol 6:788–796

Silva M, Rosado T, Teixeira D, Candeias A, Caldeira AT (2017) Green mitigation strategy for cultural heritage: bacterial potential for biocide production. Environ Sci Pollut Res 24:4871–4881

Silva M, Rosado T, Lopes da Silva Z, Nóbrega Y, Silveira D, Candeias A, Caldeira AT (2019) Green bioactive compounds: mitigation strategies for cultural heritage. Conserv Sci Culture Heritage 19:133–142

Sterflinger K (2010) Fungi: their role in deterioration of cultural heritage. Fungal Biol Rev 24:47–55

Suárez MB, Vizcaíno JA, Llobell A, Monte E (2007) Characterization of genes encoding novel peptidases in the biocontrol fungus Trichoderma harzianum CECT 2413 using the Tricho EST functional genomics approach. Curr Genet 51:331–342

Tiano P (2002) Biodegradation of cultural heritage: decay, mechanisms and control methods. Seminar article, New University of Lisbon, Department of Conservation and Restoration, 7–12 (http://www.arcchip.cz/w09/w09_tiano.pdf)

Tsuge K, Akiyama T, Shoda M (2001) Isolation of a gene essential for biosynthesis of the lipopeptide antibiotics plipastatin B1 and surfactin in Bacillus subtilis YB8. J Bacteriol 183:6265–6273

Urzì C, De Leo F (2007) Evaluation of the efficiency of water-repellent and biocide compounds against microbial colonization of mortars. Int Biodeterior Biodegrad 60:25–34

Vieira R, Nunes P, Martins S, González M, Rosado T, Pereira A, Candeias A, Caldeira AT (2014) Fluorescence in situ hybridisation for microbiological detection in mortars. In: Rogerio-

Candelera (ed) Science, technology and cultural heritage. CRC Press/Balkema Taylor & Francis Group, London, pp 257–262

Vieira R, Pazian M, González-Pérez M, Pereira A, Candeias A, Caldeira AT (2018) Detecting cells with low RNA content colonizing artworks non-invasively: RNA-FISH. In: Mosquera MJ, Gil A (eds) Conserving cultural heritage. CRC Press/Balkema, London, pp 303–305

Walker GM, Mcleod AH, Hodgson VJ (1995) Interactions between killer yeasts and pathogenic fungi. FEMS Microbiol Lett 127(3):213–222. https://doi.org/10.1111/j.1574-6968.1995.tb07476.x

Yazgan A, Ozcengiz G, Marahiel MA (2001) Tn10 insertional mutations of *Bacillus subtilis* that block the biosynthesis of bacilysin. Biochim Biophys Acta 1518:87–94

Young ME, Alakomi H, Fortune I, Gorbushina A, Krumbein W, Maxwell I, McCullagh C, Robertson P, Saarela M, Valero J, Vendrell M (2008) Development of a biocidal treatment regime to inhibit biological growths on cultural heritage: BIODAM. Environ Geol 56 (3):631–641

Zhao X, Kuipers O (2016) Identification and classification of known and putative antimicrobial compounds produced by a wide variety of Bacillales species. BMC Genomics 17(882):1–18

# Chapter 7
# New Perspectives Against Biodeterioration Through Public Lighting

**Patricia Sanmartín**

**Abstract** There is currently an increasing trend in urban centres towards the use of public outdoor lighting systems to illuminate historic and architecturally important buildings during evening hours, but for which there is no specific regulatory framework. Considering that the light is a key factor involved in regulating growth and physiological processes in photosynthetic organisms, it seems appropriate to address the effects that artificial light has on the organisms growing on the facades affected by public lighting. In this sense, despite scientific research in the fields of biological colonization of buildings surfaces and light technology has advanced greatly in recent years, the combination of both disciplines aimed at the correct handling of city public lighting remains uncharted territory with huge potential to provide innovative solutions for smart cities. Recent studies have examined how urban monuments are affected by night-time outdoor illumination in combination with natural sunlight and demonstrated that the use of suitable lighting can inhibit the development of biological colonization. In this frame, this chapter will look at ways of contribute to the long-term management of public illumination on monuments and other structures, while reducing negative impacts caused by night lighting.

**Keywords** Organism's control strategies · Cultural and natural heritage · Biodiversity · Light pollution · Public lighting · Smart cities · Stone · Biodeterioration · Urban fabric · GLOBAL change

## 1 Introduction

Light is a type of electromagnetic radiation that, via photochemical interactions, provides the main source of energy for the metabolism of photoautotrophic organisms (i.e. photosynthesis) and that is directly involved in fixing C, and in N and S

P. Sanmartín (✉)
Departamento de Edafoloxía e Química Agrícola. Facultade de Farmacia, Universidade de Santiago de Compostela, Santiago de Compostela, Spain
e-mail: patricia.sanmartin@usc.es

© The Author(s) 2021
E. Joseph (ed.), *Microorganisms in the Deterioration and Preservation of Cultural Heritage*, https://doi.org/10.1007/978-3-030-69411-1_7

metabolism (Willey 2016). Photosynthetic metabolism represents an adaptive advantage that enables photoautotrophic organisms to occupy various niches (Di Martino 2016). These organisms are therefore considered a key group in the formation of subaerial biofilms (SABs), which have been defined as "microbial communities that grow on solid surfaces exposed to the atmosphere" (Gorbushina 2007). These communities are mainly dominated by green algae and to a lesser extent by cyanobacteria in temperate climates characterized by warm summers and cool winters with high atmospheric humidity, as in many parts of Europe (Rifón Lastra 2000; Rifón Lastra and Noguerol Seoane 2001). The presence of photoautotrophic microorganisms (cyanobacteria and algae), i.e. primary colonizers, leads to the formation of more complex microbial associations, which in turn enables settlement of other species (heterotrophic microorganisms) and the concomitant growth of other colonizers, including lichens, heterotrophic bacteria, fungi, bryophytes, and vascular plants.

Because the occurrence of these pioneer algae and cyanobacteria (a phenomenon often referred to as "greening": Smith et al. 2011; Sanmartín et al. 2012) is the first step in the sequential process of colonization, it requires priority research attention. Greening can be both detrimental and beneficial, depending on the substratum and microorganisms involved. Indeed, there is some controversy regarding the impact of greening on the physical and chemical integrity of stone. Although classic studies describe a role for greening in the weathering process (Ortega-Calvo et al. 1991a, b), it is increasingly considered to have a negligible (Sanmartín et al. 2020a) or even bioprotective role (Ramírez et al. 2010; Cutler et al. 2013). In the urban fabric of a city affected by high atmospheric humidity, biofilm formation and biological colonization are unavoidable because they occur as a result of interactions between the building material and the environment (Gorbushina 2007).

One of the main challenges facing cities is the management of biological colonization on building facades (Energy & Smart cities; Smartiago project). Cleaning is an important aspect of the maintenance and rehabilitation of stone buildings and structures, which are prone to being colonized by living organisms, removed frequently to prevent biodeterioration, including the simple aesthetic impact (Warscheid and Braams 2000; Scheerer et al. 2009). Removing these organisms from building facades requires a huge financial outlay by local governments and agencies. According to the Santiago de Compostela (NW Spain) city council, the cost of cleaning the main facade (Obradoiro) of the Pazo de Raxoi amounted in 2016 to €318,000.

In the current night-time environment of historic centres with a illumination that makes the buildings more visible and showcases different features (Fig. 7.1), and known the key role of light on greening-forming organisms; it is striking that few studies have addressed to study the influence of ornamental outdoor illumination from artificial night-time sources on the biological colonization on buildings and monuments.

In this chapter, lighting-based strategies currently used against biodeterioration are presented together with progress in the public outdoor lighting systems illuminating heritage buildings, and emphasis is placed on findings of the recent regional

**Fig. 7.1** Night-time view of the Hércules Tower (A Coruña, Galicia, Spain) illuminated with purple, green, blue, and orange light in 2017. Source: Elena López

2-year Light4Heritage project (2016–2018), focused on developing lighting-based strategies to control biological colonization and to manage the chromatic integration of biofouling at laboratory scale. Finally, a critical analysis about in what form public lighting can be turned into a green method is also reported.

## 2   Lighting-Based Strategies Currently Used Against Biodeterioration

Lighting-based strategies that are currently used to control primary colonizers (algae and cyanobacteria) on the built heritage are described below (Borderie et al. 2012; Figueroa et al. 2017; Sanmartín et al. 2018a):

**Restricting the Duration of Lighting** The simplest way to restrict lampenflora, defined as "massive growth of cyanobacterial-based microbiota and of algae, mosses and other plants located near light sources in caves accessible to tourists" (Dobat 1998; Mulec and Kosi 2009; Mulec 2012), is to limit the time that caves are illuminated (during visiting hours) thus increasing the length of the dark period (Figueroa et al. 2017).

**Changing the Amount of Light** The use of low-intensity lamps reduces biological colonization. The use of low levels of illumination for short periods helps to prevent and control the growth of phototrophic biofilms (Albertano and Bruno 2003). The light must be strong enough to cause photoinhibition, considered a reduction in photosynthesis in response to increasing irradiance, although the term may refer more generally to inhibition of growth due to a change in lighting (Sanmartín et al. 2018a).

**Using Different Wavelengths of Monochromatic LED Lights** The use of LEDs with light qualities enriched in the wavelengths with low photosynthetic quantum efficiency for the target phototrophs has a biostatic effect (Sanmartín et al. 2018b). However, when LED lighting is used to control phototrophs, the ability of these organisms to photoacclimatize (MacIntyre et al. 2002) and adapt to light (Tandeau de Marsac 1977) must be taken into account, along with the presence of different pigments in photosynthetic organisms and other organisms in the same biofilm. Nonetheless, light quality is one of the most important and well-studied light parameters, and recent studies suggest that it is the main regulator of growth responses in algae and cyanobacteria (Kehoe 2010; Bussell and Kehoe 2013; Wiltbank and Kehoe 2016). Table 7.1 summarizes the most important contributions from previous studies regarding the responses of algae and cyanobacteria to LED light treatments.

**Using UV (A, B, or C) Irradiation** UV treatment is widely used to disinfect surfaces. Ultraviolet light has a short wavelength and high energy and becomes more energetic as the wavelength decreases, according to Planck's quantum theory, so that the energy increases in the order UV-A (315–400 nm) < UV-B (280–315 nm) < UV-C (190–280 nm). The sun emits radiation in all three UV ranges, but the UV radiation that reaches the Earth's surface comprises around 95% UV-A and 5% UV-B. As a physical and germicidal treatment, UV-C irradiation is a suitable alternative method of controlling or limiting biofouling, and its potential to damage phototrophic biofilms in caves has been demonstrated (Borderie et al. 2012, 2014, 2015). Although many studies have addressed the effects of UV rays not filtered by the ozone layer, i.e. UV-A and UV-B radiation, on algal and cyanobacterial cultures (Sinha et al. 1996; Wingard et al. 1997; Yakovleva and Titlyanov 2001; Mullineaux 2001; Bischof et al. 2006; Castenholz and Garcia-Pichel 2012; Moon et al. 2012; Shang et al. 2018), few studies have focused on longer-wave radiation applied to biological colonization in built heritage. Table 7.2 summarizes the most important contributions from previous studies regarding the responses of algae and cyanobacteria to UV light treatments.

All these aforementioned lighting studies are scarce in relation to poikilohydric organisms, such as lichens and bryophytes, and even more so in relation to vascular plants on cultural heritage. To the best of my knowledge, only the EU-funded ECOLIGHT (Ecological effects of light pollution) project has made major advances in discovering how terrestrial plants respond to artificial light at night. The project findings show that artificial light can prolong the retention of leaves in urban environments and initiate early onset of bud burst in the spring, thus increasing the risk of exposure to frost and pathogens.

# 3 Public Outdoor Lighting Systems Illuminating Heritage Buildings

Historic centres are tourist attractions of enormous importance to the economy of cities and regions, which today are faced with the important challenge of applying new technological and management strategies within the smart city concept. This model aspires to use technological solutions to improve the management and

**Table 7.1** Behavioural responses of phototrophs to LED lights appraised by several authors on-site and laboratory-based experiments

| Reference | Occurrence | Organism(s) | Light | Findings |
|---|---|---|---|---|
| Albertano et al. (2003) | Roman Catacombs of St. Callistus and Domitilla, Rome (Italy) | cyanobacterial and chemoorganotrophic bacterial biofilms | Blue | In 5 months, no photosynthetic activity was detected. In 10 years, the extent of phototrophic communities was reduced (Bruno et al. 2014). Bacteria increased in some treated areas, although Actinobacteria decreased by 40% over the 10 years (Urzi et al. 2014). |
| Roldán et al. (2006) | Bats' Cave, Zuheros (Spain) | *Gloeothece membranacea* (cyanobacterium) and *Chlorella sorokiniana* (alga) biofilms | Green and white | Green light may prevent the growth of photosynthetic organisms, except for those capable of modifying accessory pigments. |
| Hsieh et al. (2013) | Roman Catacombs of St. Callistus and Domitilla, Rome (Italy) | 7 cyanobacterial strains | Red | Decreased Quantum Yield of Photosystem II and photoinhibition. |
| | | | Blue | Increased photosynthetic activity and growth of biofilm. |
| Hsieh et al. (2014) | Roman Catacombs of St. Callistus and Domitilla, Rome (Italy) | 5 cyanobacterial strains | Red and white | Decreased photosynthetic pigment content. |
| | | | Blue | Increased phycocyanin and phycoerythrin content. |
| Enomoto et al. (2015) | Not covered | *Thermosynechococcus vulcanus* and *T. elongatus* (cyanobacteria) | Blue | Damage to the manganese cluster of Photosystem II. Protection by cellular aggregation. |
| Han et al. (2015) | Not covered | *Nostoc flagelliforme* (cyanobacterium) | Green and blue | Change in the C/N ratio, indicating differences in cellular components. |
| | | | Red | Negative influence on primary metabolism. Photoinhibition. |
| | | | Yellow and violet | No inhibition of growth. |

(continued)

**Table 7.1** (continued)

| Reference | Occurrence | Organism(s) | Light | Findings |
|---|---|---|---|---|
| de Mooij et al. (2016) | Not covered | *Chlamydomonas reinhardtii* CC-1690 (green alga) | Red | Low biomass. High specific growth rate. |
| | | | Blue | Low biomass. High specific growth rate. Increased Quantum Yield of Photosystem II. |
| | | | Yellow | High biomass. Improved physiological state of cells when combined with a small proportion of blue light. |
| Del Rosal Padial (2016) | Nerja Cave, Malaga (Spain) | Biofilms containing *Chroococcidiopsis* sp., other cyanobacteria taxa, red alga *Cyanidium* sp., and green algae | Green | Reduced growth of photosynthetic microorganisms. |
| Bruno and Valle (2017) | Roman Catacombs of Domitilla, Rome (Italy) | 3 cyanobacterial strains and biofilms | Red, orange and blue | In 2 months, improved biofilm growth High electron transport rate. High chlorophyll-a and carotenoids content. |
| | | | White and green | In 2 months, growth limited. |

efficiency of the urban environment, with the ultimate aim of increasing urban sustainability (Energy & Smart cities).

There is currently an increasing trend for cities to install external lighting systems to illuminate buildings and monuments. Lighting components, as floodlights and spotlights, equipped with lamps are placed on the building itself, on its facade, lighting the architectural surroundings. This trend is favoured by technological improvements in lighting installations, specifically light emitting diode (LED)-based lighting. LED lighting has rapidly become sufficiently effective for use in exterior lighting systems and has several advantages over traditional lighting (less energy demanding, longer lifetime and mercury free). LED lamps not only increase the temperature and reduce humidity on building facades, but the artificial light also generates a specific physiological response in the colonizing organisms, interrupting their natural lighting cycle. This could favour the further development of biological colonization (Fig. 7.2), as previously observed in the subterranean cultural heritage (caves, catacombs, necropolis, etc.) by Albertano and Bruno (2003) and Albertano et al. (2003), and most recently by del Rosal and colleagues (del Rosal Padial et al. 2016; Jurado et al. 2020). Indeed, the term lampenflora defined above was coined to

**Table 7.2**  Behavioural responses of phototrophs to UV lights appraised by several authors on-site and laboratory-based experiments

| Reference | Occurrence | Organism(s) | Light | Findings |
|---|---|---|---|---|
| Sinha et al. (1996) | rice paddy fields near Varanasi (India) | 4 cyanobacterial cultures | UV-B and UV-A | 30' to 5 Wm$^{-2}$ caused the death of 50% population.<br>Decrease in the uptake of $^{14}CO_2$ and protein content with increased exposure time. |
| Wingard et al. (1997) | Alkaline hot spring in the city of Rotorua (New Zealand) | 4 cyanobacterial cultures | UV-B | Increased negative effect on $^{14}C$ incorporation rate relative to the control.<br>Chlorophyll-a content decreased with increasing light intensity l. |
| Moon et al. (2012) | Not covered | *Synechocystis* sp. PCC 6803 (cyanobacterium) | UV-A | High-dose: negative phototactic response.<br>Low-dose: photoregulation. Negative phototaxis affected. |
| Borderie et al. (2012) | 3 caves of south west of France (Domme, Laugerie-Haute, Combarelles, Department of Dordogne, France) and 1 cave of northeastern of France, Moidons Cave (Department of Jura, France) | mixed algae cultures | UV-C | Chlorophyll-a, -b and carotenoids content was affected.<br>Cell viability decreased greatly.<br>Programmed cell death occurred in unicellular algae.<br>Greater effectiveness in combination with high light. |
| Borderie et al. (2014) | Moidons Cave (Department of Jura, France) | 5 biofilms (mainly algae) | UV-C | In 12 months, cleaning due to light was effective.<br>In 16 months, biological colonization continued to increase.<br>Treatment effectiveness depended on biofilm thickness. |
| Borderie et al. (2015) | Moidons Cave (Department of Jura, France) | Green algae | UV-C | Damage was greater in the planktonic than in the biofilm mode of growth. |
| Pfendler et al. (2017) | La Glacière Cave, Chaux-lès-Passavant (France) | *Chlorella* sp. (green alga) | UV-C | In situ, discoloration of biofilm.<br>In 21 months, no photosynthetic |

(continued)

**Table 7.2** (continued)

| Reference | Occurrence | Organism(s) | Light | Findings |
|---|---|---|---|---|
| | | | | microorganisms were observed. |
| Shang et al. (2018) | dry desert steppes and bare land in Mongolia (China) | *Nostoc flagelliforme* strain CCNUN1 (cyanobacterium) | UV-B | In 1.5 h, decreased the Quantum Yield of Photosystem II. In 54 h, content of mycosporine-like amino acids doubled. |

**Fig. 7.2** Occurrence of phototrophic colonization around a streetlight. Source: Patricia Sanmartín

refer to the massive biological growth near light sources in caves accessible to tourists.

In outdoor environments, the lighting of buildings in white LED has given way to the use of coloured LEDs (red, green, yellow, blue, etc.) because the chromatic performance of the installation gives the building or monument of a higher symbology. For instance, built structures illuminated in purple to mark the feminist movement and the fight for gender equality, especially the International Women's Day on March 8, or illuminated in green to celebrate actions to combat climate change. The impact of the coloured illumination from artificial night-time sources on the biological colonization on buildings and monuments largely depends on the composition and diversity of community, particularly its pigment content. Phototrophs (algae and cyanobacteria) need to live, to a greater or lesser extent, light from different parts of visible spectrum depending on the pigments that they contain. If they are rich in chlorophyll-a and -b, which absorb in the red and blue regions of spectrum, they should be more vulnerable to the effect of red and blue monochromatic lights. If their higher pigment content is in phycobiliproteins phycocyanin and allophycocyanin, which absorb in the red region, red light will have a greater effect. Finally, if total carotenoids and phycobiliprotein phycoerythrin (absorbing in the greenish-yellow and green regions respectively) are the main pigments, yellow and green lights will present the greater effect on organisms. However, this is an oversimplification. Microorganisms have several protective mechanisms that can be triggered by certain qualities of light. Account should also be taken of the light with long wavelengths (such as red and orange light) are more penetrating.

Artificial lighting on outdoor constructions cannot be considered without also considering the effects of daylight, as monuments and structures are always exposed to natural light. As indicated above, very few studies have been published regarding how urban monuments are affected by night-time outdoor illumination in combination with natural sunlight. The Light4Heritage project (2016–2018) has focused on developing lighting-based strategies to control biological colonization and to manage the chromatic integration of biofouling at laboratory scale. The first study carried out within the project involved the use of coloured cellophane films to generate different types of light (by cancelling out the spectral components in certain bands of the visible electromagnetic spectrum, thus emulating monochromatic LED lights) at different photon flux densities. The cellophane films were used to cover phototrophic cultures, derived from natural biofilm growing on a historic granitic building and mainly comprising green algae and cyanobacteria, in order to promote specific physiological responses. The blue cellophane inhibited growth of the test culture, while the yellow cellophane did not significantly decrease the biomass, pigment or EPS content, relative to uncovered, control cultures. The different coloured cellophane covers also generated colour changes in the cultures; e.g. the red cellophane produced notable greening, whereas the green cellophane enhanced the redness of the cultures (Sanmartín et al. 2017). Further experiments were carried out using phototrophs in biofilm mode of growth and LED lights. In these studies, phototrophic biofilms thrived well under blue LEDs, whereas green and red LEDs had biostatic effects (Sanmartín et al. 2018b). Phototrophs responded differently to exposure to different coloured light: the biofilms developed under blue light predominantly comprised algae, and those exposed to red and green light mainly comprised cyanobacteria (P. Sanmartín, unpublished results). Regarding the duration (number of hours daily) of LED illumination, a period of 4 hours proved sufficient to reduce colonization under red and green LED lights, while a period of 8 h proved optimal for further growth of organisms under blue LED light (P. Sanmartín, unpublished results). Finally, among cross-sectional studies, from project Light4Heritage, on the effects of UV-A and UV-B on biological colonization, another study showed UV-B irradiation to be potentially useful for eradicating green algal biofouling from granite stone (Pozo-Antonio and Sanmartín 2018).

A preliminary part of the laboratory research work has been finalized. However, not all of the laboratory-based work on biofilm study was completed in that project and it did not include studies with poikilohydric organisms, such as lichens and bryophytes, or vascular plants. It is important to establish the basis for laboratory scale evaluation in order to facilitate posterior fieldwork. In addition, experimental, laboratory-based simulation enables results to be obtained within a much shorter time than in the field. For example, in the laboratory a mature biofilm can be formed in 30 days on a membrane support (Gambino et al. 2019) and in about 45 days on an acidic, relatively non-porous substratum such as granite (Prieto et al. 2014), whereas in the field a subaerial biofilm will only begin to be formed by phototrophs after around six months on granite walls exposed to rainfall (Sanmartín et al. 2012) and more than a year in areas protected from rainfall (Sanmartín et al. 2020b).

## 4   In What Form Public Lighting Can Be Turned Into a Green Method?

The choice of the type of lighting system used to illuminate monuments in the urban fabric is completely arbitrary and is not scientifically based, nor presents a specific regulatory framework (Rodríguez Lorite 2016). The choice is not made in relation to the findings of studies on energy efficiency, environmental biodiversity or building conservation, from which the first negative data in relation to the novel night illumination systems are beginning to emerge. For instance, external lighting of cultural heritage monuments causes 5% to 20% of total light pollution (Mohar et al. 2014), which is listed among the ten main factors endangering biodiversity (Hölker et al. 2010).

There are some social and scientific concerns regarding the side effects of night-time lighting, such as urban light pollution, negative impacts on wildlife (i.e. bats, moths) and the potential new threat to pollination (Hölker et al. 2010; Mohar et al. 2014; Knop et al. 2017; ECOLIGHT project). In the framework of a LIFE project, Mohar et al. (2014) developed a lamp to improve the existing lighting of 21 pilot churches in Slovenia, reducing the negative impact of illumination on moths and bats and the energy consumption by 40% to 90%. According to these authors, blue coloured light (Fig. 7.3) interrupts melatonin (also known as sleep hormone) in humans and animals, even at low illumination levels. Likewise, reducing intensity of illumination and avoiding light with shorter wavelengths, especially in the blue, the reduction in the observed number of specimens and species results is improved. A key recommendation of this project is that lighting of cultural monuments should be omitted as much as possible, especially when they are located outside urban areas.

**Fig. 7.3** Façade of the San Fiz de Solovio Church in Santiago de Compostela (Galicia, Spain) illuminated in blue light in 2018. Source: Justo Arines

Another important recommendation is that after 23.00 hours, lighting should be switched off in order to attract fewer moths.

Furthermore, the energy consumed by the lighting systems increases the level of $CO_2$ in the atmosphere. Although the simple change of incandescence (e.g. metal halide and sodium vapour lamps) to LED technology in public lighting would be reduced by up to 90% this emission. An example is the lighting of the Puerta de Alcalá in Madrid, which until 2014 had a total consumption of 41,470 W and at present uses 4660 W of energy (Rodríguez Lorite 2016). In Spain, consumption of 1 kWh of energy is equivalent to emitting 430 g of $CO_2$ into the atmosphere (San Martín Páramo and Ferrero Andreu 2008 *apud* Rodríguez Lorite 2016), so that the illumination of this monument every night causes the release of 3440 g of $CO_2$ to the atmosphere, one tenth of what was issued until 6 years ago.

Regarding the biodiversity of higher plants, in the ECOLIGHT project mentioned above, constant artificial light was also found to increase the rates of foliar damage due to ozone in three clover species and in *Lotus pedunculatus*, which produced 10% to 25% fewer flower heads under simulated street lighting, which in turn led to reduced numbers of the aphid *Acyrthosiphon pisum* (Bennie et al. 2016, 2018). These researchers also value positively the implementation of LED technology in public lighting to achieve an environmentally sustainable system.

# 5 Current Perspectives and Future Directions

So far, it has been demonstrated at laboratory scale that the combined use of suitable lighting can promote or inhibit the development of biofilms and also shape their colour. Thus, an advancement regarding the practical application of the research findings is necessary. The objective will be to determine the criteria that would enable the use and technological implementation of outdoor lighting for effective control of biological colonization of buildings. These criteria will aim to contribute to the long-term management of public illumination on monuments and other structures, while reducing negative impacts caused by biological colonization and also preventing any further increase in light pollution. Technical solutions that will provide more energy-efficient and environmentally-sound, targeted illumination that also controls biofouling formation on buildings shall be designed. This will be achieved through the development of a pilot project and the construction of improved lighting prototype systems.

Concerning the latter, a wide range of options is offered by current light technology. It would be highly desirable, test new commercially available lights (both LED and ultraviolet lights) and examine the influence of other technological elements such as density filters, band-pass filters, and filter holders mounted on a common light source. This information will be used to identify the emissive material of the prototype lighting system. Furthermore, current recommendations regarding visual comfort, light pollution, and illumination regulations need to be taken into consideration in order to design the system. On the other hand, in the pilot project and the studies launched, a method based on the quantitative determination of colour for early detection (even before it is visible to the human eye) and real-time monitoring of epilithic phototrophic biofilms on the surface of structures (Sanmartín 2012; Sanmartín et al. 2012) could be used. The method is non-invasive, portable (can be used on-site), inexpensive, and easy to apply (enabling unskilled operators with minimal training to perform the measurements) and provides immediate results (Fig. 7.4). Likewise, the chlorophyll-a fluorescence (ChlaF) parameter Fv/Fm (maximum quantum efficiency of PSII), previously reported to be suitable for ascertaining the vitality of organisms' remains on rock surfaces (Pozo-Antonio and Sanmartín 2018) and monitoring the quantity and physiological state of the biofilm-forming phototrophs in recolonized areas (Sanmartín et al. 2020b), could be also applied (Fig. 7.4).

The development of smart lights to reduce biological colonization on monuments (Fig. 7.5) is fully consistent with smart city strategies of efficiency, applicability and adaptation of R&D&I to problems that affect heritage cities. Thus, results will be readily scalable, efficient, and replicable in cities or environments throughout the world where the historical heritage is a distinctive feature. Findings will have a significant social and economic impact, as control over biodeterioration is an important element of built heritage management worldwide, and the development of standards or regulations for managing external lighting of built heritage may help to avoid it.

**Fig. 7.4** Adrián Rodríguez (graduate student), Rafael Carballeira (expert in the field of Botany), and Patricia Sanmartín (cultural heritage conservation researcher) all involved in the Light4Heritage project, taking colour spectrophotometry and PAM fluorometry measurements on the granite-built cloister in the Monastery of San Martiño Pinario (Santiago de Compostela, Galicia, Spain). Source: Justo Arines

**Fig. 7.5** Left: Adrián Rodríguez walking on the scaffolding in which two of the system lights have been placed. Right: Scaffolding system with a commercial lighting system (acquired following the guidelines outlined by Patricia Sanmartín) provided by the company Ferrovial Servicios. Source: Patricia Sanmartín

Finally, the public response to the new lighting developed and installed should be taken into account to enable evaluation of the lighting systems from a perceptual, and not only procedural, viewpoint.

**Acknowledgements**  The author is grateful to Elena López and Adrián Rodríguez, who carried out a master thesis (MSc) and a final year (BSc) research project on this subject under her supervision. She also thanks Dr Justo Arines (Universidade de Santiago de Compostela, Spain) and Dr Rafael Carballeira (Universidade da Coruña, Spain) their collaboration on the Light4Heritage project studies. Finally, she thanks the financial support of Xunta de Galicia grant ED431C 2018/32.

# References

Albertano P, Bruno L (2003) The importance of light in the conservation of hypogean monuments. In: Saiz-Jiménez C (ed) Molecular biology and cultural heritage. Balkema, Lisse, Rotterdam, The Netherlands, pp 171–177

Albertano P, Moscone D, Palleschi G, Hermosin B, Saiz-Jiménez C, Sánchez-Moral S, Hernández-Mariné M, Urzí C, Groth I, Schroeckh V, Gallon JR, Graziottin F, Bisconti F, Giuliani R (2003) Cyanobacteria attack rocks (CATS): control and preventive strategies to avoid damage caused by cyanobacteria and associated microorganisms in Roman hypogean monuments. In: Saiz-Jiménez C (ed) Molecular biology and cultural heritage. Balkema, Lisse, Rotterdam, The Netherlands, pp 151–162

Bennie J, Davies TW, Cruse D, Gaston KJ (2016) Ecological effects of artificial light at night on wild plants. J. Ecol 104:611–620

Bennie J, Davies TW, Cruse D, Bell F, Gaston KJ (2018) Artificial light at night alters grassland vegetation species composition and phenology. J Appl Ecol 55:442–450

Bischof K, Gomez I, Molis M, Hanelt D, Karsten U, Lüder U, Roleda MY, Zacher K, Wiencke C (2006) Ultraviolet radiation shapes seaweed communities. Environ Sci Biotechnol 5:141–166

Borderie F, Alaoui-Sehmer L, Bousta F, Orial G, Rieffel D, Richard H, Alaoui-Sossé B (2012) UV irradiation as an alternative to chemical treatments: a new approach against algal biofilms proliferation contaminating building facades, historical monuments and touristic subterranean environments. In: Krueger D, Meyer H (eds) Algae: ecology, economic uses and environmental impact. Nova Science Publishers, Inc. Hauppauge, New York, pp 1–28

Borderie F, Tête N, Cailhol D, Alaoui-Sehmer L, Bousta F, Rieffel D, Aleya L, Alaoui-Sossé B (2014) Factors driving epilithic algal colonization in show caves and new insights into combating biofilm development with UV-C treatments. Sci Total Environ 484:43–52

Borderie F, Alaoui-Sossé B, Aleya L (2015) Heritage materials and biofouling mitigation through UV-C irradiation in show caves: states-of-the-art practices and future challenges. Environ Sci Pollut Res 22:4144–4172

Bruno L, Valle V (2017) Effect of white and monochromatic lights on cyanobacteria and biofilms from Roman Catacombs. Int Biodeteri Biodegr 123:286–295

Bruno L, Belleza S, Urzì C, De Leo F (2014) A study for monitoring and conservation in the Roman Catacombs of St. Callistus and Domitilla, Rome (Italy). In: Saiz-Jimenez C (ed) The conservation of subterranean cultural heritage. CRC Press/Bakelma/Taylor & Francis Group, Leiden, The Netherlands, pp 37–44

Bussell AN, Kehoe DM (2013) Control of a four-color sensing photoreceptor by a two color sensing photoreceptor reveals complex light regulation in cyanobacteria. Proc Natl Acad Sci USA 110:12834–12839

Castenholz RW, Garcia-Pichel F (2012) Cyanobacterial responses to UV-radiation. In: Whitton B (ed) Ecology of cyanobacteria II. Springer, Dordrecht, The Netherlands, pp 591–611

Cutler NA, Viles HA, Ahmad S, McCabe S, Smith BJ (2013) Algal "greening" and the conservation of stone heritage structures. Sci Total Environ 442:152–164

de Mooij T, de Vries G, Latsos C, Wijffels RH, Janssen M (2016) Impact of light color on photobioreactor productivity. Algal Res 15:32–42

del Rosal Padial Y (2016) Análisis, impacto y evolución de los biofilms fotosintéticos en espeleotemas, El caso de la Cueva de Nerja, Málaga. PhD thesis. Universidad de Málaga, Spain.

del Rosal Padial Y, Jurado Lobo V, Hernández Mariné M, Roldán Molina M, Sáiz Jiménez C (2016) Biofilms en cuevas turísticas: La Cueva de Nerja y la Cueva del Tesoro. In: Andreo B, Durán JJ (eds) El Karst y el Hombre: Las Cuevas como Patrimonio Mundial. Asociación de Cuevas Turísticas Españolas, Madrid, Spain, pp 103–114

Di Martino P (2016) What about biofilms on the surface of stone monuments. Open Conf Proc J 7:14–28

Dobat K (1998) Flore de la lumière artificielle (Lampenflora - Maladie verte). In: Juberthie C, Decu V (eds) Encyclopaedia Biospeleologica, Société Internationale de Biospéologie - International Society for Subterranean Biology. Moulis-Bucarest, France-,Romania, vol II, pp 1325–1335

ECOLIGHT Project Available online: https://www.exeter.ac.uk/esi/research/currentresearch/ecolight/ Accessed Dic 2019

Energy & Smart cities – European Comission's priority policies. Available online: https://ec.europa.eu/info/eu-regional-and-urbandevelopment/topics/cities-and-urban-development/city-initiatives/smart-cities_en Accessed Dic 2019

Enomoto G, Win NN, Narikawa RR, Ikeuchi M (2015) Three cyanobacteriochromes work together to form a light color-sensitive input system for c-di-GMP signaling of cell aggregation. Proc Natl Acad Sci USA 112:8082–8087

Figueroa FL, Álvarez-Gómez F, del Rosal Y, Celis-Plá PSM, González G, Hernández M, Korbee N (2017) In situ photosynthetic yields of cave photoautotrophic biofilms using two different pulse amplitude modulated fluorometers. Algal Res 22:104–115

Gambino M, Sanmartín P, Longoni M, Villa F, Mitchell R, Cappitelli F (2019) Surface colour: an overlooked aspect in the study of cyanobacterial biofilm formation. Sci Total Environ 659:342–353

Gorbushina AA (2007) Life on the rocks. Environ Microbiol 9:1613–1631

Han PP, Shen SG, Wang HY, Sun Y, Dai YJ, Jia SR (2015) Comparative metabolomic analysis of the effects of light quality on polysaccharide production of cyanobacterium *Nostoc* flageliforme. Algal Res 9:143–150

Hölker F, Wolter C, Perkin EK, Tockner K (2010) Light pollution as a biodiversity threat. Trends Ecol Evol 25:681–682

Hsieh P, Pedersen JZ, Albertano P (2013) Generation of reactive oxygen species upon red light exposure of cyanobacteria from Roman hypogea. Int Biodeter Biodegr 84:258–265

Hsieh P, Pedersen JZ, Bruno L (2014) Photoinhibition of cyanobacteria and its application in cultural heritage conservation. Photochem Photobiol 90:533–543

Jurado V, del Rosal Y, Gonzalez-Pimentel JL, Hermosin B, Saiz-Jimenez C (2020) Biological control of phototrophic biofilms in a show cave: the case of Nerja Cave. Appl Sci 10:3448

Kehoe DM (2010) Chromatic adaptation and the evolution of light color sensing in cyanobacteria. Proc Natl Acad Sci USA 107:9029–9030

Knop E, Zoller L, Ryser R, Gerpe C, Hörler M, Fontaine C (2017) Artificial light at night as a new threat to pollination. Nature 548:206–209

MacIntyre HL, Kana TM, Anning T, Geider RJ (2002) Photo-acclimation of photosynthesis irradiance response curves and photosynthetic pigments in microalgae and cyano- bacteria. J Phycol 38:17–38

Mohar A, Zagmajster M, Verovnik R, Skaberne BB (2014) Nature-friendlier lighting of objects of cultural heritage (churches): Recommendations. Dark-Sky Slovenia 2014, Available online: http://temnonebo.com/images/pdf/nature_friendler_lightning_churces_booklet_web.pdf. Accessed June 2020

Moon Y, Kim S, Chung Y (2012) Sensing and responding to UV-A in cyanobacteria. Int J Mol Sci 13:16303–16332

Mulec J (2012) Lampenflora. In: White WB, Culver DC (eds) Encyclopedia of caves. Elsevier/ Academic Press, Amsterdam, The Netherlands, pp 451–456

Mulec J, Kosi G (2009) Lampenflora algae and methods of growth control. J Cave Karst Stud 71:109–115

Mullineaux CW (2001) How do cyanobacteria sense and respond to light? Mol Microbiol 41:965–971

Ortega-Calvo JJ, Hernández-Mariné M, Saiz-Jimenez C (1991a) Biodeterioration of buildings materials by cyanobacteria and algae. Int Biodeter 28:165–185

Ortega-Calvo JJ, Hernández-Mariné M, Saiz-Jimenez C (1991b) Mechanical deterioration of building stones by cyanobacteria and algae. In: Roossmoore KW (ed) Biodeterioration and biodegradation. Elsevier, London, pp 392–394

Pfendler S, Einhorn O, Karimi B, Bousta F, Cailhol D, Alaoui-Sosse L, Alaoui-Sosse B, Aleya L (2017) UV-C as an efficient means to combat biofilm formation in show caves: evidence from the La Glaciere Cave (France) and laboratory experiments. Environ Sci Pollut Res 24:24611–24623

Pozo-Antonio JS, Sanmartín P (2018) Exposure to artificial daylight or UV-irradiation (A, B or C) prior to chemical cleaning: an effective combination for removing phototrophs from granite. Biofouling 34:851–869

Prieto B, Sanmartín P, Silva C, Vázquez-Nion D, Silva B (2014) Deleterious effect plastic-based biocides on back-ventilated granite facades. Int Biodeter Biodegr 86:19–24

Ramírez M, Hernández-Mariné M, Novelo E, Roldán M (2010) Cyanobacteria-containing biofilms from a Mayan monument in Palenque, Mexico. Biofouling 26:399–409

Rifón Lastra A (2000) Algas epilíticas en monumentos de interés histórico de Galicia. PhD thesis. Universidade da Coruña, Spain

Rifón Lastra A, Noguerol Seoane A (2001) Green algae associated with the granite walls of monuments in Galicia (NW Spain). Cryptogamie Algol 22:305–326

Rodríguez Lorite MA (2016) Guía de Iluminación Eficiente de Monumentos. Dirección General de Industria, Energía y Minas. DL: M-21749-2016. Available online: http://www.madrid.org/ bvirtual/BVCM015700.pdf Accessed Dic 2019

Roldán M, Oliva F, Gónzales del Valle MA, Saiz-Jimenez C, Hernández-Mariné M (2006) Does green light influence the fluorescence properties and structure of phototrophic biofilms? Appl Environ Microbiol 72:3026–3031

San Martín Páramo R, Ferrero Andreu L (2008) Los costos de la implantación y el mantenimiento de las instalaciones de alumbrado exterior. CONAMA 9. Congreso Nacional del Medio Ambiente. Cumbre del Desarrollo Sostenible. Madrid, Spain. December 1–5, 2008

Sanmartín P (2012) Color quantification in the study of biofilm formation on granite stone in historical and artistic heritage. PhD thesis. Universidade de Santiago de Compostela, Spain

Sanmartín P, Vázquez-Nion D, Silva B, Prieto B (2012) Spectrophotometric color measurement for early detection and monitoring of greening on granite buildings. Biofouling 28:329–338

Sanmartín P, Vázquez-Nion D, Arines J, Cabo-Domínguez L, Prieto B (2017) Controlling growth and colour of phototrophs by using simple and inexpensive coloured lighting: a preliminary study in the Light4Heritage project towards future strategies for outdoor illumination. Int Biodeter Biodegr 122:107–115

Sanmartín P, DeAraujo A, Vasanthakumar A (2018a) Melding the old with the new: trends in methods used to identify, monitor and control microorganisms on cultural heritage materials. Microb Ecol 76:64–80

Sanmartín P, Vázquez-Nion D, Silva B, Prieto B, Arines J (2018b) Assessing the effect of different coloured lighting in controlling biological colonization. In: Mosquera MJ, Gil A (eds) Conserving cultural heritage. CRC Press/Bakelma/Taylor & Francis Group, Leiden, The Netherlands, pp 313–318

Sanmartín P, Villa F, Cappitelli F, Balboa S, Carballeira R (2020a) Characterization of a biofilm and the pattern outlined by its growth on a granite-built cloister in the Monastery of San Martiño Pinario (Santiago de Compostela, NW Spain). Int Biodeter Biodegr 147:104871

Sanmartín P, Rodríguez A, Aguiar U (2020b) Medium-term field evaluation of several widely used cleaning-restoration techniques applied to phototrophic algal biofilm formed on a granite-built historical monument. Int Biodeter Biodegr 147:104870

Scheerer S, Ortega-Morales O, Gaylarde C (2009) Chapter 5 - Microbial deterioration of stone monuments - an updated overview. In: Laskin AI, Sariaslani S, Gadd GM (eds) Adv Appl Microbiol, Elsevier Inc., Academic Press, Cambridge, vol. 66:97–139

Shang JL, Zhang ZC, Yin XY, Chen M, Hao FH, Wang K, Feng JL, Xu HF, Yin YC, Tang HR, Qiu BS (2018) UV-B induced biosynthesis of a novel sunscreen compound in solar radiation and desiccation tolerant cyanobacteria. Environ Microbiol 20:200–213

Sinha RP, Singh N, Kumar A, Kumar HD, Häder M, Häder DP (1996) Effects of UV irradiation on certain physiological and biochemical processes in cyanobacteria. J Photochem Photobiol B: Biol 32:107–113

Smartiago Project, Available online: https://smartiago.santiagodecompostela.gal Accessed Dic 2019

Smith BJ, McCabe S, McAllister D, Adamson C, Viles HA, Curran JM (2011) A commentary on climate change, stone decay dynamics and the 'greening' of natural stone buildings: new perspectives on 'deep wetting'. Environ Earth Sci 63:1691–1700

Tandeau de Marsac N (1977) Occurrence and nature of chromatic adaptation in cyanobacteria. J Bacteriol 130:82–91

Urzi C, De Leo F, Bruno L, Pangallo D, Krakova L (2014) New species description, biomineralization processes and biocleaning applications of Roman catacombs-living bacteria. In: Saiz-Jimenez C (ed) The conservation of subterranean cultural heritage. CRC Press/Bakelma, Taylor & Francis Group, Leiden, The Netherlands, pp 65–72

Warscheid T, Braams J (2000) Biodeterioration of stone: a review. Int Biodeterior Biodegrad 46:343–368

Willey N (2016) Environmental plant physiology. Taylor & Francis, Garland Science, New York

Wiltbank LB, Kehoe DM (2016) Two cyanobacterial photoreceptors regulate photosynthetic light harvesting by sensing teal, green, yellow, and red light. MBio 7:e02130–e02115

Wingard CE, Schiller JR, Casnnholz RW (1997) Evidence regarding the possible role of c-phycoerythrin in ultraviolet-B tolerance in a thermophilic cyanobacterium. Photochem Photobiol 65:833–842

Yakovleva IM, Titlyanov EA (2001) Effect of visible and UV irradiance on subtidal *Chondrus crispus*: Stress, photoinhibition and protective mechanisms. Aquat Bot 71:47–61

# Part III
# Biocleaning and Bio-Based Conservation Methods

# Chapter 8
# Bioremoval of Graffiti in the Context of Current Biocleaning Research

**Pilar Bosch-Roig and Patricia Sanmartín**

**Abstract** Some microorganisms can be used as bioremediation agents, in biocleaning treatments, to remove undesired sulphates, nitrates and organic matter from cultural heritage surfaces. Graffiti materials (mainly spray paints) are now included in the list of materials that can be biocleaned, with studies on this topic being initiated just over 5 years ago. Research on the bioremoval of graffiti is continuing and on a promising track. This chapter reports a critical analysis of studies of the bioremoval of graffiti carried out in recent years, which are compared with similar studies of the removal of salts (mainly nitrates and sulphates) and organic matter conducted in the last thirty years. Likewise, the present challenges and ways of overcoming them are addressed towards developing a complete protocol for the use of bioremediation to remove graffiti, with particular emphasis on the use of the method for cleaning facades and buildings.

**Keywords** Bacterial strains · Biocleaning · Cultural heritage · Microbial agents · Safe methods · Stone

## 1 Introduction

The use of naturally occurring bacteria, and to lesser extent fungi, in biocleaning treatments for cultural heritage buildings and monuments is currently popular and provides an effective, non-invasive, relatively ecologically safe and inexpensive approach to cultural heritage conservation (Bosch-Roig and Ranalli 2014; Sanmartín

P. Bosch-Roig (✉)
Instituto Universitario de Restauración del Patrimonio, Universitat Politècnica de València, Valencia, Spain
e-mail: mabosroi@upvnet.upv.es

P. Sanmartín
Departamento de Edafoloxía e Química Agrícola. Facultade de Farmacia, Universidade de Santiago de Compostela, Santiago de Compostela, Spain
e-mail: patricia.sanmartin@usc.es

et al. 2018). From the beginning of the 1990s until the present, viable (non-pathogenic) cells and selected enzymes have been identified and used for the biological removal of undesired substances, such as salts (mainly nitrates and sulphates) and organic matter from stone substrates (Webster and May 2006). In the last few years graffiti paint has been added to this list as an undesired, potentially bioremovable material. The inclusion of graffiti paint in this list arose as a result of the drawbacks and environmental impacts of the methods then available (chemical and physical, including laser) for graffiti removal (Sanmartín et al. 2014). A need to develop a new inexpensive, risk-free method, especially designed for use in porous materials (such as natural stone, cement and mortar lime) was also recognized (Sanmartín et al. 2014; Sanmartín and Bosch-Roig 2019).

Every year large amounts of money and effort are expended in removing disfiguring graffiti from urban fabric surfaces (Sanmartín et al. 2014, 2016). For example, in the early 2000s about 20,000 private buildings and more than 400 municipal buildings in the city of Milan (Italy) were estimated to be disfigured every year (DDL 1607 2002). However, extrapolating the findings on bioremoval of salts and organic matter to graffiti is not an easy or direct task, and it is necessary to first identify microorganisms as potential candidates. In relation to the occurrence of microorganisms on paints, an earlier review paper concerning the biodeterioration of architectural paint by bacteria, fungi and algae indicated that studies on the microbial ecology of paints are scarce (Gaylarde et al. 2011). One study of painted walls in Latin America described microbial populations mainly formed by bacteria (including actinobacteria and cyanobacteria) as well as by populations of algae, protozoa, rotifers, nematodes and a huge variety of fungi (Gaylarde and Gaylarde 2000). The same authors also reported that biomass on external painted walls of buildings in Europe and Latin America is generally composed of fungi and to a lesser extent actinobacteria, which are more commonly found on building surfaces in Europe than in Latin America (Gaylarde and Gaylarde 2005). However, these results do not apply to other types of surfaces. Hence, some authors have reported that actinobacteria are commonly found on different types of damaged materials and buildings (e.g. ceramics and sandstone monuments) in Europe and on floor dust in damaged buildings in the USA (Hyvärinen et al. 2002; Palla et al. 2002; Macedo et al. 2009; Park et al. 2017). Allsopp et al. (2004) listed the fungi and phototrophs detected on paint films comprising refined and processed materials. Furthermore, in a review of several published studies on microbial colonization of frescoes and in regard to the presence of microorganisms and their role in damage to painted surfaces, Ciferri (1999) concluded that bacteria may be the first colonizers. After bacterial colonization chemically modifies some of the components of the paint and renders them utilizable, fungi may be able to become established and often become the principal microflora on painted surfaces (Ciferri 1999; Gaylarde et al. 2011). Whether or not they are the most abundant organisms, fungi have long been considered the major biodeteriogens in painted surfaces, producing paint-degrading metabolites (principally organic acids) that weaken and then penetrate the film (Grant et al. 1986). For example, during the 1950s, some researchers considered *Aureobasidium pullulans* (formerly *Pullularia pullulans*) to be the main biological agent involved in paint deterioration (Reynolds 1950; Klens and Lang 1956). Filamentous algae and actinobacteria can also alter painted surfaces by penetrating the paint film and releasing reactive metabolites that reduce the effectiveness and lifespan of the coating (Gaylarde et al. 2011). However, more recently, a study involving bioremediation processes showed that black fungi are not suitable for biocleaning stone artefacts, as they can damage stone treated with synthetic polymers (Troiano et al. 2014).

Most microorganisms found on paint films are actually aerial contaminants and are not involved in degradation processes (Gaylarde et al. 2011). More specifically, very few microorganisms have been shown to degrade graffiti. Indeed, in the only study to date of the microbial ecology of graffiti wall paintings, of a total of fifty-four different isolates, only nine bacterial isolates and one fungal isolate showed some capacity to degrade spray paint graffiti; of these, only three were isolated from recent or old graffiti (Sanmartín et al. 2016).

## 2   Why Can Graffiti Paint Be Removed by Biocleaning Treatments?

Graffiti spray paint contains the following main ingredients: (i) pigment, which provides colour, (ii) binding medium, in which the pigment particles are dispersed, and which hardens and binds the pigments on the painted surface and (iii) solvent, which allows the pigment/binder mixture to flow. Acrylic- and alkyd-based binders are common in current non-metallic graffiti formulations, while polyethylene-type binders predominate in metallic formulations. Semi-synthetic binders are also relatively common, although increasingly less so. Thus, although a study conducted about fifteen years ago showed that almost half of the varieties of red graffiti spray paint contain alkyd-nitrocellulose based binders (Govaert and Bernard 2004), these paints have been displaced by red alkyd paints.

Graffiti spray paints contain man-made compounds both as binders and as pigments, and they can therefore be considered xenobiotic (synthetic) substances. Although it is commonly thought that synthetic substances are more resistant to microbial attack than natural components, several studies have demonstrated the contrary (Cappitelli and Sorlini 2008; Troiano et al. 2014). Moreover, paint contains a wide variety of organic and inorganic biodegradable components, including additives such as emulsifiers, thickeners, etc., which may be utilized for growth by a large variety of microbial species (Ciferri 1999; Sanmartín et al. 2014). Most biodegradable substances, including graffiti paints, contain a mixture of carbon and elements such as oxygen, nitrogen, sulphur and phosphorus, which create charge imbalances that enzyme-containing bacteria can exploit. Thus, in previous studies, aged synthetic resins in outdoor environments (Cappitelli et al. 2007) and synthetic polymers in indoor environments (Lustrato et al. 2012) have been utilized by microorganisms as growth substrates. Likewise, cases have been reported wherein some microorganisms, such as fungi and particularly bacteria, can modify spray paint, even inside the cans (Bentley and Turner 1998). Microorganisms can enter the paint via infected raw materials (including water) or unsterile equipment. For example, the presence of bacteria has been found to cause reduced viscosity, gassing and colour drift in latex paint (Bentley and Turner 1998). In this regard, the components of spray paint affect microorganisms, either by inhibiting or stimulating their development. Thus, cellulose derivatives can act as nutrients for fungal cells (Allsopp et al. 2004), whereas organic solvents and heavy metals in pigments can adversely affect fungi (Gaylarde et al. 2011). Higher proportions of resin in gloss paint may also yield greater bioresistance (Gaylarde et al. 2011).

Paint pigments may also affect the bioremoval process. In a study of differently coloured acrylic paint, red paint was colonized more rapidly by microbes than blue paint, probably due to the presence of copper in the blue paint ($\approx 0.4\%$), which acts as a microbiocide (Breitbach et al. 2011). Zinc oxide can also act as an antimicrobial agent and has been found to exert fungicidal activity at a concentration of 2% in oil-based paint (Turner 1967). Antimicrobial compounds are added to liquid paint both to protect the paint in-can and to reduce fouling of the dry film. However, there is no universally active antimicrobial ingredient that is compatible with all paint formulations and that meets the requirements of the coating manufacturers. Thus, mixtures of antimicrobials are often used, among which isothiazolinone-based compounds are widely used for conserving paint (Gaylarde et al. 2011). In a study of 29 freshly dried synthetic resins used as paint binding media, almost all the acrylic resins proved to be resistant to fungal attack, while all alkyd resins and some polyvinyl acetates (PVAs) were found to be biodegradable (Cappitelli et al. 2005). These findings indicate that under favourable environmental conditions, and considering the same paint components, alkyd-based paint would be more readily biodegraded than acrylic-based paint. The authors of the aforementioned study also noted that knowledge of the microbial susceptibility of the binding medium under certain conditions does not necessarily enable prediction of the potential of a specific paint to undergo biodegradation. In fact, pigments and other components may be readily available sources of carbon in paint formulations.

# 3 Methodological Advances in Bioremoval of Salts, Organic Matter and Graffiti

Microorganisms have mainly been used to remove undesired inorganic salts (nitrates and sulphates) and organic substances (animal glue, casein, hydrocarbons, lipids, etc.) from inorganic surfaces (stone sculptures, building façades and frescoes). Few studies have been conducted in relation to the use of microorganisms to remove paints (included graffiti paints) produced in the twentieth and twenty-first centuries and which contain synthetic polymers.

Bioremoval strategies have in common the following basic general steps: (i) identification of the substance to be cleaned and the original material to be protected; (ii) identification, selection and growth (selection of culture media and growth protocols) of suitable microorganisms (non-spore forming, non-pathogenic, not biodeteriogens); (iii) selection of on-site application protocols (with delivery systems compatible with microorganisms and artistic surface) and conditions (treatment time and temperature) for on-site application and (iv) establishment of long-term monitoring of biocleaned substrates (Fig. 8.1).

The first step is common to all restoration-conservation strategies, and extensive research has been conducted in regard to the use of sensitive and selective techniques for identifying organic materials (gas chromatography coupled with mass

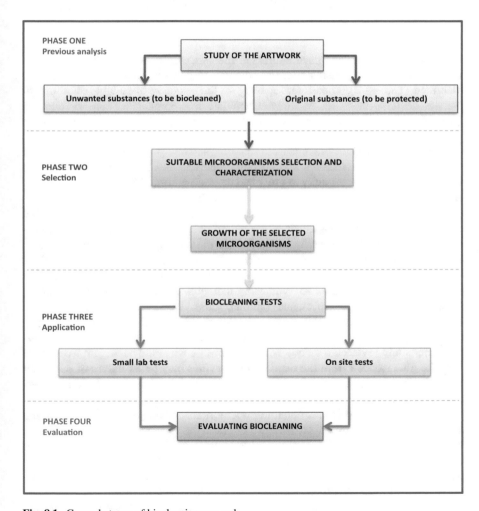

**Fig. 8.1** General stages of biocleaning research

spectrometry (GC-MS), pyrolysis-GC-MS and inorganic materials (FTIR, SEM, Ion-chromatography)) (Colombini et al. 1999, 2003; Ranalli et al. 2005; Bosch-Roig et al. 2013). As this first step is well known and used in the field of restoration, it is not considered in this chapter, which includes a comparison of the methodological analysis of the other three general steps involved in the bioremoval of salt and organic matter and of graffiti. The chapter also includes a critical analysis of the methodological advances made by the scientific community and the applicability of these in the field of graffiti removal (see Section 3 and Table 8.1).

## 3.1  Selection of Suitable Microorganisms for the Bioremoval of Graffiti

As relatively few studies have addressed the biodeterioration and biocleaning of graffiti, few microorganisms with the potential to degrade graffiti paint have been identified (Table 8.2). Therefore, focusing on the selection of suitable microorganisms is an important area of study. Candidate microorganisms can be found in international collections of microorganisms (DSMZ, ATCC, CECT, etc.) or in natural environments.

If researchers wish to use registered microorganisms, these should be selected for their particular characteristics and be purchased and assayed in the laboratory in order to evaluate their capacity to grow and degrade graffiti. Although this is a faster and probably cheaper selection method, some studies have shown that it is more difficult to obtain the most suitable microorganisms. Ranalli et al. (2005) conducted a comparative enzymatic analysis of the bioremoval capacity of a bacterium (*Pseudomonas stutzeri* A29) directly isolated from the substance to be removed (aged animal glue from frescoes) and compared with six other commercial strains of the same genera (*Pseudomonas*) and even species (*P. stutzeri*). This study revealed that all strains tested were able to synthesize inducible specific enzymes after being grown with the target substance (animal glue). However, the isolated strain (*P. stutzeri* A29) grew optimally and displayed the highest enzymatic (protease) activity (Ranalli et al. 2005). A study of the molecular mechanisms of action in this particular strain revealed a unique enzymatic profile when the bacterium was grown with the glue isolated from the fresco. The enzymes may be a mixture of particular proteases whose combined action enables the complete digestion of the hardened glue present in the frescoes (Antoniolli et al. 2005).

A logical starting point for obtaining microorganisms that can act as biocleaning agents from natural sources is to isolate them from the substance to be removed. This was done, for example, by analysing black crust biofilms, from which different researchers have isolated sulphate reducing bacteria (SRB), hydrocarbon degrading bacteria (HDB) and other microorganisms with hydrolytic activities able to degrade polycyclic aromatic hydrocarbons and fatty acids, among others (Saiz-Jimenez 1993, 1997; Fernandes 2006; Soffitti et al. 2019). In the case of graffiti paint, outdoor

**Table 8.1** Comparison of general strategies for the bioremoval of salts, organic matter and graffiti by using microorganisms from inorganic substrate

| Target substance to remove | Suitable microorganisms | | | Application protocols | | Monitoring | References |
|---|---|---|---|---|---|---|---|
| | Selected strains | Metabolic action | Growth | Delivery system | Application time/T° | | |
| Sulphate salts | *Desulfovibrio vulgaris*[a] subsp. *vulgaris* *Desulfovibrio vulgaris*[a] *Desulfovibrio desulfuricans*[a] | Sulphate reduction activity | DSMZ 63 modified medium (An) | Immersion[a] Sepiolite[a] Hydrobiogel.97[a] Carbogel[a] Mortar and algi-nate beads[a] Arbocel[a] | 12–110 h/ 17–30 °C | After treatment | Gauri et al. (1989, 1992); Heselmeyer et al. (1991); Ranalli et al. (1996a, 1997, 2000); Cappitelli et al. (2005, 2006); Polo et al. (2010); Gioventù et al. (2011); Troiano et al. (2014); Alfano et al. (2011) |
| | *Cellulosimicrobium cellulans*[a] | Carbonate solubilization | TSA medium (Ae) | Laponite micro-packs[a] | 24–48 h/ 6–37 °C | After treatment | Mazzoni et al. (2014) |
| Nitrate salts | *Pseudomonas pseudoalcaligenes*[a] *Pseudomonas aeruginosa*[a] *Pseudomonas desnitrificans*[a] | Denitrification activity | Nitrate broth medium (Ae) | Sepiolite[a] Mortar and algi-nate beads[a] Carbogel[a] Hydrobiogel-97[a] | 24 h-1 month/ 18–26 °C | After treat-ment and 8 months and 6 years | Ranalli et al. (1996b) May et al. (2008) Alfano et al. (2011) |
| | *Pseudomonas stutzeri*[a] | | | Cotton wool[a], agar[a] | 1.5–3 h/ 25 ± 4 °C | After treat-ment and 1 month | Bosch-Roig et al. (2010, 2013, 2019) |
| Organic matter | *Pseudomonas stutzeri*[a] | Collagenolytic and | M9 supplemented with 0.5–1% animal glue | Cotton wool[a], agar[a], Laponite[a] | 1.5–24 h/ 28 ± 3 °C | After treatment | Ranalli et al. (2005); Antonioli et al. (2005) Lustrato et al. (2012) |

(continued)

**Table 8.1** (continued)

| Target substance to remove | Suitable microorganisms | | Application protocols | | | Monitoring | References |
|---|---|---|---|---|---|---|---|
| | Selected strains | Metabolic action | Growth | Delivery system | Application time/T° | | |
| | | caseinolytic activity | | | | Microbial, chemical and aesthetical changes | Bosch-Roig et al. (2010) Rampazzi et al. (2018) |
| | *Stenotrophomonas maltophilia*[a] *Pseudomonas koreensis*[a] | Proteinolytic activity | TSA medium supplemented with gelatine 1% (w/v), or R2A medium supplemented with gelatine 0.4% (w/v) (Ae) | Laponite[a] | 24–48 h/ 6–37 °C | After treatment | Mazzoni et al. (2014) |
| Graffiti | *Desulfovibrio desulfuricans* | Nitroesterase activity | DSMZ 63 modified medium (An) | Immersion | 49 days/ room temperature | After treatment | Giacomucci et al. (2012) |
| | *Pantoea* sp. *Alternaria alternata Arthrobacter oryzae Pseudomonas oryzihabitans Bacillus megaterium Arthrobacter aurescens Bacillus aquimaris Pseudomonas mendocina Microbacterium* | Unknown | Tryptic soy broth (TSB), minimal salt solution Minimal salt solution modified with glucose (10 mM) (Ae) | Immersion | 25 days/ room temperature | After treatment | Sanmartín et al. (2016) |

| Microorganisms | | Medium | Application | Time/temperature | | Reference |
|---|---|---|---|---|---|---|
| oleivorans Gordonia alkanivorans | | | | | | |
| Pseudomonas stutzeri[a] Aerobacter aerogenes[a] Comamonas sp.[a] | | M9 enriched with graffiti powder (Ae) | Cotton wool[a], agar[a] | 20 days/room temperature | After treatment | Sanmartín and Bosch-Roig (2019) |
| Aerobacter aerogenes, Bacillus subtilis A mixture of Bacillus sp., Delftia lacustris, Sphingobacterium caeni and Ochrobactrum anthropi Comamonas sp. Rubellimicrobium thermophilum, Chelatococcus daeguensis, Escherichia coli Marinospirillum sp. | Unknown | Sequentially (4 steps): (1) Tryptic soy broth (TSB) during 2 days (2) 1/10 strength tryptic soy broth (TSB) during 5 days 3) Complete mineral (CM) during 10 days 4) Complete mineral (CM) during 10 days | Immersion | 27 days/30 °C | After each step | Cattò et al. (2021) |
| Aerobacter aerogenes[a] Comamonas sp.[a] | Unknown | CM medium and M9 enriched with graffiti powder[a] (Ae) | Agar[a] | 14 days/room temperature | After treatment | Sanmartín and Bosch-Roig et al. (unpublished results) |

DSMZ 63 medium: modified by eliminating any iron source; *Ae* Aerobic metabolism, *An* Anaerobic metabolism
[a]Subaereal application

**Table 8.2** Comparison of general strategies for the live microorganisms' bioremoval of graffiti

| References | Bacterial selection | | Application materials and strategy | | Analysis protocols | | | | |
|---|---|---|---|---|---|---|---|---|---|
| | Collection sources/registered microorganisms | Nature sources/isolated microorganisms | Glass slide/immersion strategy | Stone/subaerial strategy | Visual alteration | Colour variants | Surface changes[a] | Chemical modifications[b] | Enzymatic assays |
| Giacomucci et al. (2012) | • | | • | | • | • | • | • | • |
| Sanmartín et al. (2016) | | • | • | | • | | | | |
| Sanmartín and Bosch-Roig (2019) | • | | | • | • | • | • | • | |
| Cattò et al. (2021) | • | | • | | | • | • | • | |
| Sanmartín and Bosch-Roig et al. (unpublished results) | • | | | • | • | • | • | • | |

[a]Refers to microscopic observation analysis
[b]Refers to FTIR analysis

graffiti paintings and the inside of spray paint cans are candidate sources of micro-organisms (Sanmartín et al. 2016). After selection of the natural sources to be analysed, the microbial content can be examined by one of the two different methods: (i) the use of culture-dependent techniques (isolation of cultivable micro-organisms) or (ii) the use of culture-independent techniques (studying both cultivable and non-cultivable microorganisms).

The study of cultivable microorganisms is based on microbiological analysis involving isolation, growth and identification stages. Sampling can be done with sterile swabs moistened with sterile buffer solution or by contact plates with general culture media. Samples should then be cultivated in culture solid plates containing general culture media such as nutrient agar or nutrient broth (Fig. 8.2). Morphologically different colonies of microorganisms are then selected and sub-cultured to yield pure colonies. The pure colonies can be identified by classical morphological methods (by microscopic examination and dichotomous identification keys), biochemical tests (API tests, for example) and by sequencing methods based on the study of 16S rDNA (for bacteria) and ITS or 18S rDNA (for fungi) gene sequences (DNA extraction and polymerase chain reaction [PCR]) (Polo et al. 2010). In the field of graffiti biocleaning, only one molecular biological approach has been used to

**Fig. 8.2** (**a**) Sampling bacteria from new graffiti. (**b**) Example of plates initially used to screen for colonies found on graffiti and associated environments. (**c**) Different colonies of microorganisms selected, isolated and sub-cultured to test their potential use as graffiti removal agents. The images are reproduced from Sanmartín et al. (2016)

identify microorganisms from natural sources: nine cultivable bacteria and one fungus were identified as candidates for biocleaning purposes by sequencing of 16S rDNA and fungal ITS regions (Sanmartín et al. 2016).

Traditional identification techniques are sometimes tedious and time consuming. Researchers are currently focusing on new methodological advances such as matrix-assisted laser desorption ionization-time of flight mass spectrometry (MALDI-TOF-MS) to identify cultivable microorganisms. This technique captures the molecules present in the microorganisms by detecting proteins, peptides and lipid ions. It identifies a bacterial spectrum, rather than specific DNA sequences. As the samples do not require pretreatment, the method is therefore more rapid than traditional identification techniques; in addition, no reagents are required, and more than 1000 samples can be analysed per day. For identification, the bacterial spectrum obtained must then be compared with the existing databases. The main limitation of this technique is that few of the databases available include environmental microorganisms, and that many of the existing ones are costly to access (Shingal et al. 2015).

Massive DNA sequencing represents another important methodological advance in the identification of microorganisms. This approach can be used to study cultivable and non-cultivable microorganisms isolated from natural samples because it enables direct sequencing of native DNA/RNA. This next generation sequencing (NGS) technique can provide a more complete view of the microbial communities associated with an object or substrate (Vilanova and Porcar 2020). New generation single molecule nanopore technology sequencing (MinION Oxford Nanopore Technologies) is emerging and can be used for the rapid, simple identification of microorganisms as potential biocleaning agents. The huge amounts of nucleotide sequence data must be analysed and compared with existing databases in order to identify the microorganisms. The advantages of this method are that the analysis of long sequences enables more precise identification, little material is needed and the experimental process is simplified. There is no need to grow the microorganisms or to extract or amplify DNA for sequencing (Feng et al. 2015). This method provides more comprehensive information about the microbial populations colonizing a particular substrate (e.g. graffiti) because it produces data on all the microorganisms present in the sample.

A combination of culture-dependent and culture-independent techniques should be used to identify and study microorganisms as potential graffiti biocleaning agents. After identification, the microorganisms must be grown/cultivated in a particular medium supplied with the target substance as a sole carbon source. The biodegradative capacities should then be analysed, among other factors, in order to finally apply them in practical biocleaning actions (Vilanova and Porcar 2020).

## 3.2  Culture Media and Growth Protocols for the Selected Microorganisms

Once microorganisms have been purchased or isolated, they must be grown in the laboratory. The establishment of growth media and growth protocols is therefore an important methodological step to consider.

In order to use microorganisms for cleaning purposes, they must be grown in an enriched culture or minimal medium with the substrate to be degraded in order to adapt their metabolism for cleaning the substrate of interest. This is a basic biocleaning rule. For example, for removal of animal glue from frescoes, selected microorganisms were grown in the laboratory with culture media enriched with animal glue (Ranalli et al. 2005); for removal of black crust from stone materials, the selected microorganisms were grown with selected culture media enriched with sulphates (the principal component of black crusts), and for removal of salts from granite pavement stone, nitrate broth was used to grow the selected microorganisms (Bosch-Roig et al. 2019).

Following this principle, a culture media enriched with graffiti components is needed to optimally prepare/adapt the selected microorganisms for the bioremoval of graffiti. In this regard, a protocol for obtaining a minimal liquid medium (M9) enriched with powdered graffiti has recently been improved (Fig. 8.3). The established growth protocol and growth medium can help researchers to both evaluate and isolate potential new biocleaning agents and to produce large volumes of the selected microorganisms, which are required for future on-site applications (Sanmartín and Bosch-Roig 2019). On the basis of the findings of traditional microbiological sensitivity assays, a positive hydrolytic reaction will be manifested as a clear zone around the bacterial colony. Indeed, some authors have used solid media (containing calcium carbonate, phosphate and gypsum) to potentially isolate biocleaning microorganisms by the visual appearance of a halo around the colony

**Fig. 8.3** Micrograph of particles of powdered graffiti surrounded by biofilm

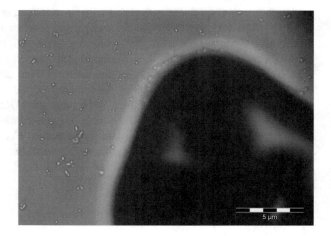

(Mazzoni et al. 2014). We are now focussing on adapting this isolation method in order to apply it to the isolation of microorganisms that could potentially be used for biocleaning graffiti. A new graffiti solid media rather than the liquid media currently used will facilitate the bacterial selection and evaluation process with visual results and prevent the need for a complementary analysis of graffiti degradation.

## 3.3 Microorganism Application Protocols

Another important aspect that should be considered in biocleaning strategies is that microorganisms require particular conditions (humidity and temperature) for growth. Therefore, different strategies have been used in the biocleaning field in order to improve application protocols (Bosch-Roig et al. 2015).

The first studies of bioremoval of graffiti used glass microscope slides painted with a graffiti layer. The microscope slides were placed in liquid cultures with the selected bacteria. Finally, the samples were incubated for long periods (25–49 days) (Giacomucci et al. 2012; Sanmartín et al. 2016). These initial studies were important because they showed that graffiti could be removed by biological means, although improvement of the application protocols was needed. The immersion protocol cannot be used for on-site graffiti biocleaning and therefore methodological advances were needed in this area. Based on the different delivery systems used in other biocleaning strategies, a more appropriate application protocol was developed by Sanmartín and Bosch-Roig (2019). This research has addressed this problem by studying two delivery systems commonly used in other biocleaning strategies: cotton wool and 2% agar gel (Bosch-Roig et al. 2015). This research has produced interesting results for bioapplication on stone, contributing to the future transfer of the biocleaning technology to real, on-site stone graffiti removal. However, further work is required, because long treatment times are required, and the graffiti layer is not completely removed.

Another important factor associated with the application protocols is the temperature required for growth of the microorganisms. Most microorganisms used in biocleaning strategies are mesophilic, i.e. they grow optimally at moderate temperatures (between 20 and 45 °C). If biocleaning is carried out at temperatures lower than 20 °C (very common in winter, for example), the effects could be altered or hampered. Different strategies (using air conditioning or heat lamps) could be used in order to reach the ideal temperature of the treatments (Ranalli et al. 2005; Bosch-Roig et al. 2013). However, some researches are focusing on the selection of microorganisms that can grow at higher temperatures. For example, Mazzoni et al. (2014) isolated biocleaning microorganisms (*Stenotrophomonas maltophilia, Pseudomonas koreensis and Cellulosimicrobium cellulans*), which have an optimal temperature range of 6–37 °C.

## 3.4   Protocols for Evaluating Graffiti Cleaning

Once microorganisms have been purchased or isolated and grown under laboratory conditions, their hydrolytic abilities and metabolic capacities must be studied in order to determine how they will contribute to graffiti removal. Most biocleaning studies use one of the two approaches: (i) an indirect approach based on study of the degradation of the target substance (graffiti) or (ii) a more direct approach based on the study of the enzymes or proteins produced by the microorganisms.

Most studies have used the indirect strategy. Few studies have investigated the enzymes and proteins produced by the microorganisms. Physico-chemical analysis is usually used for this purpose.

Considering the field of graffiti biocleaning, all the studies conducted to date have used the indirect strategy, in which microorganisms are grown with the graffiti and the graffiti is then examined in terms of visual alteration, colour variations (by reflectance colour measurements), surface change (by stereomicroscope, epifluorescence microscope and electron microscope observations) and chemical modifications, using Fourier transform infrared spectroscopy (FTIR) (Giacomucci et al. 2012; Sanmartín et al. 2016; Sanmartín and Bosch-Roig 2019) (see Table 8.2). In particular, preliminary FTIR findings on the potential biodegradation pathways induced by *Desulfovibrio desulfuricans* in nitrocellulose-based paints have been reported (Giacomucci et al. 2012). These findings include a depletion of signal related to the N-O bond, suggesting that the bacteria induce degradation of the paint via ammonification. In addition, Sanmartín and Bosch-Roig (2019)) have demonstrated bacterially induced chemical changes in graffiti paints, by using FTIR analysis. In the resulting spectra a decreasing intensity of peak bands related to alkanes, carbonyl, carbon-carbon triple bound and esters was observed. These types of indirect analysis could be complemented with other analytical strategies used in other graffiti cleaning strategies such as advanced laser-based techniques (laser-induced breakdown spectroscopy: LIBS), X-ray diffraction or RAMAN analysis (Gómez et al. 2010; Siano et al. 2012; Penide et al. 2013, Sanmartín et al. 2016).

More direct approaches based on enzymatic and proteomic analysis can also be used. In these approaches, microorganisms are grown with the target substance (graffiti) and the enzymatic activity or enzyme-protein production is then determined. In graffiti biocleaning, only one of these more direct approaches has been reported. In this study, the nitroesterase activity (nitrate and nitrite reductase activity) of a particular bacterial culture (*Desulfovibrio desulfuricans*) was evaluated by analysis of the presence of nitrate, nitrite and ammonia by use of a particular reagent that produces a colorimetric reaction (when the enzyme is expressed), which is then measured by spectrophotometry. This analysis enabled the researchers to conclude that *D. desulfuricans* was able to degrade the nitrocellulose present in paint binders, possibly by an ammonification pathway (Giacomucci et al. 2012).

In the field of biocleaning research, this type of direct analysis should inspire further studies on graffiti bioremoval. Enzymatic activity has been analysed by using biochemical (API-ZYM™ Systems BioMérieux, Rome, Italy) and colorimetric tests (based on azo dye impregnated casein) and (casein) zymograms (spectrophotometric assay) to evaluate and observe the enzymatic (protease) activity, in particular the specific caseinolytic and collagenolytic activity of the bacterium *P. stutzeri* A29 (Ranalli et al. 2005; Antonioli et al. 2005). Sulphate reducing activity has been evaluated in liquid culture medium by determining the residual sulphate content in the cultural broth medium by spectrophotometry (absorbance at 515 nm) (Ranalli et al. 1997). Some authors have used colorimetric test strips that enable the rapid semi-quantitative evaluation of bacterial enzymatic activities. For example, nitrate and nitrite test strips can be used to analyse the optimal effectiveness of different bacterial strains for nitrate removal (Bosch-Roig et al. 2013).

The protein content of bacteria can be extracted and analysed by different methods. For example, Giacomucci et al. (2012) used ultrasound-assisted extraction and the Bradford method (based on the protein interaction to Coomassie dye under acidic conditions resulting in a colour change from brown to blue) to determine the total protein content of *D. desulfuricans*.

Advanced phenotypic technologies, such as phenotype microarray systems (PM Biolog system), can also be used to obtain information about the properties of the selected strains. These systems enable rapid screening of an extensive range of growth conditions (Borglin et al. 2012). Analysis of the use of almost hundred different carbon sources by a selected microorganism can be done simultaneously (96 well microplate), to produce a "metabolic fingerprint". This type of technology has been used to study carbon and nitrogen metabolism, chemical sensitivity (osmolyte tolerance, pH tolerance and toxic compound tolerance) and the metabolic pathways sustaining the biocleaning capacity of strain *P. stutzeri* DSMZ 5190, which has been widely used for biocleaning purposes (Bosch-Roig et al. 2016).

The so-called omic techniques are increasingly used in the field of conservation and restoration research. These techniques enable complete analysis of the total pool of DNA (genomics), the total proteins (proteomics) and their metabolites (metabolomics) (Vilanova and Porcar 2020).

## 3.5  Long-Term Monitoring to Evaluate Graffiti Cleaning

The final step that should be considered in any biocleaning process, especially in the treatment of decorative surfaces, is to monitor the changes on the cleaned surface over time (Bosch-Roig and Ranalli 2014). Monitoring must focus on at least two aspects: (i) the absence of microbial growth on the surface (microbial monitoring) and (ii) the absence of aesthetic alteration of the cleaned surface over time.

All graffiti biocleaning studies include both of these monitoring strategies after the cleaning, but further long-term changes must be still determined. However, diverse biocleaning studies have focused on the changes on the cleaned surface over time. For example, in Matera cathedral, the effects of nitrate bioremoval were monitored 8 months and 6 years after the treatment. In the areas treated with bacteria, the nitrate concentration remained stable, and no notable differences in microflora or colour changes were observed in comparison with an untreated control area (Alfano et al. 2011). Most biocleaning studies include aesthetic and microbiological monitoring (by plate count method, ATP determination and SEM observation, colorimetric analysis) immediately after and/or one month after the treatment. However, some studies have also undertaken long-term monitoring (1, 2, 6 and 12 months after the treatment), revealing the absence of residual cells on the treated surface and no colour changes in the treated areas (Lustrato et al. 2012; Bosch-Roig et al. 2013; Ranalli et al. 2018, 2019).

Methodological advances in long-term monitoring are required in graffiti biocleaning, and recent interest focuses on non-invasive techniques. Digital image analysis tools can be used for this purpose. This type of analysis has recently been used for the first time to analyse a surface after a biocleaning treatment (Bosch-Roig et al. 2019).

## 4   Remaining Challenges in Graffiti Biocleaning

According to the general stages described in this chapter, one important remaining challenge in graffiti biocleaning is to reduce the treatment application times.

Graffiti biodegradation application times of up to 49 days for *Desulfovibrio desulfuricans* ATCC 13541 applied under anaerobic conditions to red alkyd-nitrocellulose-based binder spray paint have been reported (Giacomucci et al. 2012). The same authors reported aerobic treatment requiring an application time of 27 days when *Klebsiella aerogenes* ATCC 13048, ATCC 53922 (mixed culture of *Bacillus sp., Delftia lacustris, Sphingobacterium caeni* and *Ochrobactrum anthropi*) and *Comamonas* sp. ATCC 700440 were used (Giacomucci et al. 2012). This was reduced to 25 days when aerobic microorganisms (*Arthrobacter, Bacillus, Gordonia, Microbacterium, Pantoea, Pseudomonas* and *Alternaria*) were used to remove black paint (alkyd and polyester resins or varnishes) (Sanmartín et al. 2016). Important improvements have also been made towards developing a feasible on-site method, reducing the application time to 20 days in a subaerial strategy (placing the strains onto spraying graffiti paint on stone material in an aerial environment, i.e. suitable for on-site conditions) (Sanmartín and Bosch-Roig 2019).

As indicated above, until now bioremoval of graffiti has been a slow process, and the first signs of deterioration of the paint take some time to appear. Sanmartín and Pozo-Antonio (2020) have recently addressed this challenge, suggesting that prior photodegradation of the paints may improve the bioremoval treatment in terms of application time, cleaning efficiency and economy. The study evaluated the weathering of graffiti spray paint on building stones exposed to different types of UV radiation (daylight, UV-A, UV-B and UV-C) and considered a combination of UV irradiation and biological treatment for graffiti removal.

According to popular belief, graffiti should be removed promptly, as the paint is generally considered to become more difficult to remove over time (see, e.g. Weaver 1995). However, a study of accelerated ageing of graffiti paints applied to stone was the first to demonstrate that the age of graffiti (i.e. the time that the graffiti has remained on the lithic substrate) enhances the effectiveness of the cleaning process (Sanmartín and Cappitelli 2017). This finding was strengthened in subsequent studies involving the removal of graffiti from anti-graffiti surfaces, in which the same conclusion was reached, in contrast to conventional wisdom (Pozo-Antonio et al. 2018).

Another (parallel) strategy is to identify the microorganisms that are the most efficient graffiti bioremoval agents. This approach should be accompanied by study of the enzymatic reactions involved in graffiti bioremoval to better elucidate the metabolic pathways. On-site, large-scale production of microorganisms is another remaining challenge that modern biotechnologies could help to solve (Philip and Atlas 2017).

Another challenge is the study of biocleaning strategies for diverse materials and graffiti colours and types. To date, only four substrates have been analysed: glass, natural granite stone, man-made concrete stone and metal (Fig. 8.4). In addition, only three types of graffiti have been tested: red spray paint (Motip–Dupli®

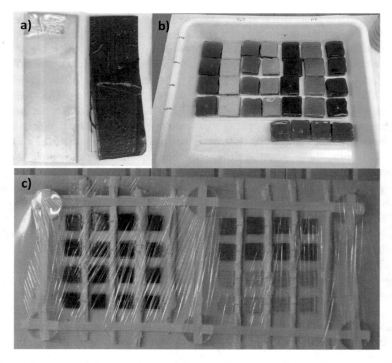

**Fig. 8.4** Laboratory-based graffiti bioremoval from diverse materials and involving coloured paints. (**a**) Remains of black and silver graffiti on glass after immersion biocleaning. (**b**) Natural granite stone and man-made concrete stone painted with black and silver graffiti paint during agar-assisted biocleaning. (**c**) Vertical metal sheet painted with black and silver graffiti paint during agar-assisted biocleaning. The image is reproduced from Sanmartín and Bosch-Roig (2019)

Autocolor, colour 5-0200), black non-metallic spray paint (*R-9011* from Montana Colors, Barcelona, Spain) and a silver metallic spray paint (*Silver Chrome* from Montana Colors).

Several authors have suggested that future research should focus on the following: (i) selection/identification of new biological formulation to effectively remove all kinds of paints and colours; (ii) the biocleaning potential of viable yeast cells (which produce few toxic metabolites); (iii) evaluation of the effectiveness of the cleaning of aged graffiti from different substrates; (iv) incomplete graffiti cleaning; (v) testing diverse delivery systems to shorten the biocleaning treatment times and resolve difficulties in application (Troiano et al. 2014; Gomes et al. 2017; Sanmartín and Bosch-Roig 2019).

# 5  Conclusion and Final Remarks

The bioremoval of graffiti is a promising green approach and risk-free method based on the use of selected microorganisms as "probiotics" (Cappitelli 2016; Sanmartín et al. 2016; Vilanova and Porcar 2020). However, it remains an emerging field in which there are still many challenges to be addressed and solved.

In this chapter, a comprehensive analysis has been conducted by focusing on comparing existing studies on the bioremoval of salts and organic matter by the use of live microorganisms in laboratory simulations or applied to real artwork. Special emphasis has been placed on the methodological advances made and on existing methodological tools and approaches used in related fields that could be applied in this particular area in order to help researchers address the remaining challenges.

**Acknowledgements**  P. Sanmartín is grateful for financial support from the Xunta de Galicia (grant ED431C 2018/32).

# References

Alfano G, Lustrato G, Belli C, Zanardini E, Cappitelli F, Mello E, Sorlini C, Ranalli G (2011) The bioremoval of nitrate and sulfate alterations on artistic stonework: the case study of Matera Cathedral after six years from the treatment. Int Biodeterior Biodegrad 65(7):1004–1011

Allsopp D, Seal K, Gaylarde C (2004) Introduction to biodeterioration. Chapter 3. Biodeterioration of refined and processed materials: paints, 2nd edn. Cambridge University Press, Cambridge, pp 78–85

Antonioli P, Zapparoli G, Abbruscato P, Sorlini C, Ranalli G, Righetti PG (2005) Art-loving bugs: the resurrection of Spinello Aretino from Pisa's cemetery. Proteomics 5:2453–2459

Bentley J, Turner GPA (1998) Introduction to paint chemistry and principles of paint technology, 4th edn. Chapman & Hall, London

Borglin S, Joyner D, DeAngelis KM, Khudyakov J, D'haeseleer P, Joachimiak MP, Hazen T (2012) Application of phenotypic microarrays to environmental microbiology. Curr Opin Biotechnol 23:41–48

Bosch-Roig P, Regidor-Ros JL, Soriano-Sancho P, Doménech-Carbó MT, Montes-Estelles RM (2010) Ensayos de biolimpieza con bacterias en pinturas murales. Arché 4–5:115–124

Bosch-Roig P, Regidor-Rosi JL, Montes-Estellés RM (2013) Biocleaning of nitrate alterations on wall paintings by *Pseudomonas stutzeri*. Int Biodeter Biodegr 84:266–274

Bosch-Roig P, Ranalli G (2014) The safety of biocleaning technologies for cultural heritage. Front Microbiol 5:155

Bosch-Roig P, Lustrato G, Zanardini E, Ranalli G (2015) Biocleaning of cultural heritage stone surfaces and frescoes: which delivery system can be the most appropriate? Ann Microbiol 65:1227–1241

Bosch-Roig P, Decorosi F, Giovannetti L, Ranalli G, Viti C (2016) Connecting phenome to genome in *Pseudomonas stutzeri* 5190: an artwork biocleaning bacterium. Res Microbiol 167:757–765

Bosch-Roig P, Allegue H, Bosch I (2019) Granite pavement nitrate desalination: traditional methods vs. biocleaning methods. Sustainability 19:4227

Breitbach AM, Rocha JC, Gaylarde CC (2011) Influence of pigments on biodeterioration of paint films. J Coating Technol Res 8:619–628

Cappitelli F (2016) Biocleaning of cultural heritage surfaces. Open Conf Proc J 7(suppl 1: M6):65–69

Cappitelli F, Sorlini C (2008) Microorganisms attack synthetic polymers in items representing our cultural heritage. Appl Environ Microbiol 74:564–569

Cappitelli F, Vicini S, Piaggio P, Abbruscato P, Princi E, Casadevall A, Nosanchuk JD, Zanardini E (2005) Investigation of fungal deterioration of synthetic paint binders using vibrational spectroscopic techniques. Macromol Biosci 5:49–57

Cappitelli F, Zanardini E, Ranalli G, Mello E, Daffonchio D, Sorlini C (2006) Improved methodology for bioremoval of black crusts on historical stone artworks by use of sulfate-reducing bacteria. Appl Environ Microbiol 72:3733–3737

Cappitelli F, Nosanchuk JD, Casadevall A, Toniolo L, Brusetti L, Florio S, Principi P, Borin S, Sorlini C (2007) Synthetic consolidants attacked by melanine producing fungi: case study of the biodeterioration of Milan (Italy) Cathedral marble treated with acrylics. Appl Environ Microbiol 73:271–277

Cattò C, Sanmartín P, Gulotta D, Troiano F, Cappitelli F (2021) Bioremoval of graffiti using novel commercial strains of bacteria. Sci Total Environ 756:144075

Colombini MP, Modugno F, Giacomelli A (1999) Two procedures for suppressing interference from inorganic pigments in analysis by gas chromatography-mass spectrometry of proteinaceous binders in painting. J Chromatogr Acta 846:101–111

Colombini MP, Bonaduce I (2003) Gas chromatography/mass spectrometry for the characterization of organic materials in frescoes of the Monumental Cemetery of Pisa (Italy). Rapid Commun Mass Spectrom 17:2523–2527

Ciferri O (1999) Microbial degradation of paintings. Appl Environ Microbiol 65:879–885

DDL 1607 del 16 luglio (2002) Disciplina del fenomeno del "graffitismo". Annunciato nella seduta ant. n. 214 del 17 luglio 2002. (In Italian)

Feng Y, Zhang Y, Ying C, Wang D, Du C (2015) Nanopore-based Fourth-generation DNA sequencing technology. Genom Proteom Bioinf 13:4–16

Fernandes P (2006) Applied microbiology and biotechnology in the conservation of stone cultural heritage materials. Appl Microbiol Biotechnol 73:291–296

Gauri KL, Chowdhury AN, Kulshreshtha NP, Punuru AR (1989) The sulfation of marble and the treatment of gypsum crusts. Stud Conserv 34:201–206

Gauri KL, Parks L, Jaynes J, Atlas R (1992) Removal of sulfated-crusts from marble using sulphate-reducing bacteria. In: Webster RGM (ed) Proceedings of the International Conference on Stone Cleaning and the Nature, soiling and decay mechanisms of stone. Donhead, Edinburgh, pp 160–165

Gaylarde PM, Gaylarde CC (2000) Algae and cyanobacteria on painted buildings in Latin America. Int Biodeter Biodegr 46:93–97

Gaylarde CC, Gaylarde PM (2005) A comparative study of the major microbial biomass of biofilms on exteriors of buildings in Europe and Latin America. Int Biodeter Biodegr 55:131–139

Gaylarde CC, Morton LHG, Loh K, Shirakawa MA (2011) Biodeterioration of external architectural paint films – a review. Int Biodeter Biodegr 65:1189–1198

Giacomucci L, Toja F, Sanmartín P, Toniolo L, Prieto B, Villa F, Cappitelli F (2012) Degradation of nitrocellulose-based paint by *Desulfovibrio desulfuricans* ATCC 13541. Biodegradation 23:705–716

Gioventù E, Lorenzi PF, Villa F, Sorlini C, Rizzi M, Cagnini A, Griffo A, Cappitelli F (2011) Comparing the bioremoval of black crusts on colored artistic lithotypes of the cathedral of Florence with chemical and laser treatment. Int Biodeterior Biodegrad 65:832–839

Gomes V, Dionísio A, Pozo-Antonio JS (2017) Conservation strategies against graffiti vandalism on cultural heritage stones: Protective coatings and cleaning methods. Prog Org Coat 113:90–109

Gómez C, Costela A, García-Moreno I, Sastre R (2010) Comparative study between IR and UV laser radiation applied to the removal of graffitis on urban buildings. Appl Surf Sci 252:2782–2793

Govaert F, Bernard M (2004) Discriminating red spray paints by optical microscopy, Fourier transform infrared spectroscopy and X-ray fluorescence. Forensic Sci Int 140:61–70

Grant C, Bravery AF, Springle WR, Worley W (1986) Evaluation of fungicidal paints. Int Biodegr 22:179–194

Heselmeyer K, Fischer U, Krumbein WE, Warsheid T (1991) Application of Desulfovibrio vulgaris for the bioconversion of rock gypsum crusts into calcite. Bioforum 1:89

Hyvärinen A, Meklin T, Vepsäläinen A, Nevalainen A (2002) Fungi and actinobacteria in moisture-damaged building materials—concentrations and diversity. Int Biodeter Biodegr 49:27–37

Klens PF, Lang JR (1956) Microbiological factors in paint preservation. J Oil Colour Chemists' Assoc 38:887–899

Lustrato G, Alfano G, Andreotti A, Colombini MP, Ranalli G (2012) Fast biocleaning of mediaeval frescoes using viable bacterial cells. Int Biodeter Biodegr 69:51–61

Macedo MF, Miller AZ, Dionísio A, Saiz-Jimenez C (2009) Biodiversity of cyanobacteria and green algae on monuments in the Mediterranean Basin: an overview. Microbiology 155:3476–3490

May E, Webster AM, Inkpen R, Zamarreño D, Kuever J, Rudolph C, Warscheid T, Sorlini C, Cappitelli F, Zanardini E, Ranalli G, Krage L, Vgenopoulos A, Katsinis D, Mello E, Malagodi M (2008) The BIOBRUSH project for bioremediation of heritage stone. In: May E, Jones M, Mitchell J (eds) Heritage microbiology and science: microbes, monuments and maritime materials. RSC Publishing, Cambridge, pp 76–93

Mazzoni M, Alisi C, Tasso F, Cecchini A, Marconi P, Sprocati AR (2014) Laponite micro-packs for the selective cleaning of multiple coherent deposits on wall paintings: the case study of Casina Farnese on the Palatine Hill (Rome-Italy). Int Biodeter Biodegr 94:1–11

Palla F, Federico C, Russo R, Anello L (2002) Identification of Nocardia restricta in biodegraded sandstone monuments by PCR and nested-PCR DNA amplification. FEMS Microbiol Ecol 39:85–89

Park JH, Cox-Ganser LM, White SK, Laney AS, Caulfield SM, Turner WA, Summer AD, Kreiss K (2017) Bacteria in a water-damaged building: associations of actinomycetes and non-tuberculous mycobacteria with respiratory health in occupants. Indoor Air 27:24–33

Penide J, Quintero F, Riveiro A, Sánchez-Castillo A, Comesaña R, del Val J, Lusquiños F, Pou J (2013) Removal of graffiti from quarry stone by high power diode laser. Opt Laser Eng 51: 64–370.

Philip J, Atlas R (2017) Chapter 4 - microbial resources for global sustainability. In: Microbial resources from functional existence in nature to applications. Ipek Kurtböke, Elsevier, Amsterdam, Netherland, pp 77–101

Polo A, Cappitelli F, Brusetti L, Principi P, Villa F, Giacomucci L, Ranalli G, Sorlini C (2010) Feasibility of removing surface deposits on stone using biological and chemical remediation methods. Microb Ecol 60:1–14

Pozo-Antonio JS, Rivas T, Jacobs RMJ, Viles HA, Carmona-Quiroga PM (2018) Effectiveness of commercial anti-graffiti treatments in two granites of different texture and mineralogy. Prog Org Coat 116:70–82

Ranalli G, Chiavarini M, Guidetti V, Marsala F, Matteini M, Zanardini E, Sorlini C (1997) The use of microorganisms for the removal of sulphates on artistic stoneworks. Int Biodeter Biodegr 40:255–261

Ranalli G, Chiavarini M, Guidetti V, Marsala F, Matteini M, Zanardini E, Sorlini C (1996a) Utilisation of microorganisms for the removal of sulphates on artistic stoneworks. 3rd Int Biodeterior Biodegrad Symposium, Sociedad Española de Microbiología, Santiago de Compostela, pp 59–60

Ranalli G, Chiavarini M, Guidetti V, Marsala F, Matteini M, Zanardini E, Sorlini C (1996b) The use of microorganisms for the removal of nitrates and organic substances on artistic stone works. In: Riederer J (ed) Proceedings of the eighth International Congress of Deterioration and Conservation of Stone. Möller, Berlin, pp 1415–1420

Ranalli G, Matteini M, Tosini I, Zanardini E, Sorlini C (2000) Bioremediation of cultural heritage: Removal of sulphates, nitrates and organic substances. In: Ciferri O, Tiano P, Mastromei G (eds) Of microbes and art - the role of microbial communities in the degradation and protection of cultural heritage. Kluwer Academic-Plenum Publisher, pp 231–245

Ranalli G, Alfano G, Belli C, Lustrato G, Colombini MP, Bonaduce I, Zanardini E, Abbruscato P, Cappitelli F, Sorlini C (2005) Biotechnology applied to cultural heritage: biorestoration of frescoes using viable bacterial cells and enzymes. J Appl Microbiol 98:73–83

Ranalli G, Zanardini E, Andreotti A, Colombini MP, Corti C, Bosch-Roig P, De Nuntiis P, Lustrato G, Mandrioli P, Rampazzi L, Giantomassi C, Zari D (2018) Hi-tech restoration by two-steps biocleaning process of triumph of death fresco at the Camposanto Monumental Cemetery (Pisa, Italy). J Appl Microbiol 125:800–812

Ranalli G, Zanardini E, Rampazzi L, Corti C, Andreotti A, Colombini MP, Bosch-Roig P, Lustrato G, Giantomassi C, Zari D, Virilli P (2019) Onsite advanced biocleaning system for historical wall paintings using new agar-gauze bacteria gel. J Appl Microbiol 126:1785–1796

Rampazzi L, Andreotti A, Bressan M, Colombini MP, Corti C, Cuzman O, d'Alessandro N, Liberatore PL, Raimondi V, Sacchi B, Tiano P, Tonucci L, Vettori S, Zanardini E, Ranalli G (2018) An interdisciplinary approach to a knowledge-based restoration: the dark T alteration on Matera Cathedral (Italy). Appl Surf Sci 458:529–539

Reynolds ES (1950) Pullularia as a cause of deterioration of paint and plastic surfaces in South Florida. Mycologia 42:432–448

Saiz-Jimenez C (1993) Deposition of airborne organic pollutants on historic buildings. Atmos Environ 27B:77–85

Saiz-Jimenez C (1997) Biodeterioration vs biodegradation: the role of microorganisms in the removal of pollutants deposited on historical buildings. Int Biodeter Biodegr 40:225–232

Sanmartín P, Cappitelli F (2017) Evaluation of accelerated ageing tests for metallic and non-metallic graffiti paints applied to stone. Coatings 7:180

Sanmartín P, Bosch-Roig P (2019) Biocleaning to remove graffiti: a real possibility? Advances towards a complete protocol of action. Coatings 9:104

Sanmartín P, Pozo-Antonio JS (2020) Weathering of graffiti spray paint on building stones exposed to different types of UV radiation. Constr Build Mater 236:117736

Sanmartín P, Cappitelli F, Mitchell R (2014) Current methods of graffiti removal: a review. Constr Build Mater 71:363–374

Sanmartín P, DeAraujo A, Vasanthakumar A, Mitchell R (2016) Feasibility study involving the search for natural strains of microorganisms capable of degrading graffiti from heritage materials. Int Biodeter Biodegr 103:186–190

Sanmartín P, Mitchell R, Cappitelli F (2016) Evaluation of cleaning methods for graffiti removal. In: Brimblecombe P (ed) Urban pollution and changes to materials and building surfaces. Imperial College Press, London, pp 291–312

Sanmartín P, DeAraujo A, Vasanthakumar A (2018) Melding the old with the new: trends in methods used to identify, monitor and control microorganisms on cultural heritage materials. Microb Ecol 76:64–80

Shingal N, Kumar M, Kanaujia PK, Virdi JS (2015) MALDI-TOF mass spectrometry: an emerging technology for microbial identification and diagnosis. Front Microbiol 6:791

Siano S, Agresti J, Cacciari I, Ciofini D, Mascalchi M, Osticioli I, Mencaglia AA (2012) Laser cleaning in conservation of stone, metal, and painted artifacts: state of the art and new insights on the use of the Nd:YAG lasers. Appl Phys A Mater Sci Process 106:419–446

Soffitti I, D'Accolti M, Lanzoni L, Volta A, Bisi M, Mazzacane S, Caselli E (2019) The potential use of microorganisms as restorative agents: an update. Sustainability 11:3853

Troiano F, Vicini S, Gioventù E, Lorenzi PF, Improta CM, Cappitelli F (2014) A methodology to select bacteria able to remove synthetic polymers. Polym Degrad Stabil 107:321–327

Turner JN (1967) The microbiology of fabricated materials. J&A Churchill, London

Vilanova C, Porcar M (2020) Art-omics: multi-omics meet archaeology and art conservation. Microb Biotechnol 13:435–441

Weaver ME (1995) Removing graffiti from historic masonry. Preservation Brief No. 38. National Park Service, Technical Preservation Services, Washington DC

Webster A, May E (2006) Bioremediation of weathered-building stone surfaces. Trends Biotecnol 24:255–260

# Chapter 9
# Ancient Textile Deterioration and Restoration: Bio-Cleaning of an Egyptian Shroud Held in the Torino Museum

**Roberto Mazzoli and Enrica Pessione**

**Abstract** Ancient textiles are fragile and several factors can affect their integrity. In the present chapter, the main agents of deterioration of old and new textiles, namely physical-chemical (light, oxygen, heat, and humidity) and biological factors as well as human erroneous interventions will be explored. As far as the biological deterioration is considered, the effects of microbial growth, primary and secondary metabolites (acids, solvents, surfactants, pigments) and enzymes (lipases, proteases, and glycosidases) on textile strength and cleanliness will be described in details. The main fungal and bacterial species involved in the damage (textile discoloration, black and green spots, cuts) will be reported. Adhesive application during restoration procedures is discussed to highlight the risk of glue thickening giving rise to dull precipitates on the fabric.

The main strategies for oil-stain and glue removal (both animal glue, such as fish collagen, and vegetal glue, *i.e.* starch) will be described in the paragraph devoted to biorestoration. Finally, a case study concerning an ancient Coptic tunic housed in the Egyptian Museum of Torino, Italy, and biocleaned by means of gellan-immobilized alpha-amylase from *Bacillus* sp. will be largely discussed by reporting historical data, adhesive characterization, methods for artificial aging of simulated sample and glue removal from the artwork.

**Keywords** Microbial and physical-chemical deterioration · Human restoration · Starch glue · Immobilized enzymes · Amylase · Wool artificial aging · Gellan · Coptic period

R. Mazzoli · E. Pessione (✉)
Structural and Functional Biochemistry, Laboratory of Proteomics and Metabolic Engineering of Prokaryotes, Department of Life Sciences and Systems Biology, University of Torino, Torino, Italy
e-mail: enrica.pessione@unito.it

© The Author(s) 2021
E. Joseph (ed.), *Microorganisms in the Deterioration and Preservation of Cultural Heritage*, https://doi.org/10.1007/978-3-030-69411-1_9

# 1 Introduction

Among artworks, textile materials constitute important human history proofs. Specimens such as Pre-Columbian and Native Indian clothes, shrouds, tapestries, carpets, soldier uniforms, ecclesiastical vestments, Olympic winner swimsuits, and spacesuits constitute a rich collection of archeological fabrics that are precious but also fragile, generally revealing a bad conservation state.

Analyzing the causes of deterioration is of primary importance in order to protect the historical items and prevent further damage. However, restoration strategies are sometimes inescapable.

In the present chapter, we will first consider mechanisms of textile aging and deterioration, then the main bio-cleaning-bio-restoration techniques applicable to textiles and finally we will report a case study involving bio-cleaning of an ancient Egyptian shroud of the Torino Egyptian Museum (Italy).

# 2 Textile Aging and Deterioration

Ancient textiles can suffer from aging, deterioration and degradation events that deeply affect their original beauty and their ethnological and economical value. The environmental conditions (temperature, humidity, light exposure, microbial contamination) these artifacts face during their life strongly affect the item value including those of the museum exhibition rooms, which are crucial for a correct conservation. However, other factors can in case account for visible damage, such as operators' interventions aimed to partially restore the item (Ferrari et al. 2017). Here we consider three main types of stresses giving rise to deterioration: (i) physical-chemical environmental stressors; (ii) microbial degradation and deterioration; (iii) human erroneous treatments.

***Physical-Chemical Factors as Cause of Textile Deterioration*** As far as physical-chemical agents are concerned, light, heat, and oxygen are among the main causes of textile alteration (Rubeziene et al. 2012). As regards solar radiation, it comprises visible, infrared (IR), and ultraviolet (UV) components. IR heats materials, but UV radiation causes most photo-chemical damages to textiles. Photo-decomposition is accelerated by the presence of heat and moisture. When also oxygen is present, oxidative reactions (such as chain scission and cross-linking) known as photo-oxidation may occur (Szostak-Kotowa 2004; Rubeziene et al. 2012). The effect of light radiation on colored textiles, such as discoloration, or generation of dark spots is easy to detect, nonetheless light also decreases tensile and tear strength of fabrics (Rubeziene et al. 2012). The effect of light can be different depending on the type of textile. Exposition of wool to radiation at 475 nm or shorter wavelength for more than 12 h decreases fiber strength (up to 20%) and causes color changes, especially in the presence of moisture (Treigiené and Musnickas 2003; Zimmermann and Hocker 1996). As regards color changes, UV irradiation generally causes photobleaching followed by yellowing, likely owing to the presence of aromatic amino acids (phenylalanine, tryptophan, tyrosine) and natural yellow pigments in wool (Nicholas and Pailthorpe 1976). Cellulose is sensitive to farther UV radiation (200–300 nm). Photo-oxidation of hydroxyl side-groups and glycosidic bonds causes changes in color, solubility and mechanical properties (increase of rigidity and brittleness) of cellulose (Timár-Balázsy and Eastop 1998). UV rays cause linen textile elongation and fabrics to become darker and slightly lose tensile strength (Abdel-Kareem 2005). Silk fibers are the most sensitive to photo-oxidation by UV irradiation (220–370 nm), leading to significant color changes and to a modification of the textile that becomes more rigid and mechanically weakened (Shubhra et al. 2011). Also synthetic fibers are sensitive to UV light exposure at different extent depending on the kind of chemical polymer, namely polyacrylonitrile is more resistant while polyamides (nylon) are more sensitive (Rubeziene et al. 2012). In this case, UV exposure causes different degrees of yellowing and decrease of mechanical strength, as well. High water content also favors disruption of the fabric structure and color loss (Gutarowska et al. 2017).

***Microbial Growth and Metabolism as Degradative and Deterioration Agents*** Humidity and relatively high temperatures favor the growth of microbial species, both bacteria and fungi, thus opening the way for release of molecules that can damage textiles (Mazzoli et al. 2018b). Among them, catabolites (surfactants, solvents, acids), secondary metabolites (pigments) and enzymes (proteases, lipases, and glycosidases) play a major role. Actually, microorganisms use the textile components (carbon, nitrogen, sulfur, phosphorous) for growth. Plant-derived fabrics host a microbial flora very different from that of animal-derived specimens, based on their different composition.

In general, the most recent proofs of human history such as synthetic fabrics are too hydrophobic to allow biodegradation and among them polyurethanes are the only suitable to bind water thus favoring microbial colonization. Exposure of polyurethane-made swimsuits from Olympic winners at museums should be carefully monitored since a huge number of fungal species bear extracellular esterases that can initiate polyurethane degradation (Rowe and Howard 2002). Other synthetic textiles, such as polypropylene, polyacrylonitrile, and polyamide, because of their hydrophobicity and the presence of ether chemical bonds (unusual in nature), undergo degradation only after light exposure since UV-induced photo-degradation generates shorter chain polymers that become available for bacteria. For this reason, textiles of historical and ethnological interest should be exposed in museums avoiding the use of intense light (Seal 1988).

Besides causing photo-oxidation, light also supports growth of phototrophic bacteria that can use textile-endogenous compounds as nitrogen source. As an example, the presence of light and a high degree of humidity can support the growth of microorganisms generating green pigments (Gutarowska et al. 2017). Other pigments (black, brown, red, orange, and yellow) are produced by non-phototrophic bacteria (*Bacillus, Corynebacterium, Achromobacter, Streptomyces*) as well as by fungi such as *Penicillium, Aspergillus,* and *Cryptococcus*. The

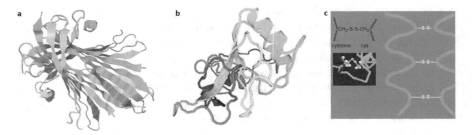

**Fig. 9.1** Molecular structure of the three main proteins present in animal-derived textiles. (**a**) silk fibroin; (**b**) silk sericin; (**c**) wool keratin

opposite phenomenon, *i.e.* discoloration, can be due to microbial production of lipases, surfactants, and solvents such as acetone. Recently, Pre-Columbian textiles made of cotton and llama- or alpaca-wool were analyzed with the aim of identifying microbial discoloration agents. The most representative genera were *Aspergillus, Penicillium,* and *Cladosporium* among fungi and *Kocuria rosea* and *Paracoccus yeei* among bacteria (Pietrzak et al. 2017). Besides pigments and discoloring agents, microbial growth and enzymatic activity can also cause depolymerization of the fabric, generating loss of strength and elasticity of the textile that can undergo fragmentation (Mazzoli et al. 2018a).

Although sensitivity to biodeterioration is mainly related to the polymerization extent of the fiber, the type of weave, the fabric thickness, as a rule, also the vegetal or animal origin of the fabric is important: actually, plant-derived textiles are more susceptible to biodegradation than wool and silk because of the peculiar chemical bonds intrinsic to keratins and fibroins/sericins (Mazzoli et al. 2018a). Among plant-derived textiles, linen and cotton undergo the major risk of deterioration since they are rich in hemicellulose and pectin that are easily degradable by microbial hydro-lases, namely endoglucanases (generating different length oligosaccharides), exoglucanases (generating di- or monosaccharides) and beta-glucosidases (converting oligosaccharides to monosaccharides) (Szostak-Kotowa 2004). Conversely, hemp and jute contain non-cellulosic components like lignin, which render the fabric more resistant to degradation, because only few organisms possess enzymes able to degrade it (Gutarowska et al. 2017). Cellulose depolymerization occurs differently in fungi and bacteria: the former directly penetrate into the fiber lumen generating a mycelium responsible for the secretion of extracellular cellu-lases, whereas the latter proceed from the fiber surface to the interior (Szostak-Kotowa 2004). However, the final effect carried on by cellulolytic enzymes is an impaired fiber strength. Silk and wool, being of animal origin, display a higher protein content (Fig. 9.1) and may undergo proteolysis. In silk, the first protein component to be used is sericin (Fig. 9.1b). However, most of the protein structure is made up of fibroin (Fig. 9.1a), a protein composed by four amino acid repeats (alanine, glycine, serine, and tyrosine) that is degraded slower. In spite of this slow biodegradation coefficient, fibroin is very sensitive to light: different bacteria

**Fig. 9.2** Glue-damaged historical carpet dating back to the Ottoman period and exhibited in the museum of the Faculty of Applied Arts, Helwan University, Egypt (modified from Ahmed and Kolisis 2011)

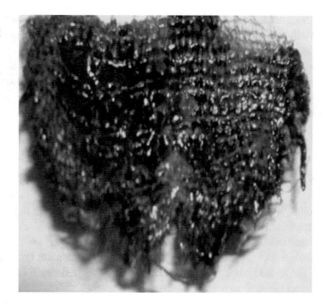

(*Bacillus, Serratia, Pseudomonas and Streptomyces*) and fungi (*Aspergillus*) can modify fibroin structure after light exposure (Seves et al. 1998; Szostak-Kotowa 2004). Wool keratin is an extra-strong structure and it is generally degraded very slowly. Unfortunately, both insects and microorganisms (both bacteria and fungi, especially *Trichophyton* and *Trichoderma*) can attack the disulfide bridges (Fig. 9.1c) that hold together keratin chains (Szostak-Kotowa 2004).

***Human Interventions as Cause of Textile Impairment*** Erroneous preservation attempts from human operators can play a role in cultural heritage item damaging (Beutel et al. 2002; Sterflinger and Pinzari 2012). As far as fabric is concerned, the use of glues has been largely employed in the effort to consolidate cuts. Actually, during aging, some parts of the textiles can be broken and both vegetal (starch) and animal (collagen) adhesives have been employed for pasting and consolidating specimens (Barbabietola et al. 2016; De La Chapelle et al. 1994; Ferrari et al. 2017). Animal-derived glue is of proteinaceous nature, namely collagen originated from fish swimming bladder or mammalian bones and cartilage. Plant-derived glues are made up of polysaccharides, namely starch (*i.e.* amylose and amylopectin) derived from rye, oat, barley, wheat, rice, corn, and potato (Barbabietola et al. 2016; Ferrari et al. 2017). Unfortunately, during time, both these adhesives can precipitate giving rise to dense and opaque material that enhances fiber distortion and fabric thickening, also forming intricate layers that cause structural fragility of the textile, besides producing a visible cloth difficult to remove (Fig. 9.2) (Blüher et al. 1995; Gostling 1989). In the case of animal glues, temperature, humidity, and light can promote protein cross-linking and peptide bond oxidation, whereas microbial attack can favor the production of unwanted pigments (Barbabietola et al. 2016).

In general, for these reasons, starch-based glues are more frequently and preferentially used for textile consolidation than collagen (Ahmed and Kolisis 2011; Whaap 2007). After aging, however, also starch paste can cause hardening, rigidity and yellowness of the textile, also constituting a nutrient-rich habitat for amylolytic bacteria and fungi, which promote textile deterioration over time. Such structural damages render ancient textiles fragile and unsuitable for exhibition. Therefore, cleaning and restoration interventions have to be applied.

## 3   Bio-Cleaning-Bio-Restoration of Textiles

*Glue Removal from Ancient Textiles*   As reported in the previous section, adhesive application on damaged or broken fabric areas is the most frequent erroneous treatment causing textile impairment. To remove glue several approaches can be followed: (i) mechanical methods; (ii) chemical methods; (iii) wet cleaning (Table 9.1). The first two strategies are sometimes too aggressive and not always applicable to precious ethnographic textiles (Mazzoli et al. 2018a). As far as wet cleaning is concerned, although it represents a milder strategy, the main constraint lies in the long application times. Since this method is only based on humidification, it is not very performant on hardened and long-aged adhesives (Ferrari et al. 2017).

A promising alternative to these techniques is a bio-based approach exploiting living microorganisms or their enzymes, or sometimes a combination of both treatments when possible (Ahmed and Kolisis 2011; Barbabietola et al. 2016; Ferrari et al. 2017). Both microbial and enzymatic glue removal occur in a water environment without use of organic solvents, hence, these treatments are environmental friendly and free of risks for the operators (Barbabietola et al. 2016). The choice between a microbiological and an enzymatic approach strongly depends on the area to be treated: a large area suggests the use of bacteria since the treatment is cheaper and requires less stringent conditions. The living microorganisms approach is also the only feasible when a mixture of complex and heterogeneous compounds have been applied to obtain consolidation (Barbabietola et al. 2016).

**Table 9.1**  Strategies for adhesive cleaning from textiles

|  | Mildness | Toxicity/ damage | Operator skill | Specificity | Cost | Time | Efficacy |
|---|---|---|---|---|---|---|---|
| Mechanical | No | Yes | No | Yes | Low | Short | Yes |
| Chemical | No | Yes | No | No | Low | Short | Yes |
| Wet cleaning | Yes | No | No | No | Low | Long | No |
| Microbial | Yes | No | No | Middle | Middle | Middle | Yes |
| Enzymatic (solution) | No | Yes | Yes | Yes | High | Long | No |
| Enzymatic (Immobilized) | Yes | No | Yes | Yes | High | Long | Yes |

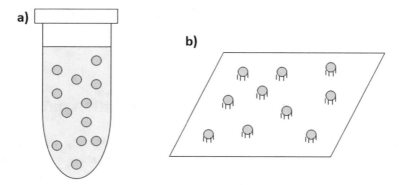

**Fig. 9.3** Enzyme immobilization strategies are necessary to treat textiles. (**a**) enzymes in solution; (**b**) immobilized enzymes

However, the limit of using microbial cells as removal agents is their low selectivity (Webster and May 2006). Therefore, more frequently, the enzymatic approach proves to be the most performant for cleaning adhesive-damaged cotton, linen and silk textiles being suitable for both protein-based and polysaccharide-based glues (Ahmed and Kolisis 2011; Ciatti et al. 2010). Actually, the very high catalytic specificity of proteases and amylases renders this method applicable on both collagen and starch. Furthermore, a wide range of microbial-derived enzymes allows the choice of the best fitting (optimum temperature and pH) according to the fabric to be treated (Germinario et al. 2017). These advantages overcome the relatively higher cost of enzymes with respect to microorganisms (Barbabietola et al. 2016). A further constraint of the protein-based strategy is the need to use aqueous solutions that can damage the textiles by excess of water (Hrdlickova Kuckova et al. 2014). During time, this humid environment can promote the growth of bacteria and fungi causing further impairment of the textile (Ahmed and Kolisis 2011). A successful alternative to bypass this bottleneck is the application of the enzyme preparations as a poultice (Bott 1990; Chapman 1986; Shibayama and Eastop 1996), using a gel as sorbent (Hrdlickova Kuckova et al. 2014) or after suitable immobilization on a solid surface (Fig. 9.3). The idea of using an enzyme-enriched poultice was developed to treat starch-damaged paper materials (Schwarz et al. 1999) and successfully translated to textiles. As far as enzyme immobilization is concerned, it has to be underlined that preliminary tests should be performed to evaluate (i) the immobilization yield (amount of enzyme really attached to the chosen surface); (ii) the catalytic activity after immobilization ($K_m$, $V_{max}$, $K_{cat}$); and (iii) the overall efficacy of the immobilized system to remove glue (Ferrari et al. 2017; Mazzoli et al. 2018b).

*Oil-Stain Removal from Historical Fabrics* Oil-stains on ancient fabrics mainly consist of unsaturated fatty acids that can undergo oxidation on the double bonds. During time, environmental oxygen can attack these double bonds generating oxidation products that can polymerize thus creating a network of molecules difficult to be removed (Blüher et al. 1997). In this case, the application of lipases (especially those derived from *Candida cylindracea*) can efficiently solve the problem without any damage on the textile (Ahmed et al. 2010).

# 4 Case Study: Adhesive-Removal by Enzymatic Approach

Gellan hydrogel-immobilized α-amylases have been originally developed for removing starch paste from ancient paper documents (Mazzuca et al. 2014). A gellan-immobilized bacterial α-amylase (Fig. 9.4) has recently been used to clean a wool shroud, dating back to the Coptic period, from starch glue that had been used in the 1950s to temporary consolidate the textile (Ferrari et al. 2017). After selection of the most suitable enzyme among those commercially available, and optimization of the conditions for enzyme immobilization, the cleaning of the back of the two fragments (about 4 m$^2$ of textile) composing the tunic was completed in 160 h of work (Ferrari et al. 2017).

## 4.1 Description of the Coptic Tunic and Its State of Conservation

Since the end of the nineteenth century, the Egyptian Museum of Turin, Italy, has housed a wool-linen tunic (inventory number INV. S. 17490) that was probably a gift by the Egyptian Museum in Cairo to the Egyptian Museum of Turin.

As shown in Fig. 9.5, the tunic (hypothetically belonging to the fifth-century A. D.) was woven in a unique rectangular piece (with an opening at the center to allow the passage of the head) and folded in two at shoulder height. In the same Figure it is also possible to appreciate that the precious textile is abundantly decorated with squares (*tabulae*) and bands (*clavi*) woven with warps and wefts of different colors (green, purple, red, and yellow).

Unfortunately, some parts of the tunic (especially on the back side, Fig. 9.6a) are damaged, fragile, and fragmented. Recent investigations have suggested the idea that the tunic was not an everyday-life cloth but rather a shroud used in the Egyptian Byzantine Period for burying the bodies after abandon of the traditional mummification procedures corresponding to the spread of Christianity. This statement is based upon the evidence that organic deposits deriving from the dead body decomposition (Fig. 9.6b) largely contaminate the fabric (Ferrari et al. 2017).

Furthermore, other parts are stiffened and deformed, due to application of adhesives to contain cuts and fragmentation (Fig. 9.6c, d). During the 1950s, the Turin Egyptian Museum restorer Erminia Caudana extensively applied glue to both papyri and textiles to consolidate them, and it is probably because of these treatments that the wool shroud appeared rigid and hard in correspondence of previous fabric cuts.

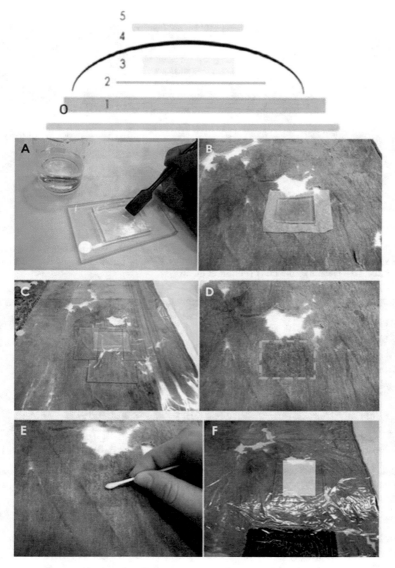

**Fig. 9.4** Immobilization of an α-amylase enzyme from *Bacillus* sp. on gellan. *Top*) Schematic representation of the enzyme poultice used. *0* Non-acidic blotter paper used to verify the amount of water released from the fabric during the cleaning procedure. *1* Coptic tunic. *2* Low basis weight (6 g/m²) Japanese paper used to prevent the gel from seeping into or migrate onto the textile. *3* Gellan-immobilized amylase tablet. *4* Melinex layer used to create a wet room to prevent water evaporation from the tablet. *5* Glass weighting to ensure uniformity of the gel-textile contact and the gradual release of water from the gel. (*Bottom*) Application phases of the enzymatic poultice to the back of the tunic (**a–f**) (Ferrari et al. 2017)

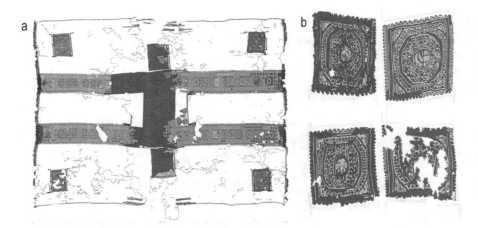

**Fig. 9.5** AutoCAD software reconstruction of the Egyptian Coptic wool-linen tunic belonging to the Egyptian Museum of Turin, Italy (inventory number INV. S. 17490) (**a**). Details of the *tabulae* of the tunic; (**b**). Details of the *clavi* (Ferrari et al. 2017)

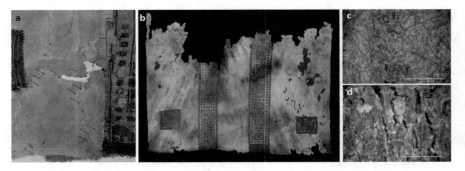

**Fig. 9.6** (**a, b**) Visible damages in the back side of the Coptic tunic of the Egyptian Museum of Turin (Italy); **c, d**) Details of the damages due to adhesive application on the fabric (modified from Ferrari et al. 2017)

Such structural damages rendered this archeological artifact not suitable for exhibition since the flexibility of the wool fibers proved to be decreased, making the textile fragile. Therefore, our research group set up a method to solve the problem.

## 4.2 Description of the Strategies to Fulfill the Objectives

First, since both protein-based and polysaccharide-based glues are employed to consolidate cuts in textile artifacts, it was necessary to establish the exact chemical nature of the adhesive to remove them by protease or glycosidase-mediated enzymatic hydrolysis, respectively. Secondarily, once established the subclass of hydrolase to be used, it was important to screen for the best-fitting enzyme, acting at the correct pH, ionic strength, and temperature. These are challenging factors when a high-value archeological textile has to be cleaned. Generally, plant and microbial enzymes are more suitable than animal enzymes that frequently display maximum activity at 37 °C, a too high temperature for textile integrity preservation. Moreover, a right immobilization approach on different matrices was to be set up, to avoid water flooding of the precious textile. Finally, before successful application of the cleaning procedure on the original Coptic shroud, a simulated sample was artificially aged and used as a test to prove the efficacy of the enzymatic treatment. The overall approach, based on bacterial α-amylases absorbed on gellan, allowed a satisfactory starch removal without any visible alteration of the fabric fibers thus resulting in a successful biorestoration of the precious artifact in study.

## 4.3    Step 1: Establishing the Nature of the Glue (Lugol Test and FT-IR)

The back of the shroud, as referred in a previous section, displayed significant damages consistent with precipitated opaque material, ascribable to a white adhesive (Fig. 9.6 c, d) because of an old restoration performed with the aim to consolidate cuts. Attempts to bio-clean the fabric with the traditional system of wet cleaning, namely vaporization-atomization, proved to be not suitable to remove the glue. In the perspective to use an enzymatic approach for removing the adhesive (Schwarz et al. 1999; Beutel et al. 2002), establishing the chemical composition of the glue was a priority. Analyses aimed to distinguish between vegetal (starch) and animal (collagen) glue. To this aim, a fabric sample of about $2 \times 5$ cm$^2$ was excised from the Coptic tunic. Based on the assumption that textile restorers most commonly use starch glue (De La Chapelle et al. 1994) a Lugol test, which employs a iodine solution (0.1% w/v KI, 1% w/v I$_2$), was firstly performed. If starch is present, the sample turns dark blue. Actually, we detected a dark-blue stain on the paste thus revealing the presence of starch on the wool textile fiber (data not shown). In parallel, a Fourier transform infrared spectroscopy (FT-IR) evaluation on a very small area of the paste-damaged textile was set up, using a FT-IR Bruker Vertex 70 spectrophotometer coupled with an infrared microscope (Bruker Hyperion 3000) at a resolution of 4 cm$^{-1}$ and 64 scans. The vibrational bands in the infrared spectra refer to different functional groups of a sample and help a general characterization of a material and, in some cases, the identification of specific compounds. This analysis furtherly confirmed that the glue was starch, by simply comparing these spectra with FT-IR spectra of standard compounds (Fig. 9.7).

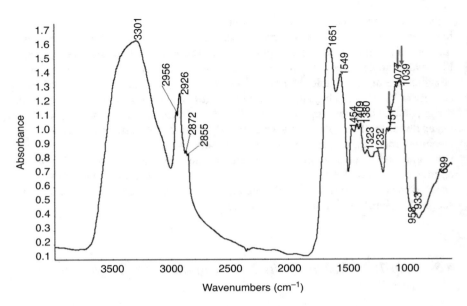

**Fig. 9.7** Fourier Transform Infrared Spectroscopy Analysis of the glue reveals the starch nature of the adhesive found on the Coptic tunic of the Egyptian Museum of Turin (Italy) (Ferrari et al. 2017)

## 4.4  Step 2: Selecting the Most Suitable Enzyme and Immobilization Strategy

Once established the starch nature of the adhesive, amylases proved to be the most suitable hydrolytic enzymes for bio-cleaning. Starch consists of two α-glucans, *i.e.* amylose (linear glucose polymer) and amylopectin (branched-chain glucose polymer), but amylopectin is generally considered the main responsible for paste-features. Both α- and β-amylases exist which hydrolyze amylose and amylopectin through different catalytic mechanisms. α-Amylases hydrolyze internal α-1-4 glycosidic bonds at random generating dextrins. β-Amylases are exoglucanases that cleave amylose and amylopectin starting from their non-reducing terminals and release maltose (Ray and Nanda 1996). Choice was addressed to microbial α-amylases because they are maltodextrin-generating endo-enzymes able to quickly eliminate the adhesive properties of starch (Schwarz et al. 1999) whereas β-amylases display a slower catalytic mechanism (Ray and Nanda 1996). Furthermore, prokaryote-derived enzymes display intrinsic higher stability in a wider pH and temperature range than animal-derived amylases (Mehta and Satyanarayana 2016). Among commercial α-amylases, that from *Bacillus* sp. was selected for starch glue removal, owing to its very high specific activity (Chand et al. 2014). More in details, lyophilized α-amylase from *Bacillus* sp. Type IIA (Sigma-Aldrich) with specific activity >1500 units/mg protein was finally chosen. A 0.1% w/v solution of the enzyme in 20 mM sodium-phosphate buffer pH 6.9 was prepared (De La Chapelle et al. 1994; Cremonesi 2009) for tunic restoration. It is worth noting that pH 6.9 corresponds to the optimal pH for *Bacillus* sp. Type IIA alpha-amylase activity.

However, as referred above, the use of water solutions of enzymes is not suitable for precious archeological artifacts since this treatment can flood the textile with a too high water load, favoring successive microbial contamination. Therefore, we designed an enzyme immobilization approach to restrict the amount of water needed for catalysis. Two immobilization matrices were tested: agar and gellan. Both are polysaccharides displaying high pH stability and highly hygroscopic properties that limit liquid transfer to the fibers, allowing a gradual release of the enzyme into the fabric. Furthermore, their un-adhesive nature and facility to form films render them suitable for application on textiles (Iannuccelli and Sotgiu 2007). Hence, one cm thick low-viscosity "soft" gels (2% w/v) agar (polygalactose and sulfurpolygalactose, Bresciani SRL) and 2% w/v Gellan gum (Phytagel, heteropolysaccharide, Bresciani SRL) in 20 mM sodium-phosphate buffer pH 6.9 were used for this purpose. Since a very good efficiency in glue removal can be obtained by adsorbing enzyme solutions at the interface between textile and support, the solution of α-amylase from *Bacillus* sp. was brush and immobilized onto the solid gel surfaces so as to have the maximum enzyme concentration at the interface and, hence, maximize efficiency of glue removal (Fig. 9.4a). However, the optimum catalytic activity for *Bacillus* sp. α-amylases is 60 °C and this temperature cannot be reached because of the risk of gel melting at temperatures higher than 40 °C. The overall enzyme immobilization procedure is illustrated in Fig. 9.4.

## 4.5   Step 3: Simulated Sample Preparation, Aging, and Damaging

To prevent any risk of damage to the original archeological tunic, a goat wool simulated sample, with similar chemical-physical properties to the original Coptic shroud fabric, was woven to compare the enzymatic cleaning and the cleaning effect of the immobilization gels alone, prior to application of the system on the real artwork. To have a simulated sample as similar as possible to the original textile, also aging should be considered. Reproducing the original aging of artifacts is a challenging procedure. In the present investigation, we tested an incubation temperature of 90 °C (maximum allowed for wool) (Quaglierini 1985) for 60 days to induce accelerated artificial aging (it has previously been established that 25 years of natural aging can be obtained by 3 days at 120 °C in silk) (Ahmed and Kolisis 2011). Dry incubation was chosen and applied here because this condition attempted to reproduce the climatic conditions characterizing the ground where the Coptic shroud was buried in Egypt. However, some events occurring in the natural aging process (contamination by burial ground and dead body decomposition liquids) cannot be reproduced. After this time, a treatment with starch glue was performed to mimic the previous inappropriate consolidation. Starch can be obtained from different plants, which cannot be distinguished by conventional diagnostic investigations. However, the use of barley, corn, oat, or rye starches in the fifties in Italy (where the restoration was made) is unlikely since these cereals were not of common use. The choice was then restricted to potato, rice, and wheat starches. For glue preparation, 3 parts of each starch were dissolved in 1 part of hot water so as to obtain a thick paste. The three glues were then applied to the simulated sample and subjected to three day-"aging" at 50 °C to further allow glue thickening, aging, and stiffening.

## 4.6   Step 4: Glue Removal Test by Immobilized α-Amylase on a Simulated Sample

After simulated sample preparation, the efficacy of agar and gellan alone was compared to the one of gellan- and agar-immobilized α-amylase in bio-cleaning the simulated samples. While the effect of gellan and agar alone was negligible, both macroscopic and microscopic evaluations have demonstrated that only after 15 min, enzymatic treatment (6 × 6 cm cold sheets) was successful in the removal of the starch glue from the simulated sample (Fig. 9.8). Furthermore, starch removal was also proved by performing the Lugol test after the enzyme application (data not shown). The liquefied glue residue was easy-to-remove and, after treatment, the wool sample was clearly more soft and flexible. Since this procedure did not cause

Fig. 9.8 Different cleaning strategies applied to simulated textile sample (Ferrari et al. 2017)

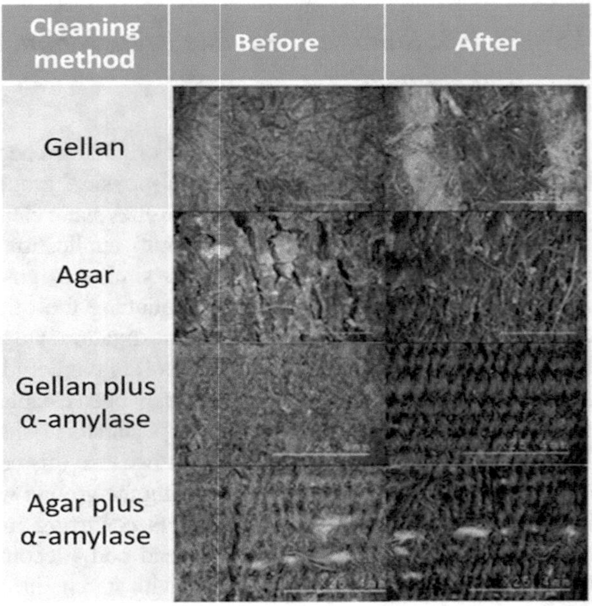

any damage to the simulated textile, it was suitable to be employed on the original Coptic tunic. However, the gellan matrix was better performing than agar also for monitoring adhesive paste removal since it is more transparent. Furthermore, the lower rigidity of gellan with respect to agar allowed a better adhesion to the discontinuity of the textile fibers. For these reasons, application of gellan-immobilized α-amylase on the original Coptic shroud was preferred.

## 4.7   Step 5: Glue Removal by Immobilized α-Amylase on the Original Archeological Shroud

Finally, the same enzymatic treatment described in the preliminary test on the simulated sample was performed on the archeological Coptic tunic. α-Amylase, immobilized on gellan sheets (that were those showing the best performances) was then placed on a thin Japanese paper sheet to avoid any seeping of the gel into the fabric (Fig. 9.4a, b). Afterwards, a "wet room" (Fig. 9.4c) was created by superimposing a Melinex sheet and a glass weight to prevent dehydration.

As shown in Fig. 9.4 e, f, 15 min were sufficient to swollen and liquefy the glue. It worth remembering that such a high efficiency was obtained although the temperature used during bio-cleaning procedure was lower than that for amylase optimum activity (60 °C) (Schwarz et al. 1999). The adhesive residue was then removed by means of a moistened cotton swab and a spatula (Fig. 9.4e). No more adhesive paste was detectable on the fabric (dried by using an absorbent paper (Fig. 9.4f), that also displayed improved elasticity and flexibility. After this bio-cleaning treatment, the needle consolidation of the textile was finally performed. The overall bio-cleaning of both sides of the textile (about 4 m$^2$) was achieved in 160 h of work.

Other advantages of immobilized amylases are the very quick time of application and the overall milder mechanical treatment required as compared to traditional cleaning methods (*i.e.* wet cleaning). It is useful mentioning that α-amylases have been already used to restore wool/linen tapestry after suitable immobilization on methylcellulose matrices (Whaap 2007) but also (both in solution and as a poultice)

**Fig. 9.9** The restored Coptic tunic-shroud exhibited at the Egyptian Museum of Turin, Italy (Ferrari et al. 2017)

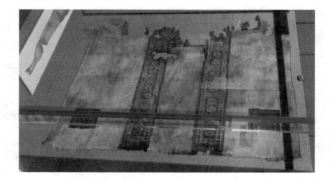

for cleaning cotton and even silk textiles from starch (Ahmed and Kolisis 2011). Furthermore, α-amylases are largely used for fabric desizing (Chand et al. 2012).

In conclusion, the specificity, mildness, and efficacy of the enzymatic treatment render this strategy especially suitable for precious archeological textiles. The Coptic shroud, "hidden" from the community for a long time is currently exhibited at the Egyptian Museum of Turin (Italy) (Fig. 9.9) without any damage to its historical value.

# 5 Conclusion

This report demonstrates that textile artifacts are fragile and undergo aging and different kinds of deterioration. However, it is possible to face damages, such as those due to wrong restoration interventions, by using purified enzymes, suitably immobilized, to contain the risk of employing aqueous solutions. Among the wide range of commercially available enzymes, those displaying a good catalytic activity at low temperatures (lower than 30 °C) are very promising since they can be applied also to fragile items. The microbial world seems to be a good candidate to be explored for finding such enzymes. This field is also promising to find future solutions to contain microbial deterioration (Mazzoli et al. 2018b) avoiding the use of acids, solvents, and surfactants (dangerous for the artworks). For the latter purpose, the use of enzyme- or bacteriocin-mediated bacterial competition seems of particular interest. Interdisciplinary approaches and collaborations between art conservators and biotechnologists, biochemists, and microbiologists are the essential requisite to preserve objects that state the immense creativity of artists and the high value of human history.

**Acknowledgements** The restoration of the Egyptian shroud has been the Master thesis work of Martina Ferrari that we would like to thank. We are also indebted with the researchers of the Restoration School of Venaria Reale, Turin, Italy, which have contributed with their respective expertizes to the achievement of the experimental part concerning adhesive characterization and sample bio-cleaning.

# References

Abdel-Kareem OMA (2005) The long-term effect of selected conservation materials used in the treatment of museum artefacts on some properties of textiles. Polym Degrad Stab 87:121–130
Ahmed HE, Kolisis FN (2011) An investigation into the removal of starch paste adhesives from historical textiles by using the enzyme α-amylase. J Cult Herit 12:169–179
Ahmed HE, Gremos SS, Kolisis FN (2010) Enzymatic removal of the oily dirt from a Coptic tunic using the enzyme lipase. J Text Apparel Technol Manag 6:1–17
Barbabietola N, Tasso F, Alisi C, Marconi P, Perito B, Pasquariello G, Sprocati AR (2016) A safe microbe-based procedure for a gentle removal of aged animal glues from ancient paper. Int Biodeter Biodegr 109:53–60

Beutel S, Klein K, Knobbe G, Königfeld P, Petersen K, Ulber R, Scheper T (2002) Controlled enzymatic removal of damaging casein layers on medieval wall paintings. Biotechnol Bioeng 80:13–21

Blüher A, Haller U, Banik G, Thobois E (1995) The application of carbopol poultices on paper objects. Restaur Int J Preserv Libr Arch Mater 16:234–247

Blüher A, Grube A, Bornscheuer U, Banik G (1997) A reappraisal of the enzyme lipase for removing drying-oil stains on paper. Pap Conserv J Inst Pap Conserv 1997:37–46

Bott G (1990) Amylase for starch removal from a set of 17th century embroidered panels. Conserv 14:23–29

Chand N, Nateri AS, Sajedi RH, Mahdavi A, Rassa M (2012) Enzymatic desizing of cotton fabric using a Ca 2+-independent α-amylase with acidic pH profile. J Mol Catal B Enzym 83:46–50

Chand N, Sajedi RH, Nateri AS, Khajeh K, Rassa M (2014) Fermentative desizing of cotton fabric using an α-amylase-producing Bacillus strain: Optimization of simultaneous enzyme production and desizing. Process Biochem 49:1884–1888

Chapman V (1986) Amylase in a viscous medium—textile applications. The Conservator 10:7–11

Ciatti M, Conti S, Fineschi C, Nelson JK, Pini S (2010) Il ricamo in or nué su disegno di Raffaellino del Garbo. Aspetti storico-stilistici, tecnici, minimo intervento e conservazione preventiva. OPD Restauro 22:81–116

Cremonesi P (2009) L'uso degli enzimi nella pulitura di opere policrome. Il Prato, Padova, Italy

De La Chapelle A, Choisy F, Gallo M, Legoy M (1994) Emploi d'amylases et de protéases pour la restauration d'arts graphiques. In: Environnement et Conservation de l'Ecrit, de l'Image et du Son: Actes des Deuxièmes Journées Internationales d'Etudes de l'ARSAG, 16 au 20 mai 1994. Association pour la recherche scientifique sur les arts graphiques, Paris, France, pp 217–229

Ferrari M, Mazzoli R, Morales S, Fedi M, Liccioli L, Piccirillo A, Cavaleri T, Oliva C, Gallo P, Borla M, Cardinali M, Pessione E (2017) Enzymatic laundry for old clothes: immobilized alpha-amylase from Bacillus sp. for the biocleaning of an ancient Coptic tunic. Appl Microbiol Biotechnol 101:7041–7052

Germinario G, van der Werf ID, Palazzo G, Regidor Ros JL, Montes-Estelles RM, Sabbatini L (2017) Bioremoval of marker pen inks by exploiting lipase hydrolysis. Prog Org Coatings 110:162–171

Gostling K (1989) Bookbinders and adhesives: part 2. New Bookbind 9:30–39

Gutarowska B, Pietrzak K, Machnowski W, Milczarek JM (2017) Historical textiles - a review of microbial deterioration analysis and disinfection methods. Text Res J 87:2388–2406

Hrdlickova Kuckova S, Crhova Krizkova M, Pereira CLC, Hynek R, Lavrova O, Busani T, Branco LC, Sandu ICA (2014) Assessment of green cleaning effectiveness on polychrome surfaces by MALDI-TOF mass spectrometry and microscopic imaging. Microsc Res Technol 77:574–585

Iannuccelli S, Sotgiu S (2007) La pulitura ad umido di opere d'arte su carta con gel rigidi di gellano: presupposti teorici, metodologia, applicazioni e verifica analitica. In: Quaderni del Cesmar7, Vol. 11. Il Prato, Padova, Italy

Mazzoli R, Giuffrida MG, Pessione E (2018a) Back to the past: "find the guilty bug—microorganisms involved in the biodeterioration of archeological and historical artifacts". Appl Microbiol Biotechnol 102:6393–6407

Mazzoli R, Giuffrida MG, Pessione E (2018b) Back to the past—forever young: cutting-edge biochemical and microbiological tools for cultural heritage conservation. Appl Microbiol Biotechnol 102:6815–6825

Mazzuca C, Micheli L, Cervelli E, Basoli F, Cencetti C, Coviello T, Iannuccelli S, Sotgiu S, Palleschi A (2014) Cleaning of paper artworks: development of an efficient gel-based material able to remove starch paste. ACS Appl Mater Interfaces 6:16519–16528

Mehta D, Satyanarayana T (2016) Bacterial and archaeal α-amylases: diversity and amelioration of the desirable characteristics for industrial applications. Front Microbiol 7:1129

Nicholas CH, Pailthorpe MT (1976) 50—primary reactions in the photoyellowing of wool keratin. J Text Inst 67:397–403

Pietrzak K, Puchalski M, Otlewska A, Wrzosek H, Guiamet P, Piotrowska M, Gutarowska B (2017) Microbial diversity of pre-Columbian archaeological textiles and the effect of silver nanoparticles misting disinfection. J Cult Herit 23:138–147

Quaglierini C (1985) Manuale di merceologia tessile. Zanichelli, Bologna, Italy

Ray RR, Nanda G (1996) Microbial β-amylases: biosynthesis, characteristics, and industrial applications. Crit Rev Microbiol 22:181–199

Rowe L, Howard GT (2002) Growth of *Bacillus subtilis* on polyurethane and the purification and characterization of a polyurethanase-lipase enzyme. Int Biodeter Biodegr 50:33–40

Rubeziene V, Varnaite S, Baltusnikaite J, Padleckiene I (2012) Effects of light exposure on textile durability. In: Annis PA (ed) Understanding and improving the durability of textiles. Woodhead Publishing, Oxford, pp 104–125

Schwarz I, Blüher A, Banik G, Thobois E, Maurer KH (1999) The development of a ready-for-use poultice for local removal of starch paste by enzymatic action. Restaurator 20:225–244

Seal KJ (1988) The biodegradation of naturally occurring and synthetic plastic polymers. Biodeter Abstr 2:296–317

Seves A, Romanò M, Maifreni T, Sora S, Ciferri O (1998) The microbial degradation of silk: a laboratory investigation. Int Biodeter Biodegr 42:203–211

Shibayama N, Eastop D (1996) Removal of flour paste residues from a painted banner with alpha-amylase. The Conservator 20:53–64

Shubhra QTH, Alam AKMM, Beg MDH (2011) Mechanical and degradation characteristics of natural silk fiber reinforced gelatin composites. Mater Lett 65:333–336

Sterflinger K, Pinzari F (2012) The revenge of time: fungal deterioration of cultural heritage with particular reference to books, paper and parchment. Environ Microbiol 14:559–566

Szostak-Kotowa J (2004) Biodeterioration of textiles. Int Biodeter Biodegr 53:165–170

Timár-Balázsy Á, Eastop D (1998) Chemical principles of textile conservation. Routledge, Taylor and Francis Group, London and New York

Treigiené R, Musnickas J (2003) Solvent pre-treated wool fabric permanent set and physical properties. Fibres Text East Eur 11:37–40

Webster A, May E (2006) Bioremediation of weathered-building stone surfaces. Trends Biotechnol 24:255–260

Whaap F (2007) The treatment of two Coptic tapestry fragments. V A Conserv J 55:11–13

Zimmermann M, Hocker H (1996) Typical fracture appearance of broken wool fibers after simulated sunlight irradiation. Text Res J 66:657–660

# Chapter 10
# Advanced Biocleaning System for Historical Wall Paintings

**Giancarlo Ranalli and Elisabetta Zanardini**

**Abstract** This chapter will focus on the potential role of safe microorganisms as biocleaning agents in the removal of altered or undesirable organic substances on historical wall paintings. Selected microbes can be adopted as biological cleaners to reduce and remove deterioration ageing phenomena, environmental pollutants and altered by-products of past intervention of restorations. The aim is to offer a comprehensive view on the role and potentiality of virtuous microorganisms pro-biocleaning of altered historical wall paintings. We also report four case studies in the CH restoration field, carried out in the last 25 years, with the innovative use of bacteria and different delivery systems, focusing the attention on the preliminary diagnosis and the monitoring of the whole process. The CH field represents a great challenge and Science and Art link together the work of conservator scientists and historians with researchers and scientists, sharing their diverse expertises and joining the knowledges to the preservation and the conservation of our artistic patrimony.

**Keywords** Conservation-Restoration · Cultural heritage · Biotechnology · Biocleaning · Organic matters · Bacteria

Cultural Heritage (CH) artworks are one of the most important elements of the identity in large part of the World, especially in the old European Continent and represent the cultural legacy from past generations defined as *tangible* (buildings, monuments, books, etc.) and *intangible* (as folklore, traditions, language, knowledge and natural as landscapes, and flora-fauna biodiversity).

The knowledge of our history has been subject to a continuous evolution over the time; that is true if we consider the modern potentialities of communications and the mobility of scientists and conservation specialists across the World. However, the

G. Ranalli (✉)
Department of Bioscience and Territory, University of Molise, Pesche, Italy
e-mail: ranalli@unimol.it

E. Zanardini
Department of Sciences and High Technology, University of Insubria, Como, Italy

© The Author(s) 2021                                                                 217
E. Joseph (ed.), *Microorganisms in the Deterioration and Preservation of Cultural Heritage*, https://doi.org/10.1007/978-3-030-69411-1_10

knowledge in the conservation field, being relatively young, remains difficult both for the differences in terminology and for the lacks in the standardization of procedures, especially considering national-international multidisciplinary sectors (trades, professionals, administrative authorities, academic lectures, conservation students, workers involved in built environment, etc.).

Three examples are here cited: the terms *restoration*, *wall painting* and *fresco*. The first still linked to possible negative aspects originated by the restoration-reconstruction methods applied in the past centuries. The second defines *wall paintings* as paintings on inorganic substrates (plaster, stone), but in other cases, with *painting on the wall* are included paintings on wooden wall (EWAGLOS 2016). The third, the term *fresco* is limited to paintings on fresh lime plaster (Italy), but used in France for all kinds of wall paintings, including combination of fresco and secco techniques.

Artworks related to CH are exposed to weather and submitted to the influence of the environmental conditions: physicals, chemicals and biological factors interact with the constitutive materials, inducing changes both in its compositional and structural characteristics. In some cases the material transformation is due to the metabolic activity associated with the growth and the activity of living organisms, from microorganisms to plants and animals.

In the last decades, an important role of CH conservation has been mostly played by physics, chemistry and material science, but actually, the scientific scenery also includes biotechnology and applied microbiology, which can contribute to the development of new methods for the detection and identification of microorganisms altering stoneworks and for the bioremediation of weathered artworks (Fernandes 2006). In the last twenty years, the advanced methodologies have contributed to the knowledge of the structure (species composition and abundance) of the microbial communities colonizing artworks and their biodeterioration potentialities (Daffonchio et al. 2000; Gurtner et al. 2000; Piñar et al. 2001; Urzì et al. 2001; Crispim and Gaylarde 2005; McNamara and Mitchell 2005; Gorbushina 2007; Abdulla et al. 2008; Portillo et al. 2009; Ranalli et al. 2009; Scheerer et al. 2009; May 2010; Alonso-Vega et al. 2011; Giacomucci et al. 2011; Saiz-Jimenez et al. 2011; Zanardini et al. 2011, 2016, 2019; Cappitelli et al. 2012).

The biodeterioration mechanism and the role played by several biological agents "biodeteriogens" in the decay of stone monuments are well known. Different kinds of organisms (primary colonization) can grow using the stone mineral components and its superficial deposits. The main consequence of their metabolic activity, such as the excretion of enzymes, inorganic and organic acids and of complex forming substances, is the dissolution of minerals of the substratum. Moreover, the growth and the swelling of some vegetative structures (e.g. roots and lichenic thalli) induce physical stresses and mechanical breaks. The growth process and vegetative development of the living species of organisms can interfere on the conservation of CHs. More, the intensity of these damages is strictly correlated with: type and dimension of the organism involved; kind of material and state of its conservation; environmental conditions, micro-climatic exposure; level and types of air pollutants (Tiano 2002).

In the last decades, the traditional microbiological culture-dependent methods based on the use of specific media for the cultivation and the identification of microorganisms have been coupled to the culture-independent methods. These allow a better understanding of the complexity of the microbial communities altering stoneworks, the detection of novel microorganisms (bacteria, fungi, algae, etc.) and their activity (Schabereiter-Gurtner et al. 2001; Portillo et al. 2009; Saiz-Jimenez et al. 2011; Villa et al. 2015; Zanardini et al. 2016, 2019).

Moreover, the molecular methods can provide an early detection of the biodeterioration process, even when the deterioration is not visible yet. So, the use of molecular techniques for a preventive diagnosis and in the conservation perspectives becomes necessary both for the scientific approach and the economical point of view. These new techniques combined with the traditional microbiological methods give a much better understanding of the number, activity, function of isolated or aggregate microorganisms, not only in the "barbarian" biodeterioration, but also in bio-restoration and bio-conservation process, for their "virtuosos" functions.

In addition, these new approaches can be useful to set up alternative biocleaning restoration methodologies and to control during the treatment the microbial growth and activity and after the treatment to prevent further colonizations (Palla 2004, 2013).

# 1   Biotechnologies Applied to CH

For these reasons, the use of biological systems, as living organisms or derivatives (e.g. enzymes), goes under the name of biotechnology in many fields including the recovery of CH artwork.

In natural environments (air, water, soil) microorganisms contribute to very important aspect of life in the Earth: the equilibrium of the "matter transformation cycle" or "carbon cycle". Some examples are the transformation of rocks in soils (pedogenesis) or the transformation of the complex substances (polymers) into simple components (monomers) until the complete degradation (mineralization). As we mentioned above, micro- and macroorganisms, ranging from microscopical bacterial cells to higher plants and animals (eukaryotes), can find as suitable habitats for their growth, the CH surfaces such as ancient archaeological remains, monumental buildings, works of arts made with different inorganic and organic materials (stone, wall paintings, metals, wood, textiles, paper, etc.).

In conservation and restoration practices, different procedures can be used and some of them require the use and application of substances (i.e. animal and vegetal glues, re-lining paste, varnishes, temperas and other materials for cleaning and soaking). These substances can contribute to the biological risk being nutrients for the microbial growth especially when the environmental conditions are favourable (high water content, air humidity, temperature, etc.).

Indeed, when in past restoration, organic matter has been applied to stone surfaces for various purposes, often stoneworks show alteration over time and, in these cases, their preservation poses serious questions (Alfano et al. 2011; Ranalli et al. 1996, 2005). The residual compounds, under certain environmental conditions, can act as an adequate growth substrate for microorganisms, as bacteria and fungi and they can deteriorate surfaces (Ranalli et al. 1997; Tiano et al. 2006).

Moreover all outdoor artworks surfaces are exposed to atmospheric inorganic and organic pollutants giving the formation of surface deposits and alterations (patinas) and these materials over the time tend to accumulate and interact with the stone surfaces.

Restoration practices, aiming the cleaning of altered surfaces, need to previously define and characterize the organic and inorganic components, the mineralogical properties and the level and extend of decay in order to understand the causes and mechanisms of the processes actually in place. In particular, CH wall paintings preservation and restoration can be complex and depends on the nature, state of conservation, environmental conditions (indoor and outdoor); therefore, an interdisciplinary technical and scientific approach is imprescindible.

Based on the chemical characterization results and their distribution on the artwork surfaces, the cleaning methods should be selective and specific in order to remove the altered materials and to respect the original materials avoiding irreversible damages (Cremonesi 2004).

During the last 20 years, the biological cleaning seems to satisfy this need of accuracy, since both cultural-dependent and molecular biology methods allow the selection of microbial strains with specific metabolic activities and safe to use. For the removal of patinas and deposits, which are the main form of alteration present in wall paintings, two approaches have been used until today: the application of purified enzymes and the use of viable microbial cells (especially bacterial strains), chosen for their selective action towards specific substrates.

Microorganisms are extremely versatile and show great potentialities to induce chemical transformations both using different substrates as energy and carbon sources and producing a large variety of enzymes. The enzymes specifically accelerate the chemical reactions, working with high selectivity; for example, amylases, proteases and lipases are widely used in CH field, for their ability to hydrolyze substrates such as starch, casein, animal glues, wax, oil, synthetic resins frequently used in past restoration practices.

In 1970, trypsin was used by Wendelbo, for the first time, for the detachment of book pages stuck with animal glue (Wendelbo 1976). Later, amylases and proteases were used by Segal and Cooper (1977) and Makes (1988) for the removal of animal glue from parchment and for the treatment of painted canvas, respectively.

The use of enzymes, pure or in mixture, has been reported for the cleaning of mural paintings, sculptures, paper, etc., evidencing the importance of the type of enzyme, operative conditions (pH, temperature, etc.) and the way of applications (Cremonesi 1999; Bonomi 1994; Wolbers 2000; Banik 2003; De la Chapelle 2003; Iannuccelli and Sotgiu 2009).

The use of the enzymes in CH requires the correct and deep detection of the materials to remove; in fact, proteases can work for the removal of proteic substances (e.g. casein and egg, animal glues and gelatines), amylases for starch and gum, and lipases for oils, waxes and fat.

If the substances to remove are complex, it can be useful to use a mixture of enzymes; so it necessary to individualize, case by case if the best choice is the use of purified enzymes or a crude mixture.

PH and temperature are important factors for the optimal enzyme activity.

In the CH restoration field, usually the enzymes are used at an optimal temperature ranging 30–37 °C, offer the possibility of a soft heating of the artworks surface, but only after a previous adequate evaluation. The pH values normally range between 4 and 9, excluding, for example, some proteolytic enzymes that require an optimal pH around 1 and obviously difficult to use.

Another important aspect is the application of the enzyme and the maintenance of the best conditions for its optimal activity, for example, the use of buffer solutions able to keep the pH stable and to guarantee the activity of purified enzymes, but the addition of some additives and/or gelling reagents could induce reduction in the effectiveness of the enzyme (Bellucci and Cremonesi 1994).

The same authors highlighted the importance of evaluating the economic impact of the use of the enzymes in the CH field, i.e. the cost depends on the grade of purity and the quantity needed. The formation of operators is also important being the use of the enzymes still not so widely accessible and optimized.

Among the biocleaning procedures, the innovative approach based on the use of viable bacterial cells has been proposed by Ranalli et al. (2005), especially when the traditional methods (solvents) cannot work for the removal of weathered and insoluble compounds.

Viable bacterial cells can produce both constitutive and inducible enzymes able to attack and to degrade several types of undesired organic substances.

The versatility and efficiency of specific groups of bacteria are well known since they are able to degrade a wide range of substrates resulting more effective compared to the use of a single enzyme that can work only specifically in the attack of certain substrates (Ranalli et al. 2005).

In the following section, different case studies are illustrated showing how in the last 20 years the use of viable bacterial cells has become an imprescindible step in the restoration practices aimed to the recover altered ancient frescoes (Table 10.1).

## 2   Case Study of the Conversion of St. Efisio and Battle (Conversione di San Efisio e Battaglia), Pisa, Italy

The restoration of one of the most important Pisa Monumental Cemetery's frescoes, *Conversione di S. Efisio e Battaglia*, a fourteenth century fresco at Monumental Cemetery at Pisa, Italy, painted by Spinello Aretino was the first pioneeristic study based on the synergic combination of bacterial metabolism with the hydrolytic

**Table 10.1** Main decay agents on organic materials (wall paintings, manuscript, others), micro-organisms and delivery systems adopted

| Decay agents | Artworks, Materials, Site | Biocleaning bacteria | Delivery systems | References |
|---|---|---|---|---|
| Animal glue | Camposanto Monumentale Pisa Frescoes, Pisa | *Pseudomonas stutzeri A29* (Ae) | *Ex-situ* Cotton Wool | Ranalli et al. (2005) Antonioli et al. (2005) Lustrato et al. (2012) |
| Proteins and inorganic compounds | Casina Farnese wall paintings (Palatine Hill, Rome) | *Stenotrophomonas maltophilia* (Ae) *Pseudomonas koreensis* (Ae) | *On-site* Laponite micro-pack | Mazzoni et al. (2014) |
| Animal glue and salt efflorescence | St. Nicholas frescoes, Valencia | *P. stutzeri* DSMZ 5190 (Ae) | *On-site* Cotton wool Agar | Bosch-Roig et al. (2010) Bosch-Roig et al. (2012, 2013) |
| Animal glue | Ancient paper | *Ochrobactrum* sp. TNS15 | Agar | Barbabietola (2012), Barbabietola et al. (2012) |
| Animal glue and efflorescence | Archaeological frescoes, ceramic and bones | *P. stutzeri* DSMZ 5190 (Ae) | Agar | Martín Ortega (2015) |
| Black organic layer (mainly lichens) | Concrete | *Thiobacillus* sp. (Ae) | Immersion | De Graef et al. (2005) De Belie et al. (2005) |
| Graffiti paintings | Glass slides Stone | *D. desulfuricans* ATCC 13541 and others (An) *P. stutzeri* DSMZ 5190 *(Ae)* | Immersion Cotton, Agar | Giacomucci et al. (2011) Sanmartín and Bosch-Roig (2019) |

*Key:* (Ae) Aerobic metabolism; (An) Anaerobic metabolism

enzymatic action for the restoration of medieval frescoes mural paintings (Ranalli et al. 2005).

The Camposanto Monumental Cemetery frescoes, a surface area of about 1500 m², were extremely damaged by a bomb in 1944 during World War II and a large number of frescoes were rapidly removed from the original walls under those extremely dangerous conditions.

As extreme intervention, the frescoes were detached using the "strappo" or "tear-off" technique; this procedure included the use of gauze and a layer of broth warm animal glue, followed by mechanical detachment by the walls. Then, by rolling them up without adding any rigid support, they were stored in deposits for about 50 years. The recovery of some frescoes became a challenge for the restorators as a consequence of humidity, atmospheric pollution and application of formaldehyde solutions to avoid microbial contamination (Antonioli et al. 2005). Most of them were also treated at the back with animal glue and casein as adhesives to the support.

After many years, the glue resulted polymerized and weathered and resistant to solubilizing agents and/or surfactants and the use of a mixture of the most performant proteolytic enzymes did not show significant effects.

In order to solve these problems and to identify suitable restoration practices and methodology to recover the frescoes, the approach required specific analyses to individualize the organic substances originally applied, as well as with those apported later in the restoration works.

For this aim, the use of sensitive, selective, and, if possible, non-invasive techniques that are minimally destructive, such as calorimetry, gas chromatography coupled with mass spectrometry (GC/MS) and pyrolysis/GC-MS, are suggested (Stassi et al. 1998; Colombini et al. 1999, 2002, 2003; Bonaduce and Andreotti 2009).

Once identified the substances to be removed, we decided that the use of viable cells could be the only way to restore the ancient frescoes.

The selection of microbial strains was a very important step, because the success of a "biocleaning" treatment depends on the efficiency of the used microorganisms. Both culture-dependent and -independent methods can help in the selection of the most effective microbial strain.

In our case, bacteria belonging to the genus *Pseudomonas*, Gram negative, asporigen, ubiquitary in nature and with extensive metabolic versatility, able in fact of using a large variety of different compounds as source of carbon and energy. Among the five species tested in vitro, *Pseudomonas stutzeri* A29 strain was selected because is not pathogenic to man and the environment and capable to grow using animal glue when supplied with it as sole organic carbon.

A direct application was carried out onto the artwork surface using viable *P. stutzeri* A29 bacterial cells in broth culture suspension ($10^6$–$10^7$ UFC/ml) embedded in sterile cotton wool, for 10–12 h at 30 °C temperature (Ranalli et al. 2005; Antonioli et al. 2005; Cappitelli et al. 2005, 2006, 2007). The costs of the bacterial suspension were quantified in about 50 euro/litre, and 100 euro/m$^2$ of fresco surface. At the end, a commercial Protease Type XIX was adopted to eliminate residues of glue on the surface.

## 3   Case Study of Stories of the Holy Fathers (Storie dei Santi Padri), Pisa, Italy

A second pioneeristic protocol based on the synergic combination of bacterial metabolism with the hydrolytic enzymatic action was performed for the restoration of medieval frescoes "*Stories of the Holy Fathers*", (size 6.10 15.65 m), painted by Buonamico Buffalmacco, in the fourteenth century (Lustrato et al. 2012).

In addition, in this case, the fresco damaged by a bomb during World War II was quickly removed from the original walls under extremely dangerous conditions.

In the 1960s, during the first restoration the fresco was cut into 17 sections (ranging from 1.3–2.0 m to 3.5–2.0 m) and attached onto asbestos-cement supports. The gauze and a large portion of the animal glue layer on the surface of the fresco were then removed using both traditional chemical and physical techniques, based on the application of ammonium carbonate solution and organic solvents.

Afterward the sections were recomposed, like a puzzle, on a metal frame at a distance of 5.0 cm from the wall in order to guarantee a better air circulation at the back of the fresco and to reduce the risk of vertical temperature gradient and condensation phenomena. However, unfortunately, alteration phenomena, such as swelling and detachment of the painting layer of the frescos, were noticed.

As for other frescoes of the Camposanto complex, degradation processes due to the synergism of the organic substances used as adhesive began and sulphation promoted by the lime putty, eternit and pollution determined the need to take again the *Stories of the Holy Fathers* frescoes down from the walls. In this case, traces of casein at the back of the fresco and glue residues ranging from a few granules (0.1 mm to about 1 mm thick) to a maximum of 20 mg/cm$^2$ were noticed on the painted surface.

Also in this case, the biotreatment with *P. stutzeri* A29 strain successfully removed animal glue and casein proteins from the altered fresco and the altered animal glue and casein were no longer detectable by visual inspection and confirmed by GC-MS and PY/GC-MS analyses (Bonaduce and Colombini 2003).

In addition, the results showed that the bioremoval procedure was quick, having a significant effect in the first 2 h after the application.

After the biocleaning treatment, the fresco treated surfaces were subjected to short- and medium-term monitoring to assess eventual microbial colonization, activity and presence of any viable *P. stutzeri* cells and the absence of viable cells was confirmed. Later, the fresco mounted on a frame was kept for one year in a confined and controlled environment in the restoration laboratory of Campaldo (Pisa) and then re-allocated in a semi-confined environment in the great hall of the Monumental Cemetery at Pisa. Periodic monitoring as well as surveillance and protection efforts continues in accordance with defined restoration protocols.

We highlight the advantages of the biocleaning process adopted: (i) the process was non-destructive and removed only extraneous substances or altered compounds; (ii) the use of safe microorganisms (not pathogenic and asporigen); (iii) safe procedure both for the operators and the environment; (iv) the biotechnological techniques were successful, low-cost, environmental friendly.

These positive results led the Technical Commission for Restoration (Pisa, Italy) to approve the use of the *P. stutzeri* strain for the biocleaning of all the ancient altered frescoes at monumental Cemetery, Pisa, that still remain to recover and the biological step has become essential in many cases.

In summary, the steps adopted were: (i) physical-chemical analysis for the characterization of the residual adhesive organic matter used in the past to detach the medieval frescoes; (ii) development and optimization of the advanced biocleaning system to remove the organic matters from the altered fresco surfaces; (iii) short- and medium-term treated surface microbial monitoring; (iv) costs–

**Fig. 10.1** (**a**, **b**) Before and after biocleaning process on *Inferno* frescoes at Monumental Cemetery Camposanto, Pisa, Italy

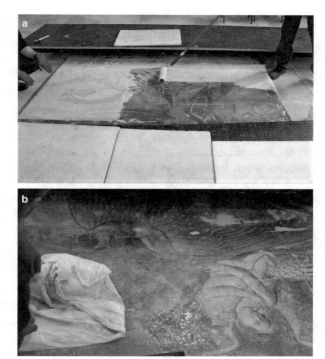

benefits analysis of the biotechnological system employed (Lustrato et al. 2012) (Fig. 10.1).

## 4   Case Study of Casina Farnese on the Palatine Hill, Rome, Italy

In this study a bio-based procedure was employed to treat several hard-coherent deposits as aged proteinaceous matter mixed with gypsum, weddellite, calcium carbonate, apatite and nitrate was present on the wall paintings of the lower loggia of the Casina Farnese (Palatine Hill, Rome) (Mazzoni et al. 2014).

The microbial screening among three asporigen bacterial strains showed that at lab scale two were able to degrade proteins (*Stenotrophomonas maltophilia*) and to degrade protein material (*Pseudomonas koreensis*). For the biotreatment, the bacterial cells were suspended in a laponite gel, easy to apply and to remove, especially from vertical walls. The *in situ* application was carried out in multiple series at temperatures ranging from 6 °C to 37 °C, with contact times of 24–48 h.

The biotreatment with colonized laponite was able to remove multiple hard-coherent deposits on the wall paintings showing no residues at the end of the restoration (Mazzoni et al. 2014).

## 5    Case Study of Animal Glue on Frescoes on the Santos Juanes Church, Valencia, Spain

Other studies showed the use of *P. stutzeri* DSMZ 5190 viable cells for the onsite removal of altered animal glue residues on frescoes of the central vault inside the Santos Juanes Church in Valencia, Spain (Bosch-Roig et al. 2010, 2012, 2013, 2015). The authors adopted in their experimental conditions, agar-gel colonized by bacteria as improved delivery systems. Agar showed to be the most appropriate carrier, both at lab scale and at onsite experiments. It was observed a good adhesive property when applied onto vertical and oblique surfaces, and an adequate water retention for bacteria without interfering with wall surfaces. Large amount of bacterial cell suspension ($10^9$ UFC/ml) was applied in an aqueous solution directly onto the fresco surface by soft brush, with short time of contact (less than 2 h) for each application. The positive results showed that the animal glue was completely removed, confirmed also by analytical pyrolysis and gas-chromatography–mass spectrometry analyses. The bacteria were able to degrade the organic matter present on the fresco compared with area of control treatments (only water).

The authors highlighted the positive results evidencing that the biotechnology is risk-free for the restorers since only non-pathogenic bacteria were used, and is not invasive for the artworks and easily monitored since these bacteria are unable to produce spores (Bosch-Roig et al. 2013).

In conclusion, the different case studies here illustrated show that the biocleaning steps are:

- Diagnosis and characterisation of the stone alteration.
- Laboratory selection of microbial strains able to remove the alteration.
- Evaluation of the degradation rate in order to individualize the most promising microbial strain (safe, efficiency, adaption to the *in situ* treatment, etc.).
- Selection of the appropriate delivery system to vehicular microbial viable cells.
- Delivery system colonization and evaluation of the better condition for the microbial metabolisms.
- Lab-scale application on specimens artificially enriched and on real altered fragments.
- Optimization of the *ex situ, onsite* application conditions.
- Evaluation of the environmental conditions at the time of the application.
- Monitoring during the application.
- Removal of the bioapplication and cleaning of the treated surfaces.
- Short– and long-term surveillance and monitoring after the biotreatment.

# 6   Final Considerations

After almost 25 years since the first use of bacteria in the CH restoration field, we believe it is important to highlight some consideration in order to individualize the aspects that still need to be improved and optimized (Fig. 10.2).

First, it is important to understand how laboratory studies have been successfully transferred *in situ* and how the evolution of technology can be further improved.

Another important aspect is the interest of the industries or the SMEs in the production and commercialization of bioformulates ready to use in the restoration field or if remains a niche sector.

A crucial question also arises: Are the conservation scientists sufficiently involved in the use of the biotechnologies? Moreover, can biodeterioration and biorecovery of cultural heritage provide a important focus for the development of informative and innovative activities in an educational setting? (Verran 2019).

Considering the frescoes at Monumental Cemetery, Camposanto (Pisa), the restorers at Campaldo Laboratories (OpaPisa) showed from the first time a great interest for these advanced and innovative methods. They decided to use the biocleaning step in many cases and in an extensively manner (for example, both for front and back of altered ancient fresco sections) to remove altered organic matters as glue, casein, etc. The restorers have been trained and prepared to recognize the potential of biotechnology but also the limits. A particular attention has also been given to the evaluation of the costs comparing the biological methods with the traditional ones and this aspect resulted positive.

Considering the scientific production in this field is worth to highlight that the number of research articles, reviews, books, proceedings of national and international conferences has greatly increased as reported in Fig. 10.3.

The research groups working in this field are also expanded over the time involving a large number of countries and consequently different environmental

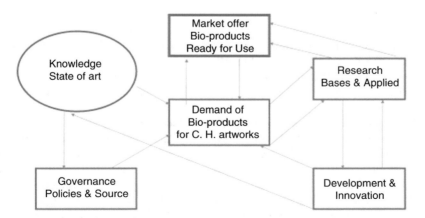

**Fig. 10.2**  Flow chart and interactions between the main actors in CH recovery

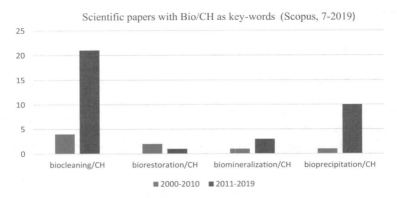

**Fig. 10.3** Representative but not exhaustive literature data from Scopus, until July 2019

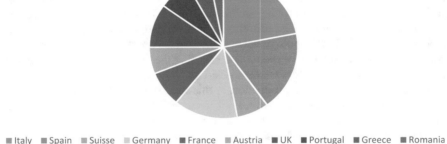

**Fig. 10.4** European scientific research groups working in cleaning/biocleaning in the cultural heritage fields (2014–2018, in %)

conditions (climate, temperature, humidity, pollution, etc.), lithotypes, restoration uses and practices, cultures, etc. (Fig. 10.4).

Despite the increase number of scientific papers and research groups, the technological transfer on larger scale and the availability of products ready to use are still a problem to solve.

In addition, the dissemination especially to stakeholders and end users can be improved through the organization of workshops and seminars specifically targeted to conservation scientists. This will help worldwide the use of the correct terminology, the understanding of the potentialities of alternative methods and the preparation of guidelines and protocols available for all the operators in this field.

As evidenced in this context, numerous studies have been carried out in order to optimize the biocleaning strategies in terms of microorganism efficiency and delivery systems, ability to maintain and guarantee the microbial activity and ready to be used also on difficult altered stone surfaces (ornamented, vertical and vaulted). Different substances can be present as stone alterations and for this reason, the choice of the best microorganism is a crucial step in the biocleaning technology.

The CH field represents a great challenge and Science and Art link together the work of conservator scientists and historians with researchers and scientists, sharing their diverse expertises and joining the knowledges to the preservation and the conservation of our artistic patrimony.

**Acknowledgements** The authors wish to thank to the restorers implicated in the before mentioned case studies because without their confidence, help and dedication none of these works could have been successfully applied on real artworks. Special thanks to Claudia Sorlini, Maria Perla Colombini, Alessia Andreotti, Laura Rampazzi and Cristina Corti for the continuous collaborations and support in the chemical aspects in our studies on CH biotechnologies. In addition to Dr. Clara Baracchini and Antonio Caleca, Soprintendenza BAAAS, Pisa, and Prof. Antonio Paolucci, Director of the Scientific Committee for the Monumental Cemetery, Opera Primaziale Pisana (OPAPisa), Italy; Ing. Giuseppe Bentivoglio and Ing. Roberto Cela, OPAPisa; Senior Restorers Gianni Caponi, Donatella Zari, Carlo Giantomassi and Gianluigi Colalucci of Conservazione and Restauro, Pisa. In addition, Stefano Lupo and Cristina Pucci as Team of Restorers, Conservazione and Restauro (Pisa).

# References

Abdulla H, May E, Bahgat M, Dewedar A (2008) Characterisation of actinomycetes isolated from ancient stone and their potential for deterioration. Pol J Microbiol 57:213–220

Alfano G, Lustrato G, Belli C, Zanardini E, Cappitelli F, Mello E, Sorlini C, Ranalli G (2011) The bioremoval of nitrate and sulfate alterations on artistic stonework: the case-study of Matera Cathedral after six years from the treatment. Int Biodeter Biodegr 65:1004–1011

Alonso-Vega P, Carro L, Martínez-Molina E, Trujillo ME (2011) *Auraticoccus monumenti* gen. nov., sp. nov., an actinomycete isolated from a deteriorated sandstone monument. Int J Syst Evol Microbiol 61:1098–1103

Antonioli P, Zapparoli G, Abbruscato P, Sorlini C (2005) Art-loving bugs: the resurrection of Spinello Aretino from Pisa's cemetery from Pisa's cemetery. Proteomics 5:2453–2459

Banik G (2003) Removal of starch paste adhesives and relinings from paper-based objects by means of enzyme poultices. In: Cremonesi P (ed) Materiali Tradizionali ed Innovativi nella Pulitura dei Dipinti e delle Opere Policrome Mobile. Il Prato, Padova, pp 33–38

Barbabietola N (2012) Development of Microbial Technologies in Conservation and Restoration. PhD Thesis. Florence University, Florence, Italy

Barbabietola N, Tasso F, Grimaldi M, Alisi C, Chiavarini S, Marconi, P, Perito B, Sprocati AR (2012) Microbe-based technology for a novel approach to conservation and restoration. Special Issue: Knowledge, Diagnostics and Preservation of Cultural Heritage. Energia, Ambiente e Innovazione, pp 69–76

Bellucci R, Cremonesi P (1994) L'uso degli enzimi nella conservazione e nel restauro dei dipinti. Kermes 21:45–64

Bonaduce I, Andreotti A (2009) Py-GC/MS of organic paint binders. In: Colombini MP, Modugno F (eds) Organic mass spectrometry in art and archaeology. John Wiley, Chichester, UK, pp 303–361

Bonaduce I, Colombini MP (2003) Gas chromatography/mass spectrometry for the characterization of organic materials in frescoes of the Monumental Cemetery of Pisa (Italy). Rapid Comm Mass Spectrom 17:2523–2527

Bonomi R (1994) Utilizzo degli enzimi per il restauro di una scultura in terracotta policroma. OPD Restauro 6:101–107

Bosch-Roig P, Regidor-Ros JL, Soriano-Sancho P, Domenech-Carbo MT, Montes-Estellés RM (2010) Ensayos de biolimpieza con bacterias en pinturas murales. Arché 4–5:115–122

Bosch-Roig P, Montes-Estellés RM, Regidor-Ros JL, Roig Picazo P, Ranalli G (2012) New frontiers in the microbial bio-cleaning of artworks. Picture Restorer 41:37–41

Bosch-Roig P, Regidor-Ros JL, Soriano-Sancho P, Montes-Estellés R (2013) Biocleaning of animal glue on wall paintings by *Pseudomonas stutzeri*. Chim Oggi 31:50–53

Bosch-Roig P, Lustrato G, Zanardini E, Ranalli G (2015) Biocleaning of cultural heritage stone surfaces and frescoes: which delivery system can be the most appropriate? Ann Microbiol 65:1227–1241

Cappitelli F, Zanardini E, Toniolo L, Abbruscato P, Ranalli G, Sorlini C (2005) Bioconservation of the marble base of the Pietà Rondanini by Michelangelo Buonarroti. Geophys Res Abstr 7:06675–06676

Cappitelli F, Zanardini E, Ranalli G, Mello E, Daffonchio D, Sorlini C (2006) Improved methodology for bioremoval of black crusts on historical stone artworks by use of sulfate-reducing bacteria. Appl Environ Microbiol 72:3733–3737

Cappitelli F, Toniolo L, Sansonetti A, Gulotta D, Ranalli G, Zanardini E, Sorlini C (2007) Advantages of using microbial technology over traditional chemical technology in removal of black crusts from stone surfaces of historical monuments. Appl Environ Microbiol 73:5671–5675

Cappitelli F, Salvadori O, Albanese D, Villa F, Sorlini C (2012) Cyanobacteria cause black staining of the National Museum of the American Indian Building, Washington DC. Biofouling 28:257–266

Colombini MP, Modugno F, Giacomelli A (1999) Two procedures for suppressing interference from inorganic pigments in analysis by gas chromatography-mass spectrometry of proteinaceous binders in painting. J Chromatogr A 846:101–111

Colombini MP, Modugno F, Fuoco R, Tognazzi A (2002) A GC-MS study on the deterioration of lipidic paint binders. Microchem J 73:175–185

Colombini MP, Bonaduce I, Gautier G (2003) Molecular pattern recognition of fresh and aged shellac. Chromatographia 58:1–8

Cremonesi P (1999) L'uso degli enzimi nella pulitura di opere policrome. In: Il Prato (ed) I Talenti, Padova, pp 5–11

Cremonesi P (2004) L'uso dei solventi organici nella pulitura di opere policrome. In: Il Prato (ed) I Talenti, Padova, pp 2–15

Crispim CA, Gaylarde CC (2005) Cyanobacteria and biodeterioration of cultural heritage: a review. Microb Ecol 49:1–9

Daffonchio D, Cherif A, Borin S (2000) Homoduplex and heteroduplex polymorphisms of the amplified ribosomal 16S-23S internal transcribed spacers describe genetic relationships in the "*Bacillus cereus* group". Appl Environ Microbiol 66:5460–5468

De Belie N, De Graef B, De Muynck W, Dick J, De Windt W, Verstraete W (2005) Biocatalytic processes on concrete: bacterial cleaning and repair. In 10th DBMC Int. Conference on Durability of Building Materials and Components, April, Lyon, pp 17–20

De Graef B, De Windt W, Dick J, Verstraete W, De Belie N (2005) Cleaning of concrete fouled by lichens with the aid of *Thiobacilli*. Mater Struct 38:875–882

De La Chapelle A (2003) Utilizzo degli enzimi nel restauro delle opere grafiche policrome. In: Cremonesi P (ed) Materiali tradizionali ed innovativi nella pulitura dei dipinti e delle opere policrome mobile. Il Prato, Padova, pp 51–53

EWAGLOS (2016) In: Weyer A, Roig Picazo P, Pop D, Carras J, Ozkose A, Vallet JM, Srsa I (eds) EwaGlos-European Illustrated Glossary of Conservation Terms for Wall Paintings and Architectural Surfaces. English definitions with translations into Bulgarian, Croatian, French, German, Hungarian, Italian, Polish, Romanian, Spanish and Turkish, vol. 17. Michael Imhof Verlag, Petersberg, Germany, pp 304–307

Fernandes P (2006) Applied microbiology and biotechnology in the conservation of stone cultural heritage materials. Appl Microb Biotechnol 73:291–296

Giacomucci L, Bertoncello R, Salvadori O, Martini I, Favaro M, Villa F, Sorlini C, Cappitelli F (2011) Microbial deterioration of artistic tiles from the Façade of the Grande Albergo Ausonia & Hungaria (Venice, Italy). Microb Ecol 62:287–298

Gorbushina AA (2007) Life on the rocks. Environ Microbiol 9:1613–1631

Gurtner C, Heyrman J, Piñar G, Lubitz W, Swings J, Rölleke S (2000) Comparative analyses of the bacterial diversity on two different biodeteriorated wall painting by DGGE and 16s rRNA sequence analysis. Int Biodeter Biodegrad 46:229–239

Iannuccelli S, Sotgiu S (2009) La pulitura superficiale di opera grafiche a stampa con gel rigidi. In: Il Prato (ed) Progetto Restauro, Padova, pp 15–24

Lustrato G, Alfano G, Andreotti A, Colombini MP, Ranalli G (2012) Fast biocleaning of mediaeval frescoes using viable bacterial cells. Int Biodeter Biodegr 69:51–61

Makes F (1988) Enzymatic consolidation of the portrait of Rudolf II as "Vertumnus" by Giuseppe Arcimboldo with a new multi-enzyme preparation isolated from Antarctic Krill (Euphausia superba). Acta Universitatis Gotheburg 23:98–110

Martín Ortega S (2015) Neteges raonades: Experimentació amb nou tractaments de neteja aplicats sobre 6 peces de material arqueologic divers. Treball final de grau ESCRBCC, Barcelona

May E (2010) Stone biodeterioration. In: Mitchell R, McNamara CJ (eds) Cultural heritage microbiology: fundamental studies in conservation science. ASM Press, Washington DC, pp 1–39

Mazzoni M, Alisi C, Tasso F, Cecchini A, Marconi P, Sprocati AR (2014) Laponite micro-packs for the selective cleaning of multiple coherent deposits on wall paintings: the case study of Casina Farnese on the Palatine Hill (Rome-Italy). Int Biodeter Biodegr 94:1–11

McNamara CJ, Mitchell R (2005) Microbial deterioration of historic stone. Front Ecol Environ 3:445–451

Palla F (2004) Le biotecnologie molecolari per la caratterizzazione e la valutazione del ruolo dei microrganismi nei processi di degrado di manufatti di interesse storico artistico. In: Lorusso S (ed) Quaderni di Scienza della Conservazione. Univerità di Bologna, Ravenna, pp 183–194

Palla F (2013) Bioactive molecules: innovative contribution of biotechnology to the restoration of cultural heritage. J Conserv Sci 13:369–373

Piñar G, Ramos C, Rölleke S, Schabereiter-Gurtner C, Vybiral D, Lubitz W, Denner EB (2001) Detection of indigenous Halobacillus populations in damaged ancient wall paintings and building materials: Molecular monitoring and cultivation. Appl Environ Microbiol 67:4891–4895

Portillo C, Saiz-Jimenez C, Gonzalez JM (2009) Molecular characterization of total and metabolically active bacterial communities of "white colonizations" in the Altamira Cave. Spain. Res Microbiol 160:41–47

Ranalli G, Chiavarini M, Guidetti V, Marsala F, Matteini M, Zanardini E, Sorlini C (1996) The use of microorganisms for the removal of nitrates and organic substances on artistic stoneworks. In: Riederer J (ed) Proceedings of the 8th Int. Congress on Deterioration and Conservation of Stone, Berlin, 30 September–4 October 1996, pp 1415–1420

Ranalli G, Chiavarini M, Guidetti V, Marsala F, Matteini M, Zanardini E, Sorlini C (1997) The use of microorganisms for the removal of sulphates on artistic stoneworks. Int Biodeter Biodegr 40:255–261

Ranalli G, Alfano G, Belli C, Lustrato G, Colombini MP, Bonaduce I, Zanardini E, Abbruscato P, Cappitelli F, Sorlini C (2005) Biotechnology applied to cultural heritage: biorestoration of frescoes using viable bacterial cells and enzymes. J Appl Microbiol 98:73–83

Ranalli G, Zanardini E, Sorlini C (2009) Biodeterioration e including cultural heritage. In: Schaechter M (ed) Encyclopedia of microbiology. Elsevier, Oxford, pp 191–205

Saiz-Jimenez C, Cuezva S, Jurado V, Fernandez-Cortes A, Porca E, Benavente D, Canaveras JC, Sanchez-Moral S (2011) Paleolithic art in peril: policy and science collide at Altamira Cave. Science 334:42–43

Sanmartín P, Bosch-Roig P (2019) Biocleaning to remove graffiti: a real possibility? Advances towards a complete protocol of action. Coatings 9:104–119

Schabereiter-Gurtner C, Gurtner G, Piñar G, Lubitz W, Rölleke S (2001) An advanced molecular strategy to identify bacterial communities on art objects. J Microbiol Methods 45:77–87

Scheerer S, Ortega-Morales O, Gaylarde C (2009) Microbial deterioration of stone monuments-an updated overview. Adv Appl Microb 66:97–139

Segal J, Cooper D (1977) The use of enzymes to release adhesives. Paper Conservator 2:47–50

Stassi A, Zanardini E, Cappitelli F, Schiraldi A, Sorlini C (1998) Calorimetric investigations on the metabolism of *Bacillus strains* isolated from artistic stoneworks. Ann Microbiol Enzim 48:111–120

Tiano P (2002) Biodegradation of cultural heritage: decay mechanisms and control methods. The Proceedings of the ARIADNE 9 WS. Institute of Theoretical and Applied Mechanics of the Academy of Sciences of the Czech Republic, Prague 4-10 February, pp 1–37

Tiano P, Cantisani E, Sutherland I, Paget JM (2006) Bioremediated reinforcement of weathered calcareous stones. J Cult Herit 7:49–55

Urzì C, Brusetti L, Salamone P, Sorlini C, Stackebrandt E, Daffonchio D (2001) Biodiversity of Geodermatophilaceae isolated from altered stones and monuments in the Mediterranean basin. Environ Microbiol 3:471–479

Verran J (2019) Mixed culture: encouraging cross-disciplinary collaboration and communication to enhance learning. Ann Microbiol 69:1107–1111

Villa F, Vasanthakumar A, Mitchell R, Cappitelli F (2015) RNA-based molecular survey of biodiversity of limestone tombstone microbiota in response to atmospheric sulphur pollution. Lett Appl Microbiol 60:92–102

Wendelbo O (1976) The use of proteolytic enzymes in the restoration of paper and papyrus. University Library of Bergen

Wolbers R (2000) Cleaning painted surface: aqueous methods. Archetype Publication Ltd, London

Zanardini E, Purdy K, May E, Moussard H, Williams S, Oakley B, Murrell C (2011) Investigation of archaeal and bacterial community structure and functional gene diversity in exfoliated sandstone at Portchester Castle (UK). In: Sterflinger K, Piñar G (eds) International Symposium on Biodeterioration and Biodegradation (IBBS-15), Vienna, September 19–24, p 90.

Zanardini E, May E, Inkpen R, Cappitelli F, Murrell JC, Purdy KJ (2016) Diversity of archaeal and bacterial communities on exfoliated sandstone from Portchester Castle (UK). Int Biodeter Biodegrad 109:78–87

Zanardini E, May E, Purdy KJ, Murrell JC (2019) Nutrient cycling potential within microbial communities on culturally important stoneworks. Environ Microbiol Rep 11:147–154

# Chapter 11
# Sustainable Restoration Through Biotechnological Processes: A Proof of Concept

**Anna Rosa Sprocati, Chiara Alisi, Giada Migliore, Paola Marconi, and Flavia Tasso**

**Abstract** An understanding of the different microbial constellations or microbiomes, which every habitat and every organism harbor, will be the key to addressing many of the challenges humanity will face in the twenty-first century. Such comprehension could launch several innovations relating to natural and cultural capital, including historical and artistic heritage. In relation to cultural heritage, microorganisms are mainly known through their role as deteriogens, but the features creating damage can be exploited positively, attaining more sustainable restoration strategies, in accordance with the principles of compatibility and retreatability deriving from reflections on the Cultural Heritage inspired by the Charter of Venice (International charter for the Conservation and restoration of monument and sites (the Venice Chart 1964). In: ICOMOS, IInd International Congress of Architects and Technicians of Historic Monuments, 1964) onwards. In this article, we show a series of case studies, using both wild-type microorganisms and plant-based extracts, providing a comprehensive proof of concept of the feasibility of biotechnological solutions for a more sustainable restoration strategy, to replace the products in use which are often dangerous for operators, aggressive for works of art and no longer compatible with the environment. The overview of the case studies presented, many of which are still unpublished, responds to the need to go beyond the state of the art and has entirely sprung from suggestions by restorers, interested in learning about potential innovations and strongly determined to introduce non-toxic products in their daily work. In this perspective, the case studies dealt with two topics: bio-cleaning and disinfection. Noteworthy results were obtained on a platform of different types of artworks and different materials with compatible, harmless and selective products.

**Keywords** Bio-cleaning · Disinfection · Bio-restoration · Microbial biotechnology

A. R. Sprocati (✉) · C. Alisi · G. Migliore · P. Marconi · F. Tasso
ENEA, Department for Sustainability, Rome, Italy
e-mail: annarosa.sprocati@enea.it; chiara.alisi@enea.it; giada.migliore@enea.it; paola.marconi@enea.it; flavia.tasso@enea.it

E. Joseph (ed.), *Microorganisms in the Deterioration and Preservation of Cultural Heritage*, https://doi.org/10.1007/978-3-030-69411-1_11

# 1  Introduction

The transition to products and practices that take environmental, social and economic aspects into account is today the challenge to be faced in order to develop a more sustainable approach to Restoration and Conservation, in accordance to European policies and the Strategic Innovation and Research Agenda (SIRA) of the Bio-Based Industry Consortium (2017). This represents a great challenge, not only for restorers and art historians, but also for researchers, politicians, scientists, entrepreneurs and also market players. Some of the products in use, recognized as harmful, are being eliminated from the market and others will be. Their replacement or reduction creates a space for new, more sustainable products in the market, with the possibility of creating a new space in the Bioeconomy chain.

Microbial biotechnologies play a full role in this process, thanks to their potential to provide new products with some advantages over traditional chemical-physical methods: selectivity thus low aggressiveness for the artifact, environmental compatibility, less danger for the operators, absence of ethical problems. The possibility of using bio-based techniques is perhaps a feasible option for the future, also in consideration of the attention that must be paid to the life cycle of a product. Microorganisms - and their derivatives - and products of plant origin are ideal candidates for the development of new bio-based products of renewable origin.

To date, however, despite numerous researches published in specialized journals, very few patents exist and very few products are available on the market. Especially, products based on microorganisms are currently absent in restoration market. The challenge must therefore point towards the transformation of research products into market products, available for restorers and ready for use.

To this end, it was necessary to broadly verify the actual potential of biotechnology, given the size and diversity of the objects involved, the enormous variety of materials used and the peculiar characteristics of the individual cases.

The case studies review described below (Tables 11.1 and 11.2) was planned by the authors to proceed beyond the state of the art (Ranalli et al. 2005; Bosch Roig et al. 2013; Gioventù et al. 2011).

The range of materials and the variety of cases involved proved that biological way is a good alternative to different products in use. Tailored solutions have been defined both to clean multiple coherent deposits and to control biodeteriogens using bacteria and plant extracts. The restorers, in order to find solutions to problems unsolved by the available methods, or to find effective alternatives to the toxic products in use, have submitted the case studies shown.

**Table 11.1** Proof of concept for bio-cleaning by using bacterial strains

| Artwork material | Deposit | Place (case number) | Bio-product | Outcome (reference) |
|---|---|---|---|---|
| Mural paintings | Overlayed layers of gypsum, carbonate, aged protein | Casina Farnese, Rome (1) | UI3, UT30, TPBF11 | Patent ENEA WO2015040647[a] |
|  | Shellac | Lab specimens (2) | CONC11, CONC18, LAM21 | Preliminary trials[b] |
| Frescoes | Vinyl glues | Lab specimens (3) | FCONT | Preliminary trials[c] |
|  | Primal resin | Galleria Carracci - Palazzo Farnese, Rome (4) | ZCONT | Positve (to be confirmed) |
|  | Lime layer | Chiesa di San Giorgio, Bitti, Nuoro (5) | CER20, TPBF11 | Negative, more research needed |
| Marble statues | Wax hydrocarbons | La lupa, Testa di Donna, Idealità e materialismo, Cleopatra- GAM, Rome (6) | LAM23, LAM30, OSS42, LAM21, SH7 | Applicable[d] |
|  | Carbonates, wax, Paraloid | Bacco col. cesto- La Venaria Reale, Turin (7) | ZCONT, TPBF11, LAM21 | Positive for Paraloid[e] |
|  | Iron oxides, copper oxides | Vatican Gardens (8) | FeIIC1, SME3.14 | Positive (to be confirmed) |
|  | Iron oxides | Lab specimens (9) | FeIIC1 | Positive (to be confirmed) |
|  | Deposits of unknown composition | Group of Michelangelo statues, Medici Chapels, Florence (10) | CONC11, ZCONT, SH7 | Positive |
| Marble portals | Iron oxides, brownish spots, lab specimens | Galleria Carracci - Palazzo Farnese, Rome (11) | FeIIC1 | Positive (to be confirmed) |
| Easel wood paintings | *Colletta* in oil | Madonna della Cintola, Vatican Museums (12) | SH7 | Positive (to be optimized)[f] |
| Ancient paper | Animal glues | Archive document XVIII century (13) | TSRNS15 | Applicable[g] |
| Calcography slabs | Zapon resin | Lab specimens (14) | LAM21 | Preliminary trials |
| Stuccoes | Copper oxides | S. Prudenziana, Rome (15) | SME1.11 | Positive |

[a]Mazzoni et al. (2014)
[b]Grimaldi (2009)
[c]Cutraro (2014)
[d]Sprocati et al. (2017)
[e]Galizzi (2015)
[f]Crisci et al. (2020)
[g]Barbabietola et al. (2016)

**Table 11.2** Proof of concept for biodeteriogens removal and control by using bio-products and bacterial strains

| Artwork material | Deposit | Place | Bio-product | Outcome (reference) |
|---|---|---|---|---|
| Mobile painting on canvas | Biofilm | Private painting | Liq | Positive |
| Frescoes | Biofilm | Chapel XIII –Orta, Novara | Liq | Positive |
| Hypogeum | Biofilm | Domus Aurea, Rome | Liq, | Underway[a] |
| Stone material | Biofilm | Vatican gardens | Liq, BioZ, SME1.11 | Positive |
| | | Diocleziano termae, Rome | Liq, BioZ, SME1.11, NopalCap | Positive[b] |
| | | Roman wall, Rome | Liq, BioZ, SME1.11, NopalCap | Underway |
| | | Outdoor statues-Vicenza | Liq | Underway |
| Mortar, bio-mortar | Bio-receptivity | Lab specimens | Nopal | Positive[c] |
| | Biofilm | Outdoor statue "Beethoven"- Naples | NopalCap | Positive |

[a]Rugnini et al. (2019)
[b]Ledda (2019)
[c]Persia et al. (2016)

## 2  Bio-Cleaning Procedures Developed in the Case Studies

The procedure involves the use of living bacteria immobilized at high concentration ($10^8$ CFU/mL) in support agents with different viscosities, to create a micro-pack, easy to apply and remove without leaving residues on the artifact.

The micro-packs are easy to apply and to remove, without dripping and leaving residues. They can be applied on vertical surfaces and ceilings; do not require strict operating conditions, in terms of temperature and pH and with no disposal problems. The micro-pack can be conveniently prepared on a support table or prepared directly on the object to be treated. In the first case a standard micro-pack is organized backwards: a plastic film, which preserves the humidity during the treatment, is first placed followed by a slightly moistened veil of paper (i.e. Japanese or English paper), a layer of bacteria in their support agent and another veil of paper. The interposition of the paper facilitates the application and the removal of the micro-pack and allows inspections to check the progress of the process. However, the opportunity to interpose paper can be evaluated from time to time, based on the specific situation. Direct contact with the original material can sometimes be more suitable. After removal, a damp swab or a sponge can easily clean the artefact.

The time of contact of the micro-pack depends on various factors, such as the nature of the substrate to be removed, the thickness of the layer; the metabolic rate of the microorganism, the nature and the state of conservation of the artifacts. The procedures developed in the case studies (Tables 11.1 and 11.2) are fairly standardized, with overnight compresses, but they can be designed according to specific needs of the case.

Competent bacteria are selected according to the ability to metabolize the deposits to be removed, without damaging the original material of the artifact. For instance, to clean paper (case 13), bacteria must not show cellulolytic activity as well as to clean limestone they must not have the ability to solubilise carbonates. In this way the cleaning is selective and safe for the artifact.

The bacteria used in the case studies belong to the laboratory in-house collection ENEA (De Vero et al. 2019). The collection gathers over 800 strains of biotechnological interest mainly isolated from harsh environments, hypogea, artworks, etc. All strains are wild type and non-pathogenic (mainly belonging to the risk group 1), their gene sequences (rDNA16S) are deposited in the GenBank database.

The support agent must be compatible with the survival of the cells, without, however, representing a source for growth. The support material is decided on the basis of the needs of the surfaces to be treated and further properties, like as the ability to retain humidity and be malleable, absorbent capacity, total inertia, etc. Often the right combination between bacteria and the support agent determines the effectiveness of the treatment. Moreover, the suspension medium to embed bacteria influences the rheological characteristics of the support agent itself and the performance of the micro-pack. The suspension medium depends on the metabolic action that bacteria have to play for the deposit removal.

Organic deposits are degraded and used by bacteria as a carbon source for growth (biodegradation), while inorganic deposits (i.e. carbonates, phosphates, oxalates, oxides, etc.) are removed through solubilisation, mediated by secondary metabolites, such as organic acids and specific molecules. To best perform specific functions, the suspension medium must comply with the specific physiological needs of the bacteria.

In the case of the removal of organic deposits, bacteria are generally separated from the culture broth and re-suspended in a mineral solution to strengthen the attack of the target deposit. Sometimes cells are starved to deplete the internal carbon reserve in order to make the attack more effective. In other cases, for instance, when it is necessary to induce the production of enzymes (i.e. lipase, protease, cellulase, etc.) the bacteria must be left in their growth media.

When cleaning inorganic deposits, after separation from the growth media, the bacteria must be re-suspended in a solution containing a very low concentration of a carbon source in order to sustain bacterial metabolism for the production of metabolites (i.e. siderophores, organic acids, etc.) responsible for the solubilisation.

In the cases listed in Table 11.1, two products were mainly selected as support agents: Laponite®RD and Vanzan®NF.

Laponite®RD (BYK-Chemie GmbH, Germany) is a colloidal clay consisting of a mixture of silicates of sodium, magnesium and lithium. In the restoration, Laponite®RD is used to remove rust stains and ferrous encrustation on frescoes. It is sometimes used in combination with citric acid and sodium citrate, but also often with tap water, to avoid further damaging to already very compromised artifacts. It must be considered that Laponite®RD can dry off and shrink in some case.

Vanzan®NF (Vanderbilt Minerals LLC, Norwolk, CT) is the commercial name of the Xanthan gum, a high molecular weight exocellular polysaccharide derived from the bacterium *Xanthomonas campestris*. Xanthan gum is widely used as a rheology control agent for aqueous systems. It increases viscosity, helps to stabilize emulsions and prevents the settling of solids in a wide variety of applications.

Vanzan®NF shows a great compatibility with cells viability, perfect adherence and plasticity also on vertical surfaces, and overall the capacity to maintain the moisture without drying off or shrinking, up to 24 h of application.

Laponite®RD and Vanzan®NF were used for the first time as support agent for microorganisms in the bio-cleaning case 1 and 12, respectively (Mazzoni et al. 2014; Crisci et al. 2020). The most appropriate concentration for a correct consistency and property of the micro-pack was 9% w/v for Laponite®RD and 6% w/v for Vanzan®NF.

# 3   Bio-Cleaning Case Studies: Review and Discussion

The series of case studies (Table 11.1) springs from the European patent EP 3046779 extended to Italy, Switzerland, Germany, Spain, France, GB (Sprocati et al. 2014) that derived from the bio-cleaning of mural painting of the external loggias in the Casina Farnese (case 1). The cases that followed (2–15) expressly ranged between different types of artifacts, materials and substrates to be removed, experimenting numerous microbial species available in the in-house collection in order to create a proof of concept of the feasibility of microbial biotechnology to contribute to a more sustainable restoration strategy.

The Casina Farnese (case 1) is situated at the top of the Palatine Hill, in Rome, the double order of open loggias representing the legend of Ercole and Caco was completed by 1593. The decoration of the lower loggia, object of the bio-cleaning tests, was ascribed to a pupil of Taddeo Zuccari and represents Ercole killing Caco. Coherent deposits composed of gypsum, weddellite, calcium carbonate, apatite, nitrate, and aged proteinaceous matter (FTIR-ATR analysis) were deemed difficult to remove from the wall paintings, especially in areas where the deposits were superimposed in layers. Following a laboratory screening, three bacterial non spore-forming strains were selected to solubilise calcium sulphate and carbonate, to degrade protein, to solubilise inorganic compounds and degrade protein material. Micro-packs with single strains were used in situ, from July 2012 to February 2013, in temperatures ranging from 6 °C to 37 °C. After experiencing some supports in use for wet compresses without satisfactory results, the living bacterial cells were eventually immobilized in a Laponite®RD gel. In this occasion, Laponite®RD was used for the first time to immobilize microbial cells, proving to be optimal both for compatibility with the vitality of the cells and for perfectly retaining the material of compress, without leaks. The strain *Cellulosimicrobium cellulans* TBF11[E] DSM 27224 (GenBank accession number: EU249577) removed the inorganic darker layer, *Stenotrophomonas maltophilia* UI3[E] DSM 27225 (JX133197) dissolved the

**Fig. 11.1** Aged-casein layer covering the architrave over the door at Casina Farnese, Rome. (**a**) before cleaning, (**b**) after bio-cleaning with a micro-pack containing UI3 strain, (**c**) after the pictorial retouching (photo from Mazzoni et al. 2014)

brownish layer (probably aged casein) and *Pseudomonas koreensis* UT30$^E$ DSM 27226 (JX133187) removed the mixed deposits. In areas with overlaid deposits, micro-packs were applied in succession, containing the strain able to remove each layer.

Another critical point threating the survival and the performance of bacteria was the low temperature in the open loggia on winter days. No significant differences were detected in the efficacy of the bio-cleaning treatment, probably the micro-pack contributes to maintain a microclimate suitable for the bacterial metabolic activity. After the bio-cleaning, the restorers successfully completed the restoration (Fig. 11.1). The microbial monitoring performed on treated surfaces up to 4 months detected only transitory population, but none of the used bacterial strains.

The procedure established by this study proved to be selective, safe for the mural painting, uncaring of the seasonal temperature changes. As a result, the cleaning was gradual, controllable and able to safeguard the surface of the painting. Laponite$^®$RD may have facilitated the cleaning process by softening the layer of deposit to be removed, as seen in Rozeik (2009), with a synergistic effect (Mazzoni et al. 2014).

This case faced three challenges: coherent overlaid deposits of different nature; choice of the right support, suitable for application on vertical surfaces and ceilings without releasing water and bacteria, and low temperatures. The identified procedure paved the way for the case studies that were addressed later.

A few cases are preliminary studies carried out only at laboratory scale on specimens reproducing the artwork. The cases 2 (shellac) and 3 (vinyl glues) deals with the removal of fixatives (natural and synthetic resins and adhesives based on proteins and hydrocarbons) from mural paintings and frescoes, used to repair the flaking or disintegration of the pigment layer; loss of cohesion or adhesion to the support.

Case 2 was focused on shellac degradation. The use of shellac as a fixative for mural paintings was very widespread in India being used for the restoration of paintings kept in the caves of Ajanta, Ellora, Bagh or Kancheepuram (Sharma et al. 1980; Singh and Balasaheb 2013). Although shellac shows good penetration features, adhesive properties and resistance to biological attacks, the resin has a tendency to yellowing and darkening, becoming hard, brittle and insoluble, because of polyesterification reactions which are accelerated by high temperatures. This makes shellac one among the undesired deposits which can be the object of cleaning, necessary to re-establish the visual perception of the paintings (Mora et al. 1984).

In the last thirty years different cleaning procedures, which have involved the use of various solvents, have been carried out: starting from aqueous solutions of sodium or potassium borate, carbonate or hydroxide, to mixtures of different solvents, such as butyl lactate, morpholine, alcohol, turpentine oil and ligroin, applied in succession and for the final cleaning. In exceptional cases, in order to make the deteriorated and harden, superficial layer of shellac susceptible to the solvent attack, a solution of formic acid 1:10 in ethyl alcohol have been used. As drawbacks, colour loss, white crusts and surfacing of soluble salts appeared after each treatment. Case 2 represented the first attempt of removing shellac through a biological approach (Barbabietola 2013).

On the molecular level, shellac is a complex mixture made of mono-and poly-esters of hydroxyl-aliphatic and sesquiterpenoid acids (Colombini et al. 2003). Comparing fresh with naturally aged shellac analysis showed that the only stable compound under pyrolysis condition was the butolic acid that is used as a marker for the molecular pattern recognition of the molecule.

In order to carry out in vivo bio-cleaning trials, specimens of wall paintings were purposely prepared. The specimens were made up of bricks covered with a plaster layer. Three different pigments, cinnabar, (red), kaolin (white) and indigo (blue), were diluted into three different media, rabbit skin glue, arabic gum and linseed oil and then were spread by brush over the plaster layer. When the paint layer was completely dried, a solution of shellac in ethanol 10% (w/v) was spread by brush over the upper surface of the wall painting specimen. In order to simulate the climatic conditions of the Ajanta caves paintings, a climatic chamber was set at 34 °C and 96% R.H, which represents the annual average values of the Maharashtra region where Ajanta caves are situated. Colorimetric measurements with the spectrophotometer Techkon sp-820λ were carried out on the samples, before they were subjected to artificial weathering. In order to monitor shellac transformation as a result of the microbial growth, gas chromatography coupled with mass spectrometry was employed.

Among the many strains tested, only three, *Pseudomonas stutzeri* CONC11 (EU275358), *Achromobacter xylosoxidans* CONC 18 (EU275351) and *Acinetobacter calcoaceticus* LAM21 (JK534248) were able to grow on shellac 0.2% as sole carbon source. A new procedure to supply shellac on a paper disk was developed, in order to achieve reproducible chromatograms for gas chromatography-mass spectrometry analysis, aimed to determine which components of shellac were able to sustain the bacterial growth. Butolic acid was reduced by 60% after 24 h of incubation and by 80% in the plateau phase and seemed to be the only component that undergoes a transformation after microbial growth. Work is actually in progress in order to tailor a microbial formula by varying the successions of the bacterial strains over time, so that to strengthen the attack to more complex compounds, once assimilated the more soluble portion.

Case 3 deals with the removal of vinyl glues from the frescoes of the Chapels of Saint Martial and Saint Jean of the Pope's Palace in Avignon, the most important

Gothic palace in the West word, built between 1335 and 1364. The frescoes, realized by an Italian painter, Matteo Giovannetti, constitute an exceptional masterpiece. The conservation of these paintings as a whole is to be considered a miracle, given the many damages to the Pope's Palace over the centuries. In a previous restoration, in the '70s, vinyl glue was applied as consolidating agent and, following ageing, proved detrimental for a correct reading of the frescos.

The removal of the resin (Vinavil K40) with products in use was critical for restorers and previous works on the biological removal of Vinavil K40 were unknown in the literature. The research aimed to find microorganisms able to attack the resin. Research started from the isolation of microorganisms inhabiting the surfaces covered with the resin. The 42 bacterial strains isolated from the frescoes together with microbial strains of the in-house collection have been tested for the ability to oxidize vinyl glue and produce esterase enzymes. Few strains showed a marked esterase activity; the best performing was the fungal strain of *Penicillium commune*, FCONT. The removal tests were performed on fresco specimens reproducing the technique used by the painter Matteo Giovannetti. He made significant changes to the traditional fresco technique, by eliminating the laying of the first wall layer and giving an unusual importance to the lights with a dry laying on the composition to improve the shine. Specimens were covered with pigments (yellow ochre, red ochre, green earth and ultramarine blue), on half of the surface a layer of Vinavil K40 dissolved in about 70–80% w/v acetone was applied. Some of the specimens have been aged in a climatic chamber for 38 days under UV light, at a constant temperature of 25 °C and 45% R.H. Laponite®RD micro-packs containing microbial suspensions were applied with a succession of 2 applications for 24 h each one. The best results were obtained with *Penicillium commune* FCONT. The efficacy of the treatment was assessed using three diagnostic techniques: Laser Induced Fluorescence (LIF) and Attenuated Total Reflectance Infrared Spectroscopy (ATR) that recorded a marked reduction in the area of the peak of the Vinavil K40 after treatment with the fungal strain ($\sim$50%). Electronic Scanning Microscopy (SEM) with Microanalysis (EDS) revealed a clear alteration of the vinyl film after the bio-cleaning treatment. The study led to define a biological process mediated by a fungal strain able to attack the vinyl glue. For bio-cleaning applications it would be better to favour the use of non-spore-forming strains. In this case-study, however, the strain is applied in the physiological state of hyphae and the contact time useful for attacking the Vinavil is less than the time required for sporulation (Cutraro 2014).

Primal resin was the problem faced by the restorers in the restoration of the frescoes in the Carracci Gallery at Palazzo Farnese, in Rome (case 4). Carracci Gallery is considered the triumph of Baroque mural painting. The frescoes were affected by widespread small/medium-sized stains, due to the darkening of the resin used in the '70s, as a consolidator of the numerous cracks that occurred following a landslide of Tiber river. The fungal strain *Penicillium commune* FCONT and the bacterial strain of *Rhodococcus* sp. ZCONT (KY697119), previously selected as candidate for the Vinavil bio-removal experiments, were tested. The strains were separately immobilized in micro-packs of Laponite®RD and applied for 24 h on the frescoes. The micro-packs containing *Rhodococcus* sp. ZCONT perfectly removed the resin layer without damaging the different pigments present in the different areas of application.

Several of the case studies concerning stone material (case 6–11) have been carried out on the artifacts. Marble lends very well to this type of treatment with wet-packs applications, as there are generally no negative implications, despite those occurring in other situations, such as on wood or canvas paintings.

Some statues of the National Gallery of Modern and Contemporary Art, in Rome, have been treated with bio-cleaning (case 6). "La Lupa" by G. Graziosi, inspired by Rodin (Count Ugolino statue) represented an interesting and unusual case. La Lupa remained outdoors in the gallery gardens about 40 years and the main problem were large areas completely blackened by deposits of urban smog, penetrated over time in the marble matrix. These alterations were impossible to clean with traditional methods without negative consequences for the statue.

Given the nature of urban contaminants, a microbial formula consisting of three bacterial strains *Exiguobacterium* sp. LAM23 (EU019988), *Bacillus cereus* LAM30 (EU019990), *Bacillus subtilis* OSS42 (EU124568), degrading oil and growing on hydrocarbons as the only carbon source, have been used for bio-cleaning. The strains were isolated from the polluted site of Bagnoli and previously used for the bioremediation of diesel and heavy metals contaminated soil (Alisi et al. 2009; Sprocati et al. 2012). These bacteria produce extracellular surfactants through which they are able to solubilise hydrocarbons and make them available for biodegradation. Large areas of the statue were treated with the bacterial formula immobilized in Laponite®RD gel, by applying in succession three micro-packs for one night each one. The synergistic effect of bacteria producing bio-surfactants and Laponite®RD with absorbent power, has allowed acting in depth, leading to a significant lightening of the marble matrix. The combination of bio-cleaning with a traditional washing, performed to remove other common deposits, allowed to obtain a satisfactory degree of cleaning for the exposition.

Other statues of the Gallery, already treated with a traditional cleaning, still had residues of waxes and fats, not further removable with usual products. The sculptures "Testa di Donna" by E. Quadrelli (Fig. 11.2), "Idealità e Materialismo" by G. Monteverde and "Cleopatra" by A. Balzico have been successfully cleaned using different strains capable of removing waxes: *Acinetobacter calcoaceticus* LAM21, (JK534248), and fats and oils: *Serratia ficaria* SH7 (KP780180).

In collaboration with the Cultural Heritage Conservation and Restoration Centre "La Venaria Reale" of Turin, it has been performed a series of bio-cleaning tests on the "Bacchus with basket", a Roman statue in Greek marble, property of the Turin Museum of Antiquities (case 7). The restoration of the statue involved the reassembly of all the fragments separated in the early twentieth century and has foreseen, in the cleaning phase, the execution of experimental tests for the removal of various deposits, the nature of which had been diagnosed by the FT-IR and XRF analysis. The bio-cleaning has been tested on a piece on right leg, left dirty after cleaning with saturated ammonium carbonate. The FT-IR analysis identified the deposit as Paraloid B72. The appearance of the piece was a black/brown and removal tests with acetone had no effect.

**Fig. 11.2** Testa di Donna- Quadrelli. The portion treated with a micro-pack containing living cells of LAM21 strain was cleaned in an overnight application

A single micro-pack containing the bacterial strain *Rhodococcus* sp. ZCONT immobilized in Vanzan®NF was applied overnight on the half of the dowel covered by Paraloid B72. After removal of the pack, a slight mechanical action with a diluted acetone-soaked swab was sufficient to completely remove the Paraloid residues, discovering the completely browned surface. FTIR spectra, before and after the treatment, confirmed the effectiveness of the treatment. Restorers reported that the noble patina of the original material was preserved unlike other more aggressive products (Galizzi 2015).

In the case of the pedestal of a statue in the Vatican gardens (case 8) copper oxides were responsible of stains. The patches have been treated with micro-packs containing the bacterial strain *Sphingopyxis macrogoltabida* SME 3.14 (MT023684) immobilized in Vanzan®NF. The strain was isolated from a mine in Sardinia and it has been selected for its capacity to produce siderophores, to have a high esterase activity and to not be able to solubilise carbonates.

The case 11 faces with brown stains on marble. During restoration of the Carracci Gallery (Palazzo Farnese, Rome), the sixteenth century portals of Carrara marble presented brown and other rust-coloured stains due to pins or brackets used for the repair of cracks due, as well as for the frescoes, to the past landslide of Tiber river. In this case the restorers put forward the hypothesis that the brown stains present on the sixteenth century portals of the Gallery may be due to an emergence of the iron contained in the marble, due to maintenance or cleaning treatments previously carried out, probably with acids and waxes.

Staining of the stone can be attributed to corrosion of inclusions of iron minerals present in the same stone or oxidation of metal parts used as elements purely aesthetic or for structural strength. The chemical removal of stains of iron compounds on marble shows the drawbacks encountered using phosphates (aggression of the limestone), oxalates (not very effective), fluorides (good solubility of iron in a saturated solution of ammonium fluoride), salicylates (unusable as it has a strong red colour). Alternatives could be phosphoric acid, phosphoric acid and EDTA disodium salt in water, glycerine and gypsum, reagents that reduce $Fe^{3+}$ to $Fe^{2+}$, and then re-oxidize producing new iron hydroxides or hydrated salts more easily removable. It has experimented the use of another complexing agent, the thioglycolic acid neutralized with ammonia, which has produced good results but it is restricted in the use because of the toxicity for the operators. Even cysteine has been proposed to treat the iron stains on marble. It is an amino acid with a high structural similarity with thioglycolic acid, but not toxic, and it has an amino group, which gives it three complexing groups. Cysteine can act both as bidentate and tridentate binder, the latter in particular with $Fe^{3+}$through the thiol and carboxylic groups. Having three donor groups and being the reactions variable according to the pH and the redox potential, it has a chemical very complex, and its chemical and kinetic behaviour is very difficult to understand. All the reactions, however, give a dark colour to the end product, which can vanish with the time.

A biological approach is based on the exploitation of bacteria producing siderophores, water-soluble molecules with low molecular weight, which can bind in a specific manner $Fe^{3+}$. Bacteria have acquired the ability to synthesize and secrete siderophores to overcome the shortage of bioavailable iron in nature. Siderophores are usually released into the extracellular environment, where they act as a captor of iron. Specific cell receptors allow the entrance into the cell of the complex Fe-siderophore. Once in the cytoplasm of the cell, the complex dissociates, with reduction of ferric iron to ferrous iron. The siderophore empty of the reduced iron can be degraded or reused releasing it into the extracellular medium. This topic has been addressed in a preliminary work with a series of tests performed on laboratory specimens prepared ad hoc, reproducing rusts on different stone materials (Pieragostini 2015).

On the Carracci Gallery portals we experimented the bacterial strain *Pseudomonas protegens* FeIIC1 (KY765342), isolated from mine drainage waters and selected for its high siderophores production.

After preliminary test on small portions, Vanzan®NF micro-packs containing the bacterial strain FeIIC1, added with a very low concentration of cysteine (0.1% w/v), were applied overnight on almost the entire surface of the four portals. The rust disappeared and the brown stains inside the crystalline matrix had lost their brown colour or were largely lightened. After the treatment, a very slight dark nuance due to cysteine, vanished with the heat of the day without leaving a trace.

The most recent study on marble was carried out on Tombs of Giuliano and Lorenzo de' Medici sculptured by Michelangelo in the Medici Chapel in Florence (case 10). It was not available a precise diagnosis of the composition of the deposits to be cleaned, as merely non-invasive analyses were performed. Based on the conservative history, restorers referred that casts have been made, the first by Vincenzo Danti around the '70s. Release agents for casts were presumably animal gelatine, oils and mixtures with soaps, used in bronzes at the time. The restoration that followed left some residues and presumably applied protective agents, such as waxes, which over time changed to a yellowish-orange tone. Withdrawals with swabs made in these areas did not give indications (personal communications by restorer M. Vincenti).

As for the conservative history of the sarcophagus of the monument to Lorenzo Duke of Urbino, the substantial traces of organic material should have been present since ancient times. The diffractometric analysis on the patina sample on the back of the sarcophagus revealed the presence of inorganic compounds (calcite, silicate, chalk). Infrared spectroscopy performed on the blackest sample suggests the presence of phosphates, which realistically could refer to the hypothesis of body fluids.

Combining historical reconstruction with the few analytical results, useful bacterial strains have been chosen from the in-house collection and tested. The marble rear floor of the altar of the chapel offered a palette to experiment with different bacterial strains for cleaning coherent generic dirty deposits. The bio-cleaning tests were performed directly on limited areas of the sculptures.

Eight different bacterial strains have been individually immobilized both in Laponite®RD and Vanzan®NF micro-packs. Depending on the specific characteristics of the portion to be treated, Laponite®RD and Vanzan®NF micro-packs were chosen. On statues, exclusively Vanzan®NF was used, for technical needs or for opportunity assessments. Vanzan®NF is a necessary choice when the surface to be treated is not smooth, such as in the case of locks of hair, ears, facial folds; moreover Vanzan®NF does not have absorbent power in itself, perfectly retains humidity and does not shrink.

Thus, on the head of Duke Giuliano a bio-cleaning was performed first using a micro-pack with Vanzan®NF without bacteria. On the head of the Night, good results have been obtained treating the hair with *Pseudomonas stutzeri* CONC11 and an earmuff with *Rhodococcus* sp. ZCONT, the deposit has been removed without interfering with the marble below.

On the sarcophagus and the altar floor, both Laponite®RD and Vanzan®NF micro-packs were tested for each strain, in order to assess which was the most effective combination between bacteria and support agent.

On the sarcophagus three strains have been tested: *Rhodococcus* sp. ZCONT with lipase activity, degradation resin, oil degradation, bio-surfactant production, calcite precipitation, no proteolytic activity, *Pseudomonas fluorescens* LAM33 very reactive for phosphates solubilisation, and *Cellulosimicrobium cellulans* TBF11, for carbonates solubilisation. The control micro-packs did not produce significant differences compared to the situation prior to treatment, nor between them. Among the tested strains, *Rhodococcus* sp. ZCONT was truly active on the coherent deposit and the best cleaning quality was obtained with the strain applied in Laponite®RD, providing a very good cleaning quality, for the degree of cleanliness, homogeneity and the respect of the proportions.

The back of the sarcophagus, where very dark, sometimes black stains were present and phosphates had been detected, was treated with the *Serratia ficaria* SH7 strain in Laponite®RD, obtaining a total removal of the deposit, except in the thickest points, where it was necessary to proceed with several applications.

By the literature, the artworks undergone bio-cleaning are mainly frescoes, mural paintings and stone (Bosch-Roig et al. 2014; Mazzoni et al. 2014). Case 12 faced a challenge, it concerned the removal of an aged darkened organic deposit from a fragile surface of a sixteenth century wood painting. The experiment was carried out in collaboration with the laboratory of paintings and the Scientific Research Cabinet of the Vatican Museums, where the painting was under restoration.

Main challenge was to meet the conflicting requirements of microorganisms - needing aqueous solutions- and painting - refractory to a long exposure to moisture. The final strategy was to couple the bio-cleaning with a chemical pre-treatment as protective of the fragile painted surfaces (Crisci et al. 2020).

The deposit consisted of an animal glue with oil, called "colletta con olio", very tenacious, altered and resistant to the usually effective chemical treatments without further damaging the painting. The original recipe includes different oils, gall and molasses.

Two original bacterial strains, *Serratia ficaria* SH7 and *Pseudomonas protegens* FeIIC1 non-pathogenic and no spore-forming, were selected following a metabolic screening, where different bacterial strains have been tested on the individual substrate composing the original recipe.

Artificial wood specimens, simulating the artistic technique and covered by a layer of "colletta con olio", were weathered and used as laboratory tests, to simulate the application on the artefact. After tests on the specimens, using different support agents and different application times (from 8 to 24 h), the best procedure resulted in applying the strains entrapped in Vanzan®NF, used for the first time to this purpose. Micro-packs allowed a complete removal of the "colletta" without altering the colours of pigments, nor raise the pictorial film and without causing swelling of the wood. The results allowed applying the procedure directly on the wood painting, where this effective combination led to the selective removal of the deposit, through a long-lasting contact of bacteria with the sensitive and hygroscopic material of the painting, without damage for the artwork.

This experiment was significant, as it demonstrated the possibility of degrading very tenacious and refractory deposits without the use of toxic products; the possibility of proceeding with the bio-cleaning method even on materials sensitive to humidity and on very fragile surfaces. In perspective, an optimisation of the procedure can be feasible by acting on microbial metabolism, to reduce the application time needed to a complete removal of tenacious deposits.

Animal glues were historically used both in paper manufacturing, during the sizing process, and in conservation as adhesive for the lining of prints or graphics and for the creation of *passe-partout*. Animal glue films are highly hygroscopic and go through degradation by ageing. The development of internal stresses affects the glue's elasticity, strength and physical stability and may lead to significant damage to the substrate. Humidity, temperature, UV radiation and pollutants can induce deterioration phenomena, such as protein cross-linking, hydrolysis of peptic bonds, oxidation, while the presence of microorganisms can lead to the production of acid metabolites and pigmented spots, which cause a strong optical-chromatic alteration. So, the removal of glue staining and the detachment of prints from the aged support become an essential step in the restoration and conservation of paper.

The chemical and mechanical methods in use, show serious drawbacks, whiles do not offer appropriate solutions, are too aggressive and use toxic products. To date, the use of enzymes is the only bio-based method. The need for skilled operators, together with the optimal application conditions required (high temperature, stable pH conditions, favourable saline concentrations), and the high costs have created difficulties in mastering the enzyme use so far.

Following a screening, a bacterial strain *Ochrobactrum* sp. TSNRS 15 (EU249585), deriving from an Etruscan hypogeum, was selected from the laboratory collection, able to grow on Cervione and rabbit glues as sole carbon source and devoid of cellulolytic activity. Original paper samples, kindly supplied by the National Institute for Graphics (case 13), representing the back supports of ancient prints from a historical volume of prints from the collection of the Institute, have been treated with the bacterial strain immobilized in an agar gel. The treatment showed its efficacy already after 4 h of incubation, allowing the complete removal of the thick layer of glue from the surface of the paper specimen. The colorimetric measurements allowed to assess the whitening of the specimens, by the increase in the L* and the decrease in the b* coordinates. SEM observations showed cellulose fibres appearing from the compact paste layer after 4 h of treatment, demonstrating the disappearance of the adhesive layer consumed by bacteria as a carbon source. To avoid undesired secondary colonization due to the remains, a final check verified that no undesirable residues were left over after the treatment.

The removal of copper oxide stains from decorative stuccos in the chapel of San Pietro, in the basilica of Santa Prudenziana (case 15, Fig. 11.3), at Rome was tested through bio-cleaning with the bacterial strain *Sphingopyxis macrogoltabida* SME 3.14 in Vanzan®NF 6% W/V. SME 3.14 is a producer of siderophores and shows esterasic activity. The stuccoes had light, medium and high intensity green stains, due to the oxidation of the "porporine" made of brass-based alloys (copper, zinc,

**Fig. 11.3** S. Prudenziana stuccoes with copper oxides blue/green spots (**a**), removed after an overnight application with a micro-pack containing SME 3.14 strain (**b**)

iron) used in previous restoration works to replace the original gilding. Nine points with different intensities were treated. The bio-cleaning tests were carried out both by applying the micro-packs by direct contact and by interposing a tissue paper. Each application was in contact for 24 h. The most intense spots were cleaned with two consecutive applications. Dino-Lite digital micro-images taken for each point before and after the treatment demonstrated the fading or an attenuation of the stains.

The application of Vanzan®NF micro-pack in direct contact with the surface of the stucco, without the interposition of paper, led to a greater adhesion of the supports to the stucco, with greater risk that the product could not be completely removed; in fact, in the portions where the product was left for 48 h the stucco absorbed part of the water of the support, making it considerably more adherent to the substrate.

As far as efficacy is concerned, we do not yet have enough data to establish which application method is preferable (by placing an interface or directly in contact) or which application time can be recommended.

Cleaning of stucco from copper oxides is therefore biologically viable, but the procedure requires further tests to be optimized.


## 4 Biodeteriogens Removal and Control

The term biodeterioration was defined about 50 years ago by Hueck (1965) as "any unwanted change in the properties of a material, caused by the vital activity of living organisms". In many natural environments, the physical and chemical transformations of materials are considered necessary and positive conditions; however, on artistic material the transformations of the substrate induced by the colonizing microflora, in association with various environmental factors, are seen as a destructive and negative phenomenon, both from a cultural and economic point of view. For a long time, abiotic factors were thought to affect physicochemical properties, facilitating subsequent microbial colonization. Furthermore, it was believed that the phenomena of biodeterioration would modify the substrate only from an aesthetic point of view, that is, inducing the appearance of patinas and coloured spots on the surface, following the release of biogenic pigments. Recently, several analytical approaches have demonstrated that microorganisms are also responsible for physical, chemical and aesthetic modifications of the colonized surface (Sterflinger and Piñar 2013). They can use the surface as growth support, using mineral components or surface deposits as metabolites necessary for their development. The speed and extent of microbial colonization are influenced not only by environmental parameters, such as the availability of water, but also by the physical-chemical properties of the material, such as the mineralogical composition. In fact, although microorganisms colonize a certain environment permanently, their harmful activity is not constant, but periodic.

Today the concept of biodeterioration is intended with a broader meaning: it is known that the phenomenon is controlled by synergistic and antagonistic relationships between colonizing microflora and environmental factors; therefore, it cannot be considered as an isolated phenomenon, as it occurs in conjunction with other phenomena of physical and chemical degradation, and it is therefore difficult to attribute any specific damage to a single cause.

Currently, to control or eliminate the phenomena of biodeterioration, restorers can intervene with mechanical (chisels, spatulas, scalpels), physical (UV rays) and chemical methods (biocides).

The biocidal products have been classified by the EU in four categories and twenty-two product-types and their use is regulated under the Biocidal Products Regulation (BPR, Regulation (EU) 528/2012, Biocidal Products Regulation (EU) 2012) (replacing the Biocidal Products Directive 98/8/EC 1998). Nowadays, many of the broad-spectrum biocides have severe limitations in their use for the toxicity and persistence in the environment. In conservation and restoration, only few of the existing biocides can be used, due to the lack of information about their interaction with the historic materials of the artworks.

Benzalkonium chloride, one of the most used substances in disinfection as part of the formulation of many commercial biocides used by restorers is being reviewed for use as a biocide in the EEA and/or Switzerland according to the ECHA site (2020).

Another aspect not to be underestimated in the use of biocides in the restoration is the onset of resistant microbial strains. As it happens in clinical microbiology, where the thoughtless use of antibiotics determines the development of resistant microorganisms, the repeated use of the same biocide or the choice of incorrect concentrations can determine the selection, in the microflora present on the artwork, of strains resistant to the biocide used. This phenomenon is very serious because it can determine the onset of attacks by biodeteriogens that become difficult to eradicate, considering that often the resistance acquired extends to the whole class of biocides to which the product used belongs. This was exemplified in Lascaux Cave, France, where the indiscriminate use of benzalkonium chloride resulted in an explosive bacterial and fungal attack (Bastian et al. 2009; Martin-Sanchez et al. 2012).

## 4.1  "Green Biocides"

There is an increasing interest in the use of naturally-produced compounds, and the need of alternative solutions led the researchers to explore the world of the natural substances such as plant extracts or bio-based products, that will be more easily degraded and environment friendly. Plants extracts are already used as biocides in food, medicine and different fields of the pharmaceutical industry. In the last years they have been tested for other applications, such as in the biodeterioration control of Cultural Heritage as a valid alternative to the traditional biocides (Rotolo et al. 2016). Many of these compounds are phenols, polyphenols, terpenoids, essential oils, alkaloids, lectins and mixtures of polypeptides derived from plants (Guiamet et al. 2006).

Oregano, thyme, clove and arborvitae essential oil have been tested for assessing the antimicrobial potential and might be used as broad-spectrum antimicrobial agents for decontaminating an indoor environment (Puškárová et al. 2017). A pilot study was carried out at the Vatican Gardens (Devreux et al. 2015), the applications of essential oil of thyme and oregano have been compared with chemical products based on quaternary ammonium salts such as benzalkonium chloride, on stones affected by biodegradation. The first results obtained showed that the synergistic use of the two essential oils allows to obtain the best biocidal efficacy. Recently an anti-musk product based on essential oils from plant became available on the market (Essenzio, Ibix Biocare). Anyway, essential oils use is not free of concern, since lavender derivatives, including essential oil, are listed in REACH registrations among the substances that cause "serious eye irritation, are harmful to aquatic life with long-lasting effects and may cause an allergic skin reaction".

Many bio-based substances have been investigated till now, as reported in a recent extensive review (Fidanza and Caneva 2019) that lists 61 natural substances, among essential oils and substances of plant and microorganism origin. Their application produces highly variable results due to a lack of a coherent assessment of the best practices, showing the need of a standard methodology in the use of natural biocides for controlling biodeterioration.

For these reasons, our effort was to develop some alternative ways to remove the biological patinas and possibly control the further microbial development, all of them based on natural products of plant and bacterial origin that are harmless for the health and the environment, using a standard procedure involving the pack application. The bio-products have been applied in the real-case studies listed in Table 11.2; here we describe briefly their origin, characteristics and range of applications performed until now.

## 4.2  BioZ

Surface active agents or "surfactants", are a group of molecules that lower the surface tension between two liquids or between a liquid and a solid. In the practical sense, surfactants may act as wetting agents, emulsifiers, foaming agents and dispersants. Bio-surfactants are synthetized by microorganisms and they are being studied for antimicrobial properties and their potential use as biocides on artefacts (Grimaldi 2009; Rivardo et al. 2009; Desai and Banat 1997). Their action is still unclear; they may inhibit infections through signals that interact with the host and/or bacterial cells or prevent the microbial adhesion and growth on various substrates (Rodrigues et al. 2006). Some bio-surfactants may show antimicrobial activities which could be applied to conservation of cultural heritage.

BioZ is a crude extract containing extracellular glycoproteins, produced by the strain MCC-Z (JF279930) from the ENEA-Lilith collection. The strain was identified by 16SRNA sequencing and belongs to the *Sphingobacteriaceae* family and *Pedobacter* genus.

The product BioZ has a good surfactant and emulsifying action which is preserved after freeze-drying: reduces the surface tension by 40 mN/m and generates an emulsion stable over time, temperature, pH and salinity, with almost constant values up to 4 months, ensuring the effectiveness in multiple application conditions (Beltrani et al. 2015).

BioZ product does not contain live bacterial cells, since its preparation involves an autoclave treatment at 121 °C for 20 min, it is applied incorporated in a suitable support to the specific situation to facilitate its application and removal without leaving residues on the surface.

The emulsifying capacity of BioZ has been tested on many hydrocarbon compounds (hexadecane, toluene, xylene, isooctane, cyclohexane and diesel) which represent common environmental contaminants and are often included in the superficial deposits of artistic artefacts. BioZ could also facilitate the removal of unwanted deposits of dust and smog particles, which often cover the surface of artifacts exposed to atmospheric agents. Its emulsifying activity was tested in the cleaning of laboratory samples from fumes of fuel oil, fats and kerosene.

BioZ antimicrobial activity was tested in-vitro on microorganisms known to be responsible for biodeterioration of Cultural Heritage, such as *Cellulomonas* sp., *Bacillus cereus*, *Bacillus pumilus*, *Bacillus megaterium*, *Acinetobacter calcoaceticus, Rhodococcus erythropolis, Rhodococcus* sp., *Paenibacillus* sp., showing an inhibitory effect on all the *Bacillus* species and on *Cellulomonas* sp. The different effectiveness was due to the different sensitivity of the single microbial strain tested. It was effective at very low concentrations (0.05% w/v) and showed an inhibitory effect on Gram-positive strains but not on Gram-negative bacteria.

## 4.3   LIQ

Licorice (*Glycyrrhiza glabra*) root extract is rich in different classes of phytocompounds, such as phenols (liquiritina, isoliquiritin, liquiritigenina, isoliquiritigenina, glabridina and glabrol), terpenoids and saponins (β-amyrin, glycyrrhizin, glycyrretol, galabrolide, licorice acid), volatiles compounds (benzaldehyde, fenchone, linalool, anethole, estragole, eugenol and hexanoic acid), vitamins (B1, B2, B3, B6, C, E, biotin, folic acid and pantothenic acid), coumarins (glycyrine, umbelliferone, ligcoumarin and herniarin) and mineral content (Öztürk et al. 2018). These are bioactive compounds which anti-bacterial, anti-fungal, anti-viral, anti-tumour, anti-inflammatory, anti-oxidant, anti-allergic, expectorant, anti-malarial and anti-convulsive activities have been widely demonstrated (Abbas et al. 2015). In addition to the innumerable properties of the root extract, phytochemical studies have recently been carried out on the extract of licorice leaves which have shown the presence of classes of phenols and acid compounds such as glabranin, licoflavanon and pinocembrin (Scherf et al. 2012). These compounds are present only in traces or completely absent in the roots, but which are also known for their antimicrobial and anti-fungal action in plants and men.

The company Trifolio-M GmbH (Lahnau, Germany) has created a formulation from the extract of licorice leaves (LIQ), commonly considered an agricultural waste. The alcoholic LIQ extract, prepared from dried and finely ground licorice leaves, is still in experimental phase and not yet commercialized. Trifolio-M is studying LIQ for its antimicrobial activity against plants pathogenic fungi for application in agriculture. In parallel, LIQ extract is under study by ENEA to verify the possibility of exploiting the antimicrobial properties of this extract towards the biodeteriogens affecting artworks and monuments. The antimicrobial effect of the extract was demonstrated in-vitro towards about 20 bacterial strains (belonging to the phyla *Firmicutes, Actinobacteria* and *Gammaproteobacteria*) and 10 fungal strains (belonging to the genera *Aspergillus, Fusarium, Epicoccum, Penicillium* and *Cladosporium*), isolated from hypogeal environments, frescoes, wall and canvas painting, and included in the ENEA collection of microbial strains. Fungal strains were less sensitive to the extract than bacterial strains; however, the intermediate sensitivity level shown by *Cladosporium* spp. and *Epicoccum nigrum* is a significant result, given that these fungal genera, widespread in hypogeal environments, are characterized by the production of melanins and other deleterious pigments.

## 4.4 SME 1.11

*Arthrobacter xylosoxidans* SME1.11 (SUB4060614) is a bacterial strain isolated from the soil of Ingurtosu mining site (Sardinia, Italy). Its cells are non-motile, non-spore forming, Gram-positive, aerobic and rod-shaped. The optimal laboratory growth conditions include 4% of salinity and pH 6 and 28 °C; SME1.11 shows no reducing power, no proteolytic and lipase activities but high pectinase activity, produces siderophores and calcite crystals, and lactate tolerance. This bacterial species belongs to the risk Group 1 that contains non-pathogenic organisms, i.e. do not cause disease in healthy adult humans.

The capacity to degrade pectin and produce siderophores are two characteristics that made this strain a good candidate for biodeterioration control. Further biochemical characterization of SME1.11 properties and its antimicrobial activity are under investigation.

## 4.5 NopalCap

A natural plant product obtained from the mucilage of *Opuntia ficus-indica* combined with the alcoholic chili extract - NopalCap- is under investigation for its potential application as additive in mortars to reduce biofilm colonization. This product has been studied and carried out in the frame of a bilateral cooperation project between Mexico and Italy, supported by the Italian Ministry of Foreign Affairs and International Cooperation (Progetto Grande Rilevanza PGR00971). According to the Mexican tradition, the addition of small amounts of *Opuntia* mucilage to lime mortar allowed to a good preservation of ancient mural paintings and other artworks. The bio-receptivity of mortars containing different percentages of mucilage was tested (Persia et al. 2016) and the positive results encouraged us to expand the study to include a further plant substance, already known for its antimicrobial property, i.e. the *Capsicum* extract. The new bio-product called NopalCap was at first tested in laboratory on bio-mortar specimens to verify the capacity to prevent a microbial attack. Then was used as additive in the production of lime mortar and hydraulic mortar for restoration intervention under the supervision of restorers (see Table 11.2).

## 4.6   Real Cases Application

The LIQ extract was tested at 3% v/v concentration on a painting on canvas for the elimination of mixed patinas including the bacterial species *Bacillus licheniformis* and *Bacillus subtilis*, and the fungal species *Arthrinium* sp., *Aspergillus* sp., *Cladosporium* sp. The treatment was proven effective both in reducing the microbial load and in preventing over time (up to 1 year later) the re-proliferation of microorganisms.

Further evidence of the potential of this formulation in relation to its efficacy towards biodeteriogens was carried out on the frescoes of the chapel XIII of the Sacro Monte di Orta (Novara), affected by a very diffused pink patina, attributable to colonization by strains of the genus *Arthrobacter* (isolated and identified by the researchers at SUPSI, The University of Applied Sciences and Arts of Southern Switzerland, Lugano). The biodeteriogen strains responsible of the pink patina were tested in laboratory for sensitivity to different natural products (NopalCap, BioZ, Liq), in comparison with benzalkonium chloride (Benz) as positive control. After a well diffusion assay LIQ extract was selected and the Minimal Biocide Concentration was quantified as 0.4% v/v, one order of magnitude higher than Benz. After a few tests on the frescoes, to verify the influence of the product on the colour, the application was carried out in the Chapel XIII, during February 2019, on a test area of about 20 square metres, intended for experimentation. The product has been applied by nebulization at the concentration of 1.5% v/v and left on the wall up to 48 h. The treatment did not cause unwanted aesthetic alterations and, after the biocide treatment, the restorers completed the cleaning of the frescoes. The first microbiological monitoring after treatment did not detected the presence of the bacterial species correlated to the pink patina. Monitoring over time is underway to evaluate the preventive or retarding properties of a new colonization, by LIQ extract.

LIQ is also under investigation in the treatment of XVIII century stone sculptures placed in the garden of a private villa near Treviso (Italy) in collaboration with a

**Fig. 11.4** Biofilm removal on the stone pedestal of a statue in the Vatican Gardens. The products were applied in cellulose pulp, (**a**) the surface after the cleaning with 1- BioZ; 2- water; 3- LIQ 3%. (**b**) the same surface 18 months after the treatments: the portion cleaned with water (2) shows a new biofilm growth, while the portions treated with BioZ and LIQ are still clean

local restoration firm (Monica Casagrande). After the application, the long-lasting effect on the delay in biofilm growth is being monitored.

BioZ and LIQ products were used in preliminary test on the basement of a statue in the Vatican Gardens. The application was performed absorbing the liquid substances with cellulose pulp before their spreading on the surface previously washed with deionized water. The packs were left in situ for 24 h and then removed (Fig. 11.4a). Comparing the effects on the microbial recolonization 18 months after the treatments (Fig. 11.4b) we observed that the portion used as control, treated with only water, was partially covered with a biofilm, while the two portions treated with BioZ and Liq were still free of colonization.

In collaboration with University of Tor Vergata (Rome) tests on biofilms sampled at the Domus Aurea (Rome) have been carried out (Rugnini et al. 2019) using alcohol extracts from *Glycyrrhiza glabra* (LIQ) and *Capsicum* (CAP), singularly or mixed. Biofilms samples were collected in one of the hypogeal rooms from an undecorated wall and were then homogenized and inoculated in BG11 agar medium. Observations showed that the biofilm was dominated by the cyanobacterium *Scytonema julianum*, often described from other hypogeal environments and known to deteriorate the substratum integrity by dissolution of minerals and the precipitation of calcium carbonate on its sheaths. Within the sampled biofilms some species belonging to *Proteobacteria*, *Actinobacteria* and *Bacteroidetes* and three fungal strains were also identified. The biofilms were treated twice with the extracts at 5 days-interval, and the photosynthetic response of the biofilm was followed for 5 days with a mini-PAM portable fluorometer. Photosynthesis is highly susceptible to this kind of treatment, so measurements of rates were used as a proxy for cell

**Fig. 11.5** Biofilm removal on marble artifact at Diocleziano Termae, Rome. (**a**) shows the scheme of the trial, with the product applied in cellulose pulp (P) and in Vanzan gel (V). From left to right: BioZ in cellulose pulp and Vanzan, LIQ in cellulose pulp and Vanzan, NopalCap in cellulose pulp and Vanzan, SME1.11 in Vanzan, C- (water in cellulose pulp) and C+ (benzalkonium chloride 1% in cellulose pulp). (**b**) the result soon after the pack removal

health. Changes in photosynthetic activity of the samples treated with the extracts were compared to control biofilms receiving no treatment. Results showed that LIQ 30% v/v had the highest photosynthesis inhibition potential, followed by LIQ extract 10% v/v. *Capsicum* extract was the least efficient. These initial results will be followed by applications on test areas on the wall and roof surfaces within the Domus Aurea site.

NopalCap was used in preparing a hydraulic mortar to fill the voids in the base of a statue, placed in the patio of the Conservatory of Naples (Italy). A year after the restoration, the mortar is free from biofilm attack, while the neighbouring areas show recolonization of the surface. This experiment had been preceded by laboratory tests on hydraulic mortar specimens also subjected to artificial ageing which gave excellent results and will be included in the Dissertation of a thesis on the restoration of the whole statue (Lorenza Cardone, Accademia delle Belle Arti di Napoli).

In collaboration with the University of Tuscia (Viterbo) and Superintendency for Roman Museum, a broad comparison of the different bio-products under investigation in our laboratory was carried out at Diocleziano's termae (Rome, Italy), on a marble artefact placed outdoor, showing a diffuse biofilm (Fig. 11.5). The surface was examined using the digital microscope Dino-Lite and the biofilm was sampled to study the microbial patina. High-throughput molecular analyses were performed on the biofilm samples: by a preliminary valuation of the data, the biofilm collected before the treatment was dominated by *Cyanobacteria*, mostly belonging to *Nostocales* (76% of the total bacterial diversity) and by fungi belonging to *Ascomycota* (90% of the total fungal diversity).

Particular attention was paid to the application process: the liquid products were immobilized in both cellulose pulp and in Vanzan®NF to test the best procedure for application. BioZ, LIQ 3%, NopalCap 10% and SME1.11 (only in Vanzan®NF) plus a negative control (water in cellulose pulp) and a positive control (benzalkonium chloride 1% in cellulose pulp) were kept in place for 2 weeks, covered with a plastic and aluminium foil to delay the dry out. As for the application procedure, it was evident that cellulose pulp is not suitable for this kind of products: during the laying of the mixture on the vertical surface, the cellulose pulp did not retain the liquid phase, while the Vanzan®NF gel was able to keep the product into its matrix. After the packs removal, the surface was washed with deionized water, let dry and the results were recorded (Fig. 11.5b): the best biofilm removal and surface cleaning was obtained with SME 1.11 in Vanzan®NF, followed by the pack with water in cellulose pulp. The other bio-products gave less effective results but better when applied in Vanzan®NF; the positive control showed a transient chromatic alteration. Bioluminometer measurement (unpublished data) allowed testing the efficacy of the bio-product on the biofilm control after 7 months. The value on the untreated area was 145,000 RLU, and the positive control (benzalkonium chloride) 8000 RLU while the treated areas show intermediate values, lower than 25,000 for the all the Vanzan®NF -packs.

The visual inspection carried out 11 months after the trial (Fig. 11.5c) showed the persistence of the clean areas for all the treatments except for NopalCap, where the biofilm seems to grow again. Monitoring of the surface is ongoing to evaluate the persistence of the biodeteriogen control.

# 5  Concluding Remarks

The demonstration cases of bio-cleaning shown and discussed in this chapter provide, together with the research studies carried out by the scientific community, a convincing proof of feasibility of using biotechnology applied to Cultural Heritage. The next step requires the transition from research to production, to make the procedures and proposed products available to the market for restorers. To this end, an effort is needed to define the repeatability of the procedures, the stability of the strains and products when transformed into a marketable product (ex: freeze-dried), to reduce the contact times and to move towards a simplification of the procedures, by selecting multifunctional and versatile microbial strains.

Our experience leads to say with certainty that investigations in the search for new products and procedures for a more sustainable strategy in cultural heritage conservation, especially with a view to reducing or replacing toxic/dangerous products, are flourishing and are widening the field of goods and the issues involved.

Around these issues numerous interests, both scientific and "policy" are gathering. That also implies interventions on the legislative, regulatory and organizational level.

In order for this process to be carried out and to take place in a virtuous way, a joint contribution from the world of Conservation, research, entrepreneurship, stakeholders, the market and public decision-makers is needed, at European and International level.

The market for new restoration products that are harmless to operators, the environment and works of art must be viewed from the illuminated perspective of sustainability, as it has a very high social, historical and cultural value, although it can be a niche market.

**Acknowledgement** We are grateful to all the restorers and Institutions that have believed in innovation, looking forward to a sustainable restoration. We especially thank the restorers Adele Cecchini and Lorenza D'Alessandro, who first involved us in the experimentation in their restoration sites and who have spread the interest among their colleagues. Special thanks also go to Maria Bartoli of the Domus Aurea, to Professor Santamaria, head of the GRS of the Vatican Museums and to his collaborators, to Marina Vincenti and Daniela Manna, to Rodolfo Corrias of National Gallery GAM, to Carla Tomasi and Paolo Pastorello, to the IscR and to a long list of people who collaborated in each of the case studies, which cannot be contained here.

The NopalCap studies are supported by the Italian Ministry of Foreign Affairs and International Cooperation [Progetto Grande Rilevanza PGR00971 and 000784].

# References

Abbas A, Zubair M, Rasool N, Rizwan K (2015) Antimicrobial potential of *Glycyrrhiza glabra*. J Drug Design Med Chem 1:17–20

Alisi C, Musella R, Tasso F, Ubaldi C, Manzo S, Cremisini C, Sprocati AR (2009) Bioremediation of diesel oil in a co-contaminated soil by bioaugmentation with a microbial formula tailored with native strains selected for heavy metals resistance. Sci Total Environ 407:3024–3032

Barbabietola N (2013) Innovative microbe-based technology for Conservation and Restoration. PhD thesis dissertation, Science for conservation of cultural heritage, University of Florence

Barbabietola N, Tasso F, Alisi C, Marconi P, Perito B, Pasquariello G, Sprocati AR (2016) A safe microbe-based procedure for a gentle removal of aged animal glues from ancient paper. Int Biodeterior Biodegradation 109:53–60

Bastian F, Alabouvette C, Jurado V, Saiz-Jimenez C (2009) Impact of biocide treatments on the bacterial communities of the Lascaux cave. Naturwissenschaften 96:863–868

Beltrani T, Chiavarini S, Cicero DO, Grimaldi M, Ruggeri C, Tamburini E, Cremisini C (2015) Chemical characterization and surface properties of a new bioemulsifier produced by *Pedobacter* sp. strain MCC-Z. Int J Biol Macromol 72:1090–1096

Bio-Based Industry Consortium (2017) Strategic Innovation and Research Agenda (SIRA). https://www.bbi-europe.eu/sites/default/files/sira-2017.pdf

Biocidal Products Directive 98/8/EC (1998). https://eurlex.europa.eu/LexUriServ/LexUriServ.do?uri=OJ:L:1998:123:0001:0063:EN:PDF

Biocidal Products Regulation (EU) 528/2012. https://eurlex.europa.eu/LexUriServ/LexUriServ.do?uri=OJ:L:2012:167:0001:0123:EN:PDF

Bosch Roig P, Regidor Ros JL, Montes Estelles R (2013) Bio-cleaning of nitrate alteration on wall paintings by *Pseudomonas stutzeri*. Int Biodeterior Biodegradation 84:266–274

Bosch-Roig P, Lustrato G, Zanardini E, Ranalli G (2014) Bio-cleaning of cultural heritage stone surfaces and frescoes: which delivery system can be the most appropriate? Ann Microbiol 65:1227–1241

Colombini MP, Bonaduce I, Gautier G (2003) Molecular pattern recognition of fresh and aged shellac. Chromatographia 58:357–364

Crisci L, Alisi C, Pratelli M, Santamaria U, Sprocati AR (2020) Il Biorestauro: messa a punto di una procedura su misura per l'opera. In: Tra cielo e terra. La Madonna della Cintola di Vincenzo Pagani. Edizioni Musei Vaticani (in press)

Cutraro S (2014) Ricerca di una via biologica per la rimozione di colle viniliche da affreschi: il caso studio degli affreschi delle cappelle di Saint Jean e Saint Martial (Palazzo dei Papi, Avignone). Master's thesis dissertation, Sapienza University, Rome

De Vero L, Boniotti MB, Budroni M, Buzzini P, Cassanelli S, Comunian R, Gullo M, Logrieco AF, Mannazzu I, Musumeci R, Perugini I, Perrone G, Pulvirenti A, Romano P, Turchetti B, Varese GC (2019) Preservation, characterization and exploitation of microbial biodiversity: the perspective of the Italian network of culture collections. Microorganisms 7:685

Desai JD, Banat IM (1997) Microbial production of surfactants and their commercial potential. Microbiol Mol Biol Rev 61:47–64

Devreux G, Santamaria U, Morresi F, Rodolfo A, Barbabietola N, Fratini F, Reale R (2015) Fitoconservazione. Trattamenti alternativi sulle opere in materiale lapideo nei giardini vaticani. In: Abstract, 13th Congresso Nazionale IGIIC -Lo Stato dell'Arte; Turin, Italy. 22–24 October 2015, pp 199–206

ECHA Site (2020). https://echa.europa.eu/substance-information/-/substanceinfo/100.063.544

Fidanza MR, Caneva G (2019) Natural biocides for the conservation of stone cultural heritage: a review. J Cult Herit 38:271–286

Galizzi V (2015) Studio e riassemblaggio della scultura lapidea di Bacco del Museo di Antichità di Torino: ricerca di un equilibrio tra fruizione e conservazione. Master's degree dissertation, Conservation and restoration of cultural heritage, Torino University

Gioventù E, Lorenzi PF, Villa F, Sorlini C, Rizzi M, Cagnini A, Griffo A, Cappitelli F (2011) Comparing the bioremoval of black crusts on colored artistic lithotypes of the cathedral of Florence with chemical and laser treatment. Int Biodeterior Biodegradation 65:832–839

Grimaldi M (2009) Search for new surface-active compounds of microbial origin in view of the development of biorestoration techniques. PhD thesis dissertation, University of Florence

Guiamet PS, Gómez de Saravia SG, Arenas P, Pérez ML, de la Paz J, Borrego SF (2006) Natural products isolated from plants used in biodeterioration control. Pharmacologyonline 3:537–544

Hueck HJ (1965) The biodeterioration of materials as a part of hylobiology. Mater Org 1:5–34

Ledda M (2019) Prodotti biotecnologici per il restauro sostenibile: un caso studio alle terme di Diocleziano, Roma. Bachelor thesis, University of Cagliari, Italy

Martin-Sanchez PM, Nováková A, Bastian F, Alabouvette C, Saiz-Jimenez C (2012) Use of biocides for the control of fungal outbreaks in subterranean environments: the case of the Lascaux cave in France. Environ Sci Technol 46:3762–3770

Mazzoni M, Alisi C, Tasso F, Cecchini A, Marconi P, Sprocati AR (2014) Laponite micro-packs for the selective cleaning of multiple coherent deposits on wall paintings: the case study of Casina Farnese on the Palatine Hill (Rome-Italy). Int Biodeterior Biodegrad 94:1–11

Mora P, Mora L, Philippot P (1984) Conservation of wall paintings. Butterworths, Sevenoaks

Öztürk M, Altay V, Hakeem KR, Akçiçek E (2018) Pharmacological activities and phytochemical constituents. In: Öztürk M, Altay V, Hakeem KR, Akçiçek E (eds) Liquorice: from botany to phytochemistry. Springer, Cham, pp 45–60

Persia F, Alisi C, Bacchetta L, Bojorquez E, Colantonio C, Falconieri M, Insaurralde M, Meza-Orozco A, Sprocati AR, Tatì A (2016) Nopal as organic additive for bio-compatible and eco-sustainable lime mortars. In: Campanella L, Piccioli C (eds) Diagnosis for conservation and valorization of cultural heritage. AIES, Naples, pp 245–251

Pieragostini E (2015) Comparative evaluation of chemical and biological methods for the removal of iron oxide stains from calcareous surfaces. Master's thesis dissertation, Sapienza, University of Rome, Italy

Puškárová A, Bučková M, Kraková L, Pangallo D, Kozics K (2017) The antibacterial and antifungal activity of six essential oils and their cyto/genotoxicity to human HEL 12469 cells. Sci Rep 7:8211

Ranalli G, Alfano G, Belli C, Lustrato G, Colombini MP, Bonaduce I, Zanardini E, Abbruscato P, Cappitelli F, Sorlini C (2005) Biotechnology applied to cultural heritage: biorestoration of frescoes using viable bacterial cells and enzymes. J Appl Microbiol 98:73–83

Rivardo F, Turner RJ, Allegrone G, Ceri H, Martinotti MG (2009) Anti-adhesion activity of two biosurfactants produced by *Bacillus* spp. prevents biofilm formation of human bacterial pathogens. Appl Microbiol Biotechnol 83:541–553

Rodrigues L, Banat IM, Teixeira J, Oliveira R (2006) Biosurfactants: potential applications in medicine. J Antimicrob Chemother 57:609–618

Rotolo V, Barresi G, Di Carlo E, Giordano A, Lombardo G, Crimi E, Costa E, Bruno M, Palla F (2016) Plant extracts as green potential strategies to control the biodeterioration of cultural heritage. Int J Conserv Sci 7:839–846

Rozeik C (2009) The treatment of an unbaked mud statue from ancient Egypt. J Am Inst Conserv 48:69–81

Rugnini L, Ellwood NTW, Sprocati AR, Migliore G, Tasso F, Alisi C, Bruno L (2019) Plant products as a green solution in the fight against biodeterioration of stone monuments. In: Poster 114° congresso S.B.I. (IPSC) - Padova, 4–7 September 2019, p 34

Scherf A, Treutwein J, Kleeberg H, Schmitt A (2012) Efficacy of leaf extract fractions of Glycyrrhiza glabra L. against downy mildew of cucumber (Pseudoperonospora cubensis). Eur J Plant Pathol 134:755–762

Sharma RK, Yenkateswaran NS, Singh SK (1980) Solvent used for the removal of shellac and other accretionary deposits. In: Symposium on mural paintings, archaeological survey of India-ICCROM, Dehra Dun, pp 10–15

Singh M, Balasaheb A (2013) Chemistry of preservation of the Ajanta murals. Int J Conserv Sci 4 (2):161–176

Sprocati AR, Alisi C, Tasso F, Marconi P, Sciullo A, Pinto V, Chiavarini S, Ubaldi C, Cremisini C (2012) Effectiveness of a microbial formula, as a bioaugmentation agent, tailored for bioremediation of diesel oil and heavy metal co-contaminated soil. Process Biochem 47:1649–1655

Sprocati AR, Alisi C, Tasso F (2014) Biotechnology process for the removal of coherent deposits of organic and inorganic origin from materials and works of artistic historical interest. European patent WO2015040647

Sprocati AR, Alisi C, Tasso F, Marconi P, Migliore G (2017) Formule microbiche per l'arte. Kermes 100: 23–26. Nardini Editore

Sterflinger K, Piñar G (2013) Microbial deterioration of cultural heritage and works of art — tilting at windmills? Appl Microbiol Biotechnol 97:9637–9646

Venice (1964) International charter for the Conservation and restoration of monument and sites (the Venice Chart 1964). In: ICOMOS, IInd International Congress of Architects and Technicians of Historic Monuments, Venice

# Chapter 12
# The Role of Microorganisms in the Removal of Nitrates and Sulfates on Artistic Stoneworks

**Giancarlo Ranalli and Elisabetta Zanardini**

**Abstract** This chapter will focus on the role of microorganisms in the removal of nitrates and sulfates on artistic stoneworks. The main groups of microbes and their metabolisms involved in bioremoval methods for the preservation and protection of cultural artifacts are reported. The aim is to offer a comprehensive view on the role and potentiality of virtuous microorganisms in the biocleaning and bioremoval of black crusts and salts altering CH stoneworks. We highlight the importance of the use of the selected microorganisms and the adoption of adequate carriers for the anaerobic metabolism of nitrate and sulfate reducers to be applied on the altered stone surfaces. The following characteristics of the delivery system are of great importance: the ability to guarantee water content for microbes, the absence of toxicity for the environment, no negative effects to the stone surfaces, easy to prepare, to apply, and to remove from different stone surfaces at the end of the treatment. We report an overview of the last 30 years on the biocleaning processes including diagnostic studies of the alterations, the assessment of associated risks, the effectiveness and efficacy of the proposed method, and the evaluation in terms of economic and environmental sustainability.

**Keywords** Cultural Heritage · Biocleaning · Black crust · Nitrates · Sulfates · Biotechnology · Bacteria

It is well known that microorganisms have a crucial role in their natural habitats being involved in the biogeochemical cycles and in the transformation of inorganic and organic compounds as energy and carbon sources. However, in the Cultural

G. Ranalli (✉)
Department of Biosciences and Territory, University of Molise, Pesche, Italy
e-mail: ranalli@unimol.it

E. Zanardini
Department of Sciences and High Technology, University of Insubria, Como, Italy
e-mail: elisabetta.zanardini@uninsubria.it

Heritage (CH) field microorganisms have a double role in relation to the biodeterioration processes and the biorecovery potentialities.

In this chapter, we briefly describe the role of microorganisms in the biodeterioration of stoneworks and, in particular, we focus the attention on the use of viable bacterial cells for the removal of nitrates and sulfates from CH altered stone surfaces.

Outdoor archeological and monumental CH stoneworks are subjected to deterioration damages due to physical, chemical, and biological processes causing structural, mechanical, chemical, and aesthetic alterations of the original material and leading to stone decay (Sorlini et al. 1987; Ortega-Calvo et al. 1995; Mansch and Bock 1998; McNamara and Mitchell 2005; Gorbushina 2007; Pinna and Salvadori 2008; Ranalli et al. 2009; Scheerer et al. 2009; May 2010).

Climate change can indirectly affect stone deterioration due to fluctuation of temperature, rainfall and humidity determining higher salt crystallization processes (Brimblecombe and Grossi 2007; Grossi et al. 2007; Duthie et al. 2008).

Salts, deriving from various sources (e.g., air pollution, soil, wind from the sea or the desert, de-icing salt, cleaning materials, garden fertilizers, etc.) can seriously damage the CH stone surfaces due to the stresses generated by crystal growth in the stone pores (Scherer 2000; Doehne and Price 2010).

All the deterioration processes depend on the stone physical and chemical properties since different lithotypes have diverse strength, porosity, hardness, absorption, etc. These aspects strongly affect the decay of monuments and building materials and can also favor the surface colonization of microorganisms and this concept is named bioreceptivity (Viles and Moses 1998; Herrera and Videla 2009; Doehne and Price 2010). Indeed, together with physical and chemical agents, biological agents are also important in the deterioration of historical and artistic stoneworks acting in different ways: mechanically, chemically, and aesthetically (Warscheid and Braams 2000; Dornieden et al. 2000; Saiz-Jimenez 2001; Ranalli et al. 2009; Bhatnagar et al. 2010; Zanardini et al. 2011, 2016, 2019; De Leo et al. 2012).

The growth of the biodeteriogens on stone mainly occurs with the formation of biofilm that can produce mechanical stresses to the mineral structure leading to changes in the stone pore size and in the moisture circulation patterns and temperature response (Gorbushina 2007). However, it has also evidenced that biological patinas and biofilms in some situations do not cause stone decay but can act as a protective layer of the stone surfaces (Caneva et al. 2008; De Muynck et al. 2010).

Among the principal groups of microorganisms involved in CH biodeterioration processes, diverse bacteria have been detected and isolated from altered stonework surface and/or at the depth of few millimeters; chemolithoautotrophs such as bacteria involved in sulfur and nitrogen cycles that can strongly acidify by the production of inorganic acid (sulfuric and nitric acids, respectively). In the past, great attention was focused on acidifying bacteria such as sulfur oxidizers (*Thiobacillus* spp.) and nitrifiers (*Nitrosomonas* spp.) which can cause the transformation of calcium carbonate giving calcium sulfate and calcium nitrate leading to material corrosion (Mansch and Bock 1998; Tiano et al. 1999, 2006; Abeliovich 2006).

# 1   Stone Sulfation and Nitration Mechanisms

Air pollution has been recognized in numerous studies as a strong factor involved in the outdoor stoneworks deterioration for causing material transformation and surface deposits (Camuffo 1998; Warscheid and Braams 2000; Zanardini et al. 2000; Saiz-Jimenez 2003; Doehne and Price 2010). Inorganic atmospheric pollutants such as air sulfur oxides ($SO_2$ and $SO_3$), nitrogen oxides ($NO_x$), and carbon oxides, in the presence of water, produce acidic solutions, which react with the calcareous materials and transform calcium carbonate into calcium sulfate dihydrate and calcium nitrate giving consequently decay phenomena of "sulfation" and "nitration" and the formation of damaging crusts namely "black crust" (Gauri et al. 1989; Rivadeneyra et al. 1991, 1994; Orial et al. 1992; Salvadori and Realini 1996; Saiz-Jimenez 2003, 2004; Doehne and Price 2010).

In the case of the black crust formation, the main mechanism is the sulfation reaction involving calcium carbonate as reported below:

$$2SO_2 + O_2 \rightarrow 2SO_3$$
$$SO_3 + H_2O \rightarrow H_2SO_4$$
$$H_2SO_4 + CaCO_3 \rightarrow CaSO_4 \cdot (2H_2O) + CO_2$$

The corrosion processes operated by acidic pollutants on stone can differently happen based on the environment where monuments are located: exposed and sheltered areas can show in fact different levels of stone surfaces deterioration. When the products from chemical reactions can accumulate, black crusts on stone surfaces are often formed and the blackness mainly depends on the air pollutants from fossil fuel combustion, while in exposed areas the rainfall normally inhibits the formation of these kinds of alterations as shown in Fig. 12.1a and b. (Doehne and Price 2010).

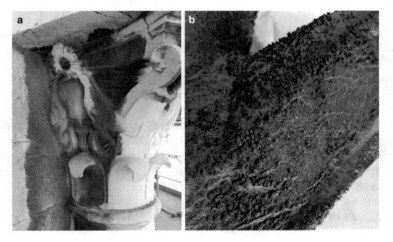

**Fig. 12.1**   (**a–b**) Typical black crusts on ancient altered marble by exposition to outdoor conditions, at Pisa Cathedral (Credits: Photos by OpaPisa)

Restoration practices, aiming the cleaning of altered stone surfaces, need to previously define and characterize the stone mineralogical properties and the level and extend of decay in order to understand the causes and mechanisms of the processes actually in place. Preservation and conservation of CH stoneworks is not simple, as material decay is a natural process, and, therefore, can be only slowed down (Fassina 1994).

Therefore, the conservation of stone surfaces needs a technical and scientific approach. As we said above, outdoor stone surfaces accumulate different kinds of materials, i.e. atmospheric pollutants, salts and residues from past restorative interventions giving the formation of surface deposits.

In the case of the black crusts, stone surfaces appear extremely damaged in their aesthetic, chemical, and physical aspects; furthermore, stone properties are also affected with the formation of fissures, fractures, exfoliation, disintegration up to loss of original material.

## 2 Removal of Black Crusts, Sulfates, Nitrates, and Deposits

The removal of these kinds of alteration is a serious issue for the conservators-restorers and the different treatments aim to be not too much aggressive and to avoid the alteration of the original stone surfaces.

The first step is the accurate characterization of the materials (diagnosis) that must be removed and the conservation state of the stoneworks followed by the individualization of the best methodology to use, showing high efficiency, selective cleaning, absence of aesthetic alteration, and durability (Vergès-Belmin 1996; Gulotta et al. 2014).

Conventional cleaning methods are mechanical (brushing and rubbing, washing and steaming, wet and dry abrasives, etc.) and chemical (alkaline treatments, acidic treatments, or organic solvents, etc.). These methods can be too aggressive and not selective and therefore the effort to find an alternative has been recognized as a priority.

Methodologies based on laser and biological cleaning can give a relevant contribution in the treatment of such stone alterations (Ramirez et al. 2005; Doehne and Price 2010; Junier and Joseph 2017). Lasers can discriminate between the soiling and the substratum, are less intrusive, more easily controlled, allowing a cleaning method with high selectivity. In some cases, the use of laser does not permit the complete removal of the deposits and can cause color changes. Moreover, studies report problems in the treatment of polychrome sculptures and in the use of laser on large superficial areas since the cost considerably increases (Salimbeni et al. 2003).

Until now, we cited the conventional cleaning methods, considering the following definition of the term cleaning reported in EWAGLOS project book: *"Action performed to remove dirt deposits, foreign matters and products of alteration present on artworks surface that can be a source of decay or deterioration or aesthetic disturbance"* (EWAGLOS 2016). The main cleaning techniques are mechanical, physico-chemical and also biological; the last (the biocleaning) includes the use of living organisms and/or the enzymes as cleaning agents by controlled reproduction of their metabolic processes.

The biological methods based on the use of microorganisms and enzymes as cleaning agents in the "biorestoration" of artworks represented in the last decades an attractive alternative to the mechanical and chemical methods (Bellucci et al. 1999; Cremonesi 1999). The key idea of using living cells in the conservation and preservation of works of art is supported by the fact that microorganisms (mainly bacteria) are the most versatile and ubiquitous organisms found on earth, and they are able of colonizing almost any environment (Maier et al. 2000; Sorlini and Cappitelli 2008).

Even if we know that some microorganisms have "negative effects," many of them are responsible for "positive effects" such as the degradation and transformation processes of pollutants (Sorlini et al. 1987).

The microorganisms involved in the biorestoration can be isolated from natural environments (soil, water, etc.) and/or selected among the autochthonous microbial communities inhabiting historical artworks. Once verified their ability in the removal of the undesired substances, after an adequate increased cellular biomass, they can be used in the recovery applications (bioaugmentation) showing in this case not a negative factor, but a new and positive conservation perspective.

Therefore, the methods based on the use of specific microorganisms can positively help in the cleaning of stone surfaces exploiting the microbial versatility, their different metabolic activities and role in the biogeochemical cycles in their natural habitat. (Atlas et al. 1988; Ranalli et al. 2003). Safe living microbial cells under optimal controlled conditions reproduce the same processes that occur in nature, softly remove certain substances and represent an eco-friendly solution without health risks for the conservator-restorers and the environment (Boquet et al. 1973; Atlas et al. 1988; Ferrer et al. 1988; Heselmeyer et al. 1991; Saiz-Jimenez 1997; Castanier et al. 2000; Maier et al. 2000; Rodriguez-Navarro et al. 2000; Ranalli et al. 2003, 2005; Tiano et al. 2006; Biavati and Sorlini 2008; Valentini et al. 2010; Sasso et al. 2015).

In the last decades biocleaning technologies have been applied in the removal of organic and inorganic unwanted substances on CH stone surfaces (marble, tuff, sandstone, limestone, etc.), on ceramic material (brick-work), on paper materials, and on concrete using specific bioformulations containing *Desulfovibrio* sp. and *Pseudomonas* sp. cells (Gauri et al. 1989, 1992; Heselmeyer et al. 1991; Delgado Rodrigues and Valero 2003; Ranalli et al. 2005; De Graef et al. 2005; De Belie et al. 2005; Cappitelli et al. 2006, 2007; Alfano et al. 2011; Gioventù et al. 2011; Barbabietola et al. 2012; Troiano et al. 2013).

Table 12.1 reports the biotreatment studies carried out specifically for the removal of salts from CH stoneworks, together with the bacterial strains and the delivery systems adopted.

Sulfate and nitrate reducers (see Prokaryotes organisms involved in sulfur and nitrogen cycles and reported in Table 12.2 a and b) have been used for the removal of

**Table 12.1** Biotreatment studies carried out specifically for the removal of salts from CH stoneworks

| Main decay agents | Type of materials | Biocleaning bacteria | Delivery systems | References |
|---|---|---|---|---|
| Sulfates | Marble (Georgia) stone and statue | *Desulfovibrio desulfuricans* (An) | Immersion | Gauri et al. (1989, 1992) |
| | Marble and sandstone | *D. vulgaris* (An) | Immersion | Heselmeyer et al. (1991) |
| | Marble | *D. desulfuricans,* (An) *D. vulgaris* (An) | Sepiolite | Ranalli et al. (1996, 1997) |
| Black crusts | Marble (Candoglia stone), Milan | *D. vulgaris* subsp. *vulgaris* (An) | Sepiolite Hydrobiogel 97 | Cappitelli et al. (2005) |
| | Marble sculpture, Milan | *D. vulgaris* subsp. *vulgaris* (An) | Carbogel | Cappitelli et al. (2006) |
| | Limestone sculpture, Trento | *D. vulgaris* subsp. *vulgaris* (An) | Carbogel | Polo et al. (2010) |
| | Colored lithotypes, Firenze | *D. vulgaris* subsp. *vulgaris* (An) | Carbogel | Gioventù et al. (2011) |
| Black crusts and grey deposits | Marble column and statue, Cemetery of Milan | *D. vulgaris* subsp. *vulgaris* (An) | Arbocel | Troiano et al. (2013) |
| Nitrates | Brickworks and calcareous stones, (marble and Vicenza stones) | *Pseudomonas stutzeri* (Ae) | Sepiolite | Ranalli et al. (1996) |
| | Sandstone walls, Matera | *P. pseudoalcaligenes* (An) | Mortar and alginate beads | May et al. (2008) |
| Nitrates and Sulfates | Sandstone walls, Matera | *P. pseudoalcaligenes* (Ae) *D. vulgaris* (An) | Carbogel | Alfano et al. (2011) |
| Calcium sulfate and carbonate | Casina Farnese wall paintings (Palatine Hill, Rome) | *Cellulosimicrobium cellulans* (Ae) | Laponite | Mazzoni et al. (2014) |
| Saline efflorescence | Frescoes, Valencia | *P. stutzeri* DSMZ 5190 (Ae) | Cotton wool Agar | Bosch-Roig et al. (2010, 2012, 2013) |
| Efflorescence | Archaeological frescoes, ceramic and bones | *P. stutzeri* DSMZ 5190 (Ae) | Agar | Martín Ortega (2015) |

*Ae* Aerobic metabolism, *An* Anaerobic metabolism

sulfates and nitrates from CH stoneworks being able to reduce, in anaerobic conditions, sulfates to $H_2S$ and nitrates to $N_2$, respectively.

In the case of the sulfate-reducing bacteria, the mechanism of action occurs through gypsum dissociation into $Ca^{2+}$ and $SO_4^{2-}$ ions where the sulfates are

**Table 12.2**  **a** and **b** Prokaryotes organisms involved in sulfur and nitrogen cycles and reported

| Processes | Reactions | Microorganisms | Metabolic pathways |
|---|---|---|---|
| Nitrification $[NH_4^+ \rightarrow NO_3^-]$ | $[NH_4^+ \rightarrow NO_2^-]$ $[NO_2^- \rightarrow NO_3^-]$ | *Nitrosomonas Nitrobacter* | Aerobic Aerobic |
| Nitrate reduction | $[NO_3^- \rightarrow N_2]$ | *Bacillus, Pseudomonas* | Anaerobic |
| $N_2$- fixation | Free-living cells | *Azotobacter* | Aerobic |
| $[N_2 + 8H^+ \rightarrow NH_3 + H_2]$ | | *Clostridium* | Anaerobic |
| | Symbiotic cells | *Rhizobium, Frankia* | Anoxic |
| Ammonification | $[Organic\text{-}N \rightarrow NH_4^+]$ | Several microorganisms | Anaerobic |
| Ammonium oxidation | $[NH_4^+ NO_2^- \rightarrow N_2 + 2H_2O]$ | *Brocadia, Kuenenia, Scalindua* | Anaerobic |
| Processes | Reactions | Microorganisms | Metabolic pathways |
| Sulfate reduction | $[SO_4^{2-} \rightarrow H_2S]$ | *Desulfovibrio, Desulfobacter* | Anaerobic |
| Sulfur reduction | $[S^0 \rightarrow H_2S]$ | *Desulforomonas,* Extreme thermophilic Archaea | Anaerobic |
| Sulfide/sulfur oxidation | $[H_2S \rightarrow S^0 \rightarrow SO_4^{2-}]$ | *Thiobacillus, Beggiatoa,* others | Aerobic |
| | | Phototrophic bacteria (purple and green), chemolithotrophs | Anaerobic |
| Organic S ox/red | $[Org\text{-}S \rightarrow H_2S]$ | Several microorganisms | Aerobic/ Anaerobic |

reduced by the bacteria into S, and the $Ca^{2+}$ ions react with $CO_2$ to originate new calcite, following this reaction:

$6CaSO_4 + 4H_2O + 6CO_2 \rightarrow 6\ CaCO_3 + 4H_2S + 11O_2 + 2S$

With regard to the use of these bacteria for the removal of black crusts and sulfates, the first applied studies were performed by immersion of marble fragments in broth anaerobic cultures of strains of *Desulfovibrio desulfuricans* and *Desulfovibrio vulgaris,* for a period of 60–84 hours obtaining a removal efficiency of 40–100% (Gauri et al. 1989, 1992; Heselmeyer et al. 1991).

After these studies, it has been recognized that the immersion technique in liquid culture can have many limitations for large, fragile artworks; therefore, in order to reduce the water absorption to stone due to this methodology, the use of a delivery system has been considered in the successive studies. For this reason, biotreatments have been performed, always in anaerobic conditions, using an inorganic delivery system such as sepiolite colonized with *D. desulfuricans* and/or *D. vulgaris* for the removal of sulfates. Lab-scale tests were carried out both using artificial enriched specimens and real altered stone fragments; these studies evidenced the possibility of obtaining a relevant sulfate removal percentage (around 80%) in a short period of time (36 hours) (Ranalli et al. 1996, 1997). Sepiolite was also tested for the removal of nitrates on brick-works and calcareous stones (marble and Vicenza-stones) using a *Pseudomonas stutzeri* strain in anaerobic conditions (Ranalli et al. 1996). However, the use of sepiolite gave in some cases problems of dark precipitates (for the presence of iron in trace) requiring a pre-treatment of the sepiolite to remove the iron. In addition, sepiolite as delivery system did not guarantee the right amount of water for the metabolic activity of bacterial cells during the application on stone surface.

The inorganic delivery system was therefore substituted by an organic carrier named hydrobiogel-97 for the immobilization of *D. vulgaris* subsp. *vulgaris* ATCC 29579 bacterial cells. The biotreatments were carried out both under lab-scale trials on artificially specimens (different lithotypes) and real altered stoneworks, and in situ trials to remove black crusts present on Candoglia stone at Milan Cathedral, Italy. The in situ bioapplication has been carried out using carbogel (CTS, Vicenza, Italy), covered by a thin film to create microaerophilic condition, and repeated 3 times (15 hours each) giving a removal of 98% of sulfates. A comparison of the biological methods with a chemical one (using ammonium carbonate and EDTA) was carried out in order to define the results in terms of removal, but also to determine the effects on the stone surface; in this case the biotreatment was very promising (Cappitelli et al. 2006, 2007).

Another case study where the sulfate crust was efficiently biocleaned by *D. vulgaris* subsp. *vulgaris* ATCC 29579 and carbogel was the Pietà Rondanini marble base sculpture by Michelangelo Buonarroti (Milan, Italy). In order to eliminate secondary iron black deposits, bacteria were grown in a modified broth medium free of iron. Treatment times were of 24–30 hours in anaerobiosis (Cappitelli et al. 2005).

In another applied studies on altered stone sculptures carried out by Polo et al. (2010) the presence of black crusts and the microbial growth causing discoloration were evidenced on the *Demetra* and *Cronos* sculptures, two of 12 stone statues decorating the courtyard of the Buonconsiglio Castle in Trento (Italy). *Desulfovibrio vulgaris* subsp. *vulgaris*, grown in DSMZ 63 medium without iron source and carbogel as delivery system, was applied to remove the black crusts but preserved the original stone and the patina noble. The altered surfaces were treated with three 12-hours applications for a total duration of 36 hours. This case study, combining traditional and biomolecular methods, showed that conservators can benefit from an integrated biotechnological approach aimed at the biocleaning of the alterations together with the abatement of biodeteriogens (Polo et al. 2010).

Other positive results involved the colored lithotypes of Florence Cathedral external walls. Green serpentine, red marlstone, and Carrara white marble black crust were biocleaned. This work compared three different cleaning treatments: chemical poultice, laser, and biocleaning. Results shows that the better cleaning method was biocleaning using *D. vulgaris* subsp. *vulgaris* ATCC 29579 strain and carbogel as a carrier, but long-time treatments and numerous applications were needed (Gioventù and Lorenzi 2013).

In another recent study, the combination of chemical and biological (arbocel nonionic detergent treatment colonized with *D. vulgaris* subsp. *vulgaris*) methods allowed efficient cleaning of grey deposits and black crusts on marble columns and statues at the Cemetery in Milan, Italy, with a notable reduction in the treatment duration (Troiano et al. 2013).

Carbogel alone together with mortar and alginate beads have been also used on tuff stone walls at Matera Cathedral, Italy, for the removal of both sulfates and nitrates (the latter present at higher concentration) using different microorganisms such as *Pseudomonas pseudoalcaligenes* and *D. vulgaris* (May et al. 2008; Alfano et al. 2011).

At Matera Cathedral the main problem arises from the rising of salts from the groundwater rich in nitrates and sulfates causing reduction in cohesion of the tuff stone, loss of materials, and the darkening of the lower external walls. The nitrates originate from the oxidation of various N organic compounds (including proteins, amino acids, and urea) from bodies buried in the ground when the area was used as a cemetery (until eighteenth century). At the same way, sulfates derived mainly by amino acids such as methionine and cysteine rather than environmental depositions. In addition, chemical analyses detected egg residues as previous consolidant treatments of the tuff surface. The biotreatment was efficient, showing a removal of surface deposits and salt alterations after 24 hours (55% of the nitrate and 85% of the sulfate deposits) (Alfano et al. 2011; Rampazzi et al. 2018).

Many of the reported studies were performed during a 3 years EU project named BIOBRUSH project (*BIOremediation for Building Restoration of the Urban Stone Heritage*). The main aims of the project were the selection of microorganisms and delivery system for the removal of black crusts, sulfates, nitrates and also the possibility of using bio-calcifying bacteria for the stone bio-consolidation (Webster and May 2006; May et al. 2008).

The EU BIOBRUSH project gave the opportunity to perform studies under different scale (lab and in situ experiments) giving promising results and showing the aspects to improve and optimize. For example, in some cases, the delivery systems adopted resulted not appropriate to treat large surface areas or decorative elements.

In another study, the combination of chemical and biological (arbocel nonionic detergent treatment colonized with *D. vulgaris* subsp. *vulgaris*) methods allowed efficient cleaning of grey deposits and black crusts on marble columns and statues at the Cemetery in Milan, Italy, with a notable reduction in the treatment duration (Troiano et al. 2013).

More recent studies showed the use of improved delivery systems as agar-gel colonized by *Pseudomonas stutzeri* DSMZ 5190 strain for the removal of salt efflorescence on frescoes on the Santos Juanes Church in Valencia, Spain, as reported in Fig. 12.2 (Bosch-Roig et al. 2010, 2012, 2013) and on archeological frescoes, ceramic, and bones (Martín Ortega 2015).

In the first case, Bosch-Roig et al. (2013) demonstrated that short term (90 min) application of *P. stutzeri* DSMZ 5190 and the use of agar as delivery system provided an efficient cleaning of insoluble nitrate efflorescence deposited on wall painting surfaces confirmed by Ion Chromatography (92% in nitrate efflorescence reduction). Bosch-Roig et al. (2013) underlined the positive properties of agar supporting the microorganisms as follows: (i) it can provided an adequate water

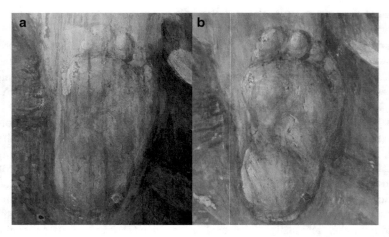

**Fig. 12.2 a–b** Before (*left*)/after (*right*) comparison picture of one detail of the biocleaning of saline efflorescence from the frescos at *St. Nicholas* church in Valencia, Spain (Gentle concession of the pictures by J.L. Regidor Ros & P. Bosch-Roig)

supply to the bacterial viability, (ii) can release water only on the surface, in homogeneous and controlled way to avoid damages on the artwork; (iii) absence of stains or residues on the artwork; (iv) easy to apply and remove; (v) non-toxic and eco-friendly. Furthermore, the use of agar is based both on the small volume of water and on shorter time of contact; this means a reduction of risks if compared to others systems that require extended contact periods (Lustrato et al. 2012).

Recently, hydrobiogel-97 colonized with *D. vulgaris* was also used for the removal of black crusts from weathered stone surfaces in Failaka Island, Kuwait (Elhagrassy and Hakeem 2018).

Laponite colonized with *Cellulosimicrobium cellulans* was used instead for the removal of calcium sulfate and carbonate at Casina Farnese wall paintings (Palatine Hill, Rome, Italy) as reported by Mazzoni et al. (2014).

Furthermore, advanced biocleaning system using new agar-gauze bacteria gel was carried out at *onsite* historical wall paintings (Ranalli et al. 2019).

A recent study reports the use of agar-gel colonized with extremophilic bacteria (*Halomonas campaniensis*) for the biocleaning of nitrate crusts from stoneworks (Romano et al. 2019).

# 3  Conclusions

Numerous studies have been carried out in order to optimize the biocleaning strategies in terms of microorganism efficiency and delivery systems able to maintain and guarantee the microbial activity and ready to be used also on difficult altered stone surfaces (ornamented, vertical, and vaulted). Different substances can be

## Biotechnologies for Cultural Heritage *(BioteCH)*

(Risk free, Selective performance,
Easy to apply and remove,
Effective, Environmentally sustainable)

Bio-cleaning, Bio-removal
Bio-treatment, Bio-consolidation,
Bio-restoration
with
Microorganisms/Enzymes
+ delivery systems

before                                                                                              after

**Fig. 12.3** Advantages and strengths of biotechnologies for Cultural Heritage (BioteCH) based on the use of selected microorganisms (before and after comparison pictures of the biocleaning of black crusts on one detail of statue *St. Benedetto*, at Montecassino Abbey, Italy (Pontone 2014)

present as stone alterations and for this reason, the choice of the best microorganism is a crucial step in the biocleaning technology.

Microbial diversity includes *Bacteria, Archaea*, and *Eukarya* (prokaryotic and eukaryotic cells) able to live and adapt to different natural habitats (soil, rocks, hot springs, oceans, etc.), to perform a wide range of metabolic strategies in terms of energy and carbon sources, in presence/absence of oxygen (aerobic and anaerobic conditions) and using light for their photometabolism.

Therefore, a careful selection of the most performant microorganism in the removal of the undesired substances present on stone surfaces as deposits and crusts (in this case mainly represented by salts such as nitrates and sulfates) is the first step in the biorestoration strategies.

The application of the selected microorganism is also important since the carrier, used to apply the viable cells on the altered stone surfaces, must have specific properties and provide them an adequate microenvironment to optimize their activity.

The ideal delivery system should have the following characteristics: (i) able to guarantee the appropriate water content for the microbial activity; (ii) not toxic both for the microorganisms and for the environment; (iii) do not cause any color variation to the stone surface; (iv) applicable to all types of surfaces (horizontal, vertical, oblique, vaulted, rough, smooth, etc.), and finally, (v) easy to prepare and also to apply and to remove at the end of the treatment.

All these aspects about the carrier properties and characteristics were evaluated in a review in order to individualize the best choice for the different situations (Bosch-Roig et al. 2015).

The biocleaning process also requires a deep diagnostic study of the work of art to be cleaned, an assessment of any associated risks, an evaluation of the effectiveness and efficiency of the process, and last but not least, an evaluation in terms of economic and environmental sustainability (Bosch-Roig and Ranalli 2014) (Fig. 12.3).

**Acknowledgements** The authors wish to thank to the restorers involved in the before mentioned case studies because without their confidence, help, and dedication none of this works could have been successfully applied on real artworks. Special thanks to Claudia Sorlini, Maria Perla Colombini, Alessia Andreotti, Laura Rampazzi and Cristina Corti for the continuous collaborations and support in the chemical aspects in our studies on CH biotechnologies. Special thanks to Arch. B. Lafratta, Superintendence for Restoration, The Technical and Scientific Committee (Matera, Italy); Mons. D. Lionetti, ArciDiocesi di Matera-Irsina; Restorer G. D'Alessandro, Impresa D'Alessandro Restauri, Matera, Italy. Moreover, we are very grateful to the Veneranda Fabbrica del Duomo for assistance and support of the diagnostic work on the Milan Cathedral (Milan, Italy).

# References

Abeliovich A (2006) The nitrite-oxidising bacteria. In: Dworkin M, Falkow S, Rosenberg E, Schleifer K-H, Stackebrandt E (eds) The prokaryotes. Springer, New York, pp 861–871

Alfano G, Lustrato G, Belli C, Zanardini E, Cappitelli F, Mello E, Sorlini C, Ranalli G (2011) The bioremoval of nitrate and sulfate alterations on artistic stonework: the case-study of Matera cathedral after six years from the treatment. Int Biodeterior Biodegradation 65:1004–1011

Atlas RM, Chowdhury AN, Gauri KL (1988) Microbial calcification of gypsum-rock and sulfated marble. J Stud Conserv 33:149–153

Barbabietola N, Tasso F, Grimaldi M, Alisi C, Chiavarini S, Marconi P, Perito B, Sprocati AR (2012) Microbe-based technology for a novel approach to conservation and restoration. EAI Speciale II. Knowledge, diagnostics and preservation of cultural heritage, pp. 69–76

Bellucci R, Cremonesi P, Pignagnoli G (1999) A preliminary note on the use of enzymes in conservation: the removal of aged acrylic resin coatings with lipase. Stud Conserv 44:278–281

Bhatnagar P, Khan AA, Jain SK, Rai MK (2010) Biodeterioration of archaeological monuments and approach for restoration. In: Jain SK, Khan AA, Rai MK (eds) Geomicrobiology. CRC Press, Enfield, pp 255–302

Biavati B, Sorlini C (2008) Microbiologia agroambientale. Ambrosiana, Milano, p 684

Boquet E, Boronat A, Ramos-Cormenzana A (1973) Production of calcite (calcium carbonate) crystals by soil bacteria is a general phenomenon. Nature 246:527–528

Bosch-Roig P, Lustrato G, Zanardini E, Ranalli G (2015) Biocleaning of cultural heritage stone surfaces and frescoes: which delivery system can be the most appropriate. Ann Microbiol 65:1227–1241

Bosch-Roig P, Montes-Estellés RM, Regidor-Ros JL, Roig Picazo P, Ranalli G (2012) New frontiers in the microbial bio-cleaning of artworks. The Picture Restorer 41:37–41

Bosch-Roig P, Ranalli G (2014) The safety of biocleaning technologies for cultural heritage. Front Microbiol 5:155

Bosch-Roig P, Regidor-Ros JL, Montes-Estellés R (2013) Biocleaning of nitrate alterations on wall paintings by *Pseudomonas stutzeri*. Int Biodeterior Biodegradation 84:266–274

Bosch-Roig P, Regidor-Ros JL, Soriano-Sancho P, Domenech-Carbo MT, Montes-Estellés RM (2010) Ensayos de biolimpieza con bacterias en pinturas murales. Arché 4–5:115–122

Brimblecombe P, Grossi CM (2007) Damage to buildings from future climate and pollution. APT Bull 38:13–18

Camuffo D (1998) Microclimate for cultural heritage. Elsevier, Amsterdam, p 415

Caneva G, Nugari MP, Salvadori O (2008) Plant biology for cultural heritage: biodeterioration and conservation. The Getty Conservation Institute, Los Angeles, CA

Cappitelli F, Toniolo L, Sansonetti A, Gulotta D, Ranalli G, Zanardini E, Sorlini C (2007) Advantages of using microbial technology over traditional chemical technology in removal of black crusts from stone surfaces of historical monuments. Appl Environ Microbiol 73:5671–5675

Cappitelli F, Zanardini E, Ranalli G, Mello E, Daffonchio D, Sorlini C (2006) Improved methodology for bioremoval of black crusts on historical stone artworks by use of sulfate-reducing bacteria. Appl Environ Microbiol 72:3733–3737

Cappitelli F, Zanardini E, Toniolo L, Abbruscato P, Ranalli G, Sorlini C (2005) Bioconservation of the marble base of the Pietà Rondanini by Michelangelo Buonarroti. Geophys Res Abstr 7:06675–00676

Castanier S, Le Metayer-Levrel G, Orial G, Loubiere JF, Perthuisot JP (2000) Carbonatogenesis and applications to preservation and restoration of historic property. In: Ciferri O, Tiano P, Mastromei G (eds) Of microbes and art. Kluwer Academic/Plenum Publishers, Amsterdam, pp 203–218

Cremonesi P (1999) L'uso degli enzimi nella pulitura di opere policrome. In: Cremonesi P (ed) I Talenti. Il Prato, Padova, pp 5–11

De Belie N, De Graef B, De Muynck W, Dick J, De Windt W, Verstraete W (2005) Biocatalytic processes on concrete: bacterial cleaning and repair. In: 10th DBMC international conference on durability of building materials and components, Lyon, 17–20 April

De Graef B, De Windt W, Dick J, Verstraete W, De Belie N (2005) Cleaning of concrete fouled by lichens with the aid of *Thiobacilli*. Mater Struct 38:875–882

De Leo F, Iero A, Zammit G, Urzì CE (2012) Chemoorganotrophic bacteria isolated from biodeteriorated surfaces in cave and catacomb. Int J Speleol 41:125–136

De Muynck W, De Belie N, Verstraete W (2010) Microbial carbonate precipitation in construction materials: a review. Ecol Eng 36:118–136

Delgado Rodrigues J, Valero J (2003) A brief note on the elimination of black stains of biological origin. Stud Conserv 48:17–22

Doehne EF, Price CA (2010) Stone conservation: an overview of current research. The Getty Conservation Institute, Los Angeles

Dornieden T, Gorbushina AA, Krumbein WE (2000) Biodecay of cultural heritage as a space/time-related ecological situation an evaluation of a series of studies. Int Biodeterior Biodegradation 46:261–270

Duthie L, Hyslop E, Kennedy C, Phoenix V, Lee MR (2008) Quantitative assessment of decay mechanisms in Scottish building sandstones. In: Lukaszewicz JW, Niemcewicz P (eds) Proceedings of the 11th int. congress on deterioration and conservation of stone, Torun, pp 73–80

Elhagrassy AF, Hakeem A (2018) Comparative study of biological cleaning and laser techniques for conservation of weathered stone in Failaka Island, Kuwait. Sci Cult 4:43–50

EWAGLOS (2016) Weyer A, Roig Picazo P, Pop D, Carras J, Ozkose A, Vallet JM, Srsa I (Eds.). In: EwaGlos-European illustrated glossary of conservation terms for wall paintings and architectural surfaces. English definitions with translations into Bulgarian, Croatian, French, German, Hungarian, Italian, Polish, Romanian, Spanish and Turkish, vol. 17. Michael Imhof Verlag, Petersberg, Germany, pp. 304–307

Fassina V (1994) General criteria for the cleaning of stone: theoretical aspects and methodology of application. In: Zezza F (ed) Stone material in monuments: diagnosis and conservation. Scuola Universitaria C.U.M. Conservazione dei Monumenti, Heraklion, pp 131–138

Ferrer MR, Quevedo-Sarmiento J, Rivadeneira MA, Bejar V, Delgado R, Ramos-Cormenzana A (1988) Calcium carbonate precipitation by two groups of moderately halophilic microorganisms at different temperatures and salt concentrations. Curr Microbiol 17:221–227

Gauri KL, Chowdhury AN, Kulshreshtha NP, Punuru AR (1989) The sulfation of marble and the treatment of gypsum crusts. Stud Conserv 34:201–206

Gauri KL, Parks L, Jaynes J, Atlas R (1992) Removal of sulfated-crust from marble using sulfate-reducing bacteria. In: Proceedings of the int. conference on stone cleaning and the nature, soiling and decay mechanisms of stone, Edinburgh, 14–16 April, pp. 160–165

Gioventù E, Lorenzi P (2013) Bio-removal of black crust from marble surface: comparison with traditional methodologies and application on a sculpture from the Florence's English cemetery. Procedia Chem 8:123–129

Gioventù E, Lorenzi PF, Villa F, Sorlini C, Rizzi M, Cagnini A, Griffo A, Cappitelli F (2011) Comparing the bioremoval of black crusts on colored artistic lithotypes of the Cathedral of Florence with chemical and laser treatment. Int Biodeterior Biodegradation 65:832–839

Gorbushina AA (2007) Life on the rocks. Environ Microbiol 9:1613–1631

Grossi CM, Brimblecombe B, Esbert RM, Alsono FJ (2007) Color changes in architectural limestone from pollution and cleaning. Color Res Appl 32:320–331

Gulotta D, Saviello D, Gherardi F, Toniolo L, Anzani M, Rabbolini A, Goidanich S (2014) Setup of a sustainable indoor cleaning methodology for the sculpted stone surfaces of the Duomo of Milan. Heritage Science 2:6–19

Herrera LK, Videla HA (2009) Surface analysis and materials characterization for the study of biodeterioration and weathering effects on cultural property. Int Biodeterior Biodegradation 63:813–822

Heselmeyer K, Fischer U, Krumbein KE, Warscheid T (1991) Application of *Desulfovibrio vulgaris* for the bioconversion of rock gypsum crusts into calcite. BIOforum 1(/2):89

Junier P, Joseph E (2017) Microbial biotechnology approaches to mitigating the deterioration of construction and heritage materials. Microb Biotechnol 10:1145–1148

Lustrato G, Alfano G, Andreotti A, Colombini MP, Ranalli G (2012) Fast biocleaning of mediaeval frescoes using viable bacterial cells. Int Biodeterior Biodegrad 69:51–61

Maier RM, Pepper I, Gerba CP (2000) Environmental microbiology. Elsevier, San Diego, p 585

Mansch R, Bock E (1998) Biodeterioration of natural stone with special reference to nitrifying bacteria. Biodegradation 9:47–64

Martín Ortega S (2015) Neteges raonades: Experimentació amb nou tractaments de neteja aplicats sobre 6 peces de material arqueologic divers. *Treball final de grau*, ESCRBCC, Barcelona

May E (2010) Stone biodeterioration. In: Mitchell R, McNamara CJ (eds) Cultural heritage microbiology: fundamental studies in conservation science. ASM Press, Washington, DC, pp 1–39

May E, Webster AM, Inkpen R, Zamarreno D, Kuever J, Rudolph C, Warcheid T, Sorlini C, Cappitelli F, Zanardini E, Ranalli G, Kgrage L, Vgenopoulos A, Katsinis D, Mello E, Malagodi M (2008) The BIOBRUSH project for bioremediation of heritage stone. In: May E, Jones M, Mitchell J (eds) Heritage microbiology and science. Microbes, monuments and maritime materials. RCS Publishing, Cambridge, pp 76–93

Mazzoni M, Alisi C, Tasso F, Cecchini A, Marconi P, Sprocati AR (2014) Laponite micro-packs for the selective cleaning of multiple coherent deposits on wall paintings: the case study of Casina Farnese on the Palatine Hill (Rome-Italy). Int Biodeterior Biodegradation 94:1–11

McNamara CJ, Mitchell R (2005) Microbial deterioration of historic stone. Front Ecol Environ 3:445–451

Orial G, Castanier S, Le Métayer G, Loubiere JF (1992) The biomineralisation: a new process to protect calcareous stone applied to historic monuments. In: Toishi K, Arai H, Kenjo T, Yamano K (eds) Proceedings of 2nd int. conference on biodeterioration of cultural property, Yokohama, pp. 98–116

Ortega-Calvo JJ, Arino X, Hernandez-Marine M, Saiz-Jimenez C (1995) Factors affecting the weathering and colonisation of monuments by phototrophic microorganisms. Sci Total Environ 167:329–341

Pinna D, Salvadori O (2008) Biodeterioration processes in relation to cultural heritage materials. Stone and related materials. In: Caneva G, Nugari MP, Salvatori O (eds) Plant biology for cultural heritage. Biodeterioration and conservation. Getty Publications, Los Angeles, pp 128–149

Polo A, Cappitelli F, Brusetti L, Principi P, Villa F, Giacomucci L, Ranalli G, Sorlini C (2010) Feasibility of removing surface deposits on stone using biological and chemical remediation methods. Microb Ecol 60:1–14

Pontone M (2014) Biorestauro: applicazione di batteri alle opere d'arte - Il caso dell'abbazia di Montecassino. Thesis biology degree, University of Molise, Italy

Ramirez JL, Santana MA, Galindo-Castro I, Gonzalez A (2005) The role of biotechnology in art preservation. Trends Biotechnol 23:584–588

Rampazzi L, Andreotti A, Bressan M, Colombini MP, Corti C, Cuzman O, D'Alessandro N, Liberatore L, Palombi L, Raimondi V, Sacchi B, Tiano P, Tonucci L, Vettori S, Zanardini E, Ranalli G (2018) An interdisciplinary approach to a knowledge-based restoration: the dark alteration on the Matera cathedral (Italy). Appl Surf Sci 458:529–539

Ranalli G, Alfano G, Belli C, Lustrato G, Colombini MP, Bonaduce I, Zanardini E, Abbruscato P, Cappitelli F, Sorlini C (2005) Biotechnology applied to cultural heritage: biorestoration of frescoes using viable bacterial cells and enzymes. J Appl Microbiol 98:73–83

Ranalli G, Belli C, Baracchini C, Caponi G, Pacini P, Zanardini E, Sorlini C (2003) Deterioration and bioremediation of frescoes: a case-study. In: Saiz-Jimenez C (ed) Molecular biology and cultural heritage. Balkema Publishers, Lisse, pp 243–246

Ranalli G, Chiavarini M, Guidetti V, Marsala F, Matteini M, Zanardini E, Sorlini C (1996) The use of microorganisms for the removal of nitrates and organic substances on artistic stoneworks. In: Riederer J (ed) Proceedings of the 8th int. congress on deterioration and conservation of stone, Berlin, 30 September–4 October 1996

Ranalli G, Chiavarini M, Guidetti V, Marsala F, Matteini M, Zanardini E, Sorlini C (1997) The use of microorganisms for the removal of sulphates on artistic stoneworks. Int Biodeterior Biodegradation 40:255–261

Ranalli G, Zanardini E, Rampazzi L, Corti C, Andreotti A, Colombini MP, Bosch-Roig P, Lustrato G, Giantomassi C, Zari D (2019) Onsite advanced biocleaning system on historical wall paintings using new agar-gauze bacteria gel. J Appl Microbiol 126:1785–1796

Ranalli G, Zanardini E, Sorlini C (2009) Biodeterioration—including cultural heritage. In: Schaechter M (ed) Encyclopedia of microbiology. Elsevier, Oxford, pp 191–205

Rivadeneyra MA, Delgado R, del Moral A, Ferrer MR, Ramos-Cormenzana A (1994) Precipitation of calcium carbonate by *Vibrio* spp. from an inland saltern. FEMS Microbiol Ecol 13:197–204

Rivadeneyra MA, Delgado R, Quesada E, Ramos-Cormenzana A (1991) Precipitation of calcium carbonate by *Deleya halophila* in media containing NaCl as sole salt. Curr Microbiol 22:185–190

Rodriguez-Navarro C, Rodriguez-Gallego M, Ben Chekroun K, Gonzales-Munoz MT (2000) Carbonate production by Myxococcus xanthus: a possible application to protect/consolidate calcareous stones. In: Proceeding of the int. congress quarry, laboratory, Monument, Pavia, pp. 493–498

Romano I, Abbate M, Poli A, D'Orazio L (2019) Bio-cleaning of nitrate salt efflorescence on stone samples using extremophilic bacteria. Sci Rep 9:1668

Saiz-Jimenez C (1997) Biodeterioration vs. biodegradation: the role of microorganisms in the removal of pollutants deposited on historic buildings. Int Biodeterior Biodegradation 24:225–232

Saiz-Jimenez C (2001) The biodeterioration of building materials. In: Stoecker J (ed) A practical manual on microbiologically influenced corrosion, vol 2. NACE, Houston, pp 4.1–4.20

Saiz-Jimenez C (2003) Organic pollutants in the built environment and their effect on the microooorganisms. In: Brimblecombe P (ed) The effects of air pollution on the built environment. Air pollution reviews, vol 2. Imperial College Press, London, pp 183–225

Saiz-Jimenez C (2004) Air pollution and cultural heritage. In: Saiz-Jimenez C (ed) Proceedings of the int. workshop on air pollution and cultural heritage, Balkema, Seville

Salimbeni R, Pini R, Siano S (2003) A variable pulse width Nd: YAG laser for conservation. J Cult Herit 4:72–76

Salvadori O, Realini M (1996) Characterization of biogenic oxalate films. In: Realini M, Toniolo L (eds) Proceedings of the 2nd int. symposium: the oxalate films in the conservation of works of art, Editream, Milan, pp. 335–351

Sasso S, Miller AZ, Rogerio-Candelera MA, Cubero B, Scrano L, Bufo SA, Saiz-Jimenez C (2015) Non-destructive testing of stone biodeterioration and biocleaning effectiveness. In: Non-destructive and microanalytical techniques in art and cultural heritage. TECHNART 2015. Catania

Scheerer S, Ortega-Morales O, Gaylarde C (2009) Microbial deterioration of stone monuments-an updated overview. Adv Appl Microbiol 66:97–139

Scherer GW (2000) Stress from crystallization of salt pores. In: Fassina V (ed) Proceedings of the 9th int. congress on deterioration and conservation of stone, Elsevier, Amsterdam, pp. 187–194

Sorlini C, Cappitelli F (2008) The application of viable bacteria for the biocleaning of cultural heritage surfaces. Coalition N 15:18–20

Sorlini C, Sacchi M, Ferrari A (1987) Microbiological deterioration of Gambaras' frescos exposed in open at Brescia, Italy. Int Biodeterior 23:167–179

Tiano P, Biagiotti L, Mastromei G (1999) Bacterial bio-mediated calcite precipitation for monumental stone conservation: methods of evaluation. J Microbiol Methods 36:139–145

Tiano P, Cantisani E, Sutherland I, Paget JM (2006) Bioremediated reinforcement of weathered calcareous stones. J Cult Herit 7:49–55

Troiano F, Gulotta D, Balloi A, Polo A, Toniolo L, Lombardi E, Daffonchio D, Sorlini C, Cappitelli F (2013) Successful combination of chemical and biological treatments for the cleaning of stone artworks. Int Biodeterior Biodegradation 85:294–304

Valentini F, Diamanti A, Palleschi G (2010) New bio-cleaning strategies on porous building materials affected by biodeterioration event. Appl Surf Sci 256:6550–6563

Vergès-Belmin V (1996) Towards a definition of common evaluation criteria for the cleaning of porous building materials: a review. Sci Tech Cult Heritage 5:69–83

Viles HA, Moses CA (1998) Experimental production of weathering nanomorphologies on carbonate stone. Q J Eng Geol 31:347–357

Warscheid T, Braams J (2000) Biodeterioration of stone: a review. Int Biodeterior Biodegradation 46:343–368

Webster A, May E (2006) Bioremediation of weathered-building stone surface. Trends Biotechnol 24:255–260

Zanardini E, Abbruscato P, Ghedini N, Realini M, Sorlini C (2000) Influence of atmospheric pollutants on the biodeterioration of stone. Int Biodeterior Biodegradation 45:35–42

Zanardini E, May E, Inkpen R, Cappitelli F, Murrell JC, Purdy KJ (2016) Diversity of archaeal and bacterial communities on exfoliated sandstone from Portchester Castle (UK). Int Biodeterior Biodegradation 109:78–87

Zanardini E, May E, Purdy KJ, Murrell JC (2019) Nutrient cycling potential within microbial communities on culturally important stoneworks. Environ Microbiol Rep 11:147–154

Zanardini E, Purdy K, May E, Moussard H, Williams S, Oakley B, Murrell C (2011) Investigation of archaeal and bacterial community structure and functional gene diversity in exfoliated sandstone at Portchester Castle (UK). In: Sterflinger K, Piñar G (eds) Int. Symposium on biodeterioration and biodegradation (IBBS-15), Vienna, September 19–24, p. 90

# Chapter 13
# Protection and Consolidation of Stone Heritage by Bacterial Carbonatogenesis

Fadwa Jroundi, Maria Teresa Gonzalez-Muñoz, and
Carlos Rodriguez-Navarro

**Abstract** For millennia, artists and architects around the world used natural stone for the carving of sculptures and the construction of monuments, such as Roman, Greek, and Maya temples, the European cathedrals, and the Taj Mahal, just to name a few. Currently, the survival of these irreplaceable cultural and historical assets is under threat due to their continued degradation caused by various biotic and abiotic weathering processes that affect not only the aesthetic appearance of these structures, but also their durability and survival. The natural precipitation of calcium carbonate minerals by bacteria has been proposed for conservative interventions in monument restoration. This chapter reviews the application of biomineralization by (indigenous) bacterial carbonatogenesis as a novel technology for the protection and consolidation of altered ornamental materials. Carbonatogenesis is based on the ability of some bacteria to induce calcium carbonate precipitation. Laboratory and in situ results support the efficacy of bacterial carbonatogenesis, since remarkable protection and consolidation are achieved on the surface and in depth, without alterations in color or porosity, and without fostering the development of microbiota that could be harmful to the stone material. A discussion on the advantages of this novel biotechnology is provided. Challenges and future work on bioconsolidation of stone artifacts are also outlined.

**Keywords** Consolidation · Stone · Biomineralization · Self-inoculation strategy · Indigenous bacteria · Cultural heritage · Conservation

F. Jroundi · M. T. Gonzalez-Muñoz
Department of Microbiology, Faculty of Science, University of Granada, Granada, Spain
e-mail: fadwa@ugr.es; mgonzale@ugr.es

C. Rodriguez-Navarro (✉)
Department of Mineralogy and Petrology, Faculty of Science, University of Granada, Granada, Spain
e-mail: carlosrn@ugr.es

© The Author(s) 2021
E. Joseph (ed.), *Microorganisms in the Deterioration and Preservation of Cultural Heritage*, https://doi.org/10.1007/978-3-030-69411-1_13

# 1   Introduction

The stone built heritage undergoes severe deterioration and damage resulting in the irreversible loss of priceless cultural assets (Crispim et al. 2003; Negi and Sarethy 2019). This alteration is a consequence of complex weathering phenomena in which physical, chemical, and biological processes are involved, which typically occur simultaneously, thus being difficult to distinguish the independent role played by each one (Strzelczyk 1981; Sebastián and Rodríguez-Navarro 1995; Saiz-Jimenez 1997; Warscheid and Braams 2000). Carbonate stones (limestone, marble, and dolostone) made up of calcium (and magnesium) carbonate minerals are since ages among the most commonly used in artworks and monuments. Despite their apparent strength, these materials are not highly durable; what is astonishing is that they are prone not only to chemical and physical damage but also to microbial attack. Carbonate stones are indeed highly susceptible to weathering factors, involving the interaction of several phenomena that include changing environmental conditions, such as temperature, relative humidity, pH, and sun light, which altogether contribute to chemical, physical, and biological weathering processes (Warscheid and Braams 2000; Rodriguez-Navarro et al. 2011; Mihajlovski et al. 2017). One of the most remarkable phenomena in this regard is the effect of acid attack (for example, by the so-called acid rain) promoted by emissions of acid pollutant gases to the atmosphere from industries, heating and traffic (Rodriguez-Navarro and Sebastian 1996; Urosevic et al. 2012), and the formation of sulfate compounds (i.e., gypsum). But these phenomena of acid attack can also be a consequence of biodeterioration, being in many cases practically impossible to distinguish if the origin of the alteration is chemical or biological. In fact, the deterioration caused by biological agents (biodeterioration), which is conditioned by the development of living organisms, can cause weathering phenomena spanning from patina, which mask the surfaces, to severe damage derived from the production of very aggressive acids for the calcareous material such as sulfuric acid, produced by sulfur-oxidizing bacteria. The action of organic acids produced by a multitude of heterotrophic bacteria and fungi that are favored by organic pollutants is also deleterious for carbonate stones. For example, deterioration of marble at the cathedral of Milan has been associated with this type of organic acids (Strzelczyk 1981). All these types of alteration processes cause a gradual increasing of the material porosity, degrading its mechanical characteristics, causing the loss of cohesion and disintegration, and finally leading to their complete destruction. It is only in the last decades that this concern has received serious attention from scientists including archaeologists, geologists, chemists, biologists, and conservators aiming for the search of preventive and remedial strategies to preserve these historical monuments. This task is, however, challenging. Many attempts were directed towards the conservation and consolidation of such structures by using conventional methods including the use of inorganic and organic products that can act as protective coatings and/or consolidant agents (Lazzarini and Laurenzi Tabasso 2010; Rodriguez-Navarro et al. 2011).

## 2 Conventional Methodologies for Stone Conservation

Organic products have been frequently used for their versatility, easy synthesis, and application methodology. This is the case of acrylic and epoxy resins or copolymers of a mix of diverse type of polymers. Among the non-polymeric products, there are organic ones such as oxalates which aim at mimicking the formation of natural protective patinas or inorganic ones such as calcium hydroxide solutions (lime water), which is a well-known treatment used since Roman times for in situ consolidation of objects made of calcium carbonate. Other family of products are alkoxysilanes such as ethyl silicate that generate a silica gel, frequently used for friable limestone and wall paintings consolidation, or alkyl alkoxysilanes which impart a dual protective and consolidation effect (Doehne and Price 2010). More recently, phosphates have emerged as protective coatings for carbonate stones (Sassoni et al. 2011). Some of these treatments have protective and sacrificial qualities, while others have more specific consolidating effects for limestone and lime plaster (Hansen et al. 2003). Although these traditional organic and inorganic consolidants resulted to be efficient in some cases, providing the stone with some hydrophobicity and/or consolidation effect, most of them, however, prove to have various disadvantages. They were prone to form superficial films that block the porous system or they caused alteration of the color and brightness of the treated stone, in addition to their limited resistance to alteration and their physico-chemical incompatibility with the original substrate. In some cases, very limited consolidation was achieved, as it is the case of the lime water technique which revealed ineffective, since precipitation of $CaCO_3$ often led to the formation of a superficial, micrometer-thick, friable aggregate of submicron-size calcite crystals, as a powder without cohesion that had an insufficient protection and/or consolidation effect (Price et al. 1988; Hansen et al. 2003; Doehne and Price 2010). These shortcomings have in part been overcome by the use of nanolimes, which are alcohol dispersions of $Ca(OH)_2$ nanoparticles (Rodriguez-Navarro and Ruiz-Agudo 2018). In general, however, the use of conventional protection and consolidation treatments, sometimes indiscriminate, has led to the production of more drawbacks than those intended to be corrected (Rodriguez-Navarro and Ruiz-Agudo 2018). Additionally, large quantities of volatile organic solvents are commonly used during their application, which contribute to pollution (Rodriguez-Navarro et al. 2003; Doehne and Price 2010). Thus, most of the conventional treatments used so far in the conservation of decayed stone have revealed to be only partially successful for the preservation and consolidation of deteriorated monuments (Price et al. 1988; Hansen et al. 2003; Wheeler 2005; Rodriguez-Navarro et al. 2013). Moreover, some of them have been even harmful because they accelerated the deterioration of the treated stone (Rodriguez-Navarro et al. 2003; Doehne and Price 2010). To address the shortcomings of these conventional strategies, numerous efforts have been dedicated to the search and development of new more effective and compatible conservation treatments for the consolidation and protection of the stone cultural heritage.

# 3 New Methodologies for Stone Conservation

In the last decades, bacterial biomineralization has been proposed as a compatible and an environmentally friendly technology for the protection of decayed ornamental stones, particularly those having a carbonate composition (such as limestone and marble) (Adolphe et al. 1990; Orial et al. 1993; Tiano et al. 1999; Castanier et al. 2000). This strategy takes advantage on the capacity of bacteria to induce the formation of calcium carbonate, which cements carbonate rocks and is a biogeochemical phenomenon that commonly occurs in a wide range of natural environments (Boquet et al. 1973). Although often bacteria have been related to harmful effects on stone structures, affecting the integrity of minerals or exacerbating powerful physico-chemical deterioration processes, there is increasing evidence that carbonate-producing bacteria (i.e., carbonatogenic bacteria) could be used to reverse or ameliorate the effects of weathering processes affecting stone objects of historical and artistic interest (Le Métayer-Levrel et al. 1999; Rodriguez-Navarro et al. 2003; Jimenez-Lopez et al. 2007, 2008; Dhami et al. 2014; Seifan and Berenjian 2019). The bacterial conservation method is based on the bacterially induced precipitation of a compatible calcium carbonate mineral cement on the stone substrate, and unlike most conventional treatments, the precipitate seems to be highly coherent (Le Métayer-Levrel et al. 1999; Rodriguez-Navarro et al. 2003).

Below, an attempt is made to provide an outline of the current methods based on bacterially induced carbonate mineralization for the conservation of stone works as a promising approach for the bioremediation of historical and important building materials.

# 4 Bacterial Biomineralization of Calcium Carbonates

Biomineralization refers to the process by which organisms synthesize minerals. It can be either biologically induced mineralization (BIM) or biologically controlled mineralization (BCM) (Lowenstan and Weiner 1989; Mann 1995; González-Muñoz et al. 2010). BCM commonly results in the formation of complex and specialized structures like skeletons, shells, and teeth, and occurs mainly in tissue-forming multicellular eukaryotes. These biogenic minerals are synthesized either intracellularly or extracellularly in a way that is unique to each species, independently of environmental conditions. The process is thought to be completely regulated under specific metabolic and genetic control (Bazylinski and Moskowitz 1997; Bäuerlein 2003; Baeuerlein 2004). Although rare, BCM also occurs in the case of microorganisms: well-known examples are magnetite mineralization by magnetotactic bacteria, as well as calcite precipitation and silica deposition by algae and diatoms, respectively (Bäuerlein 2003; Baeuerlein 2004; Bazylinski and Schübbe 2007). In contrast, in BIM, as it is the case of bacterial carbonatogenesis, the organisms do not control the biomineralization process directly. Due to their metabolism, they alter the

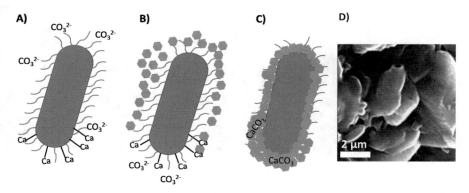

**Fig. 13.1** Representation of the process of bacterially induced calcium carbonate precipitation. Calcium ions in the solution are attracted to the bacterial cell wall due to the negative charge of the latter. Oxidative deamination of amino acids results in the production of ammonia and $CO_2$, which is converted into $HCO_3^-$ or $CO_3^{2-}$ at higher pH in the microenvironment surrounding the bacteria (**A**). In the presence of calcium ions, this can result in the heterogeneous precipitation of calcium carbonate on the bacterial cell wall (**B**). The whole cell is finally entombed with time (**C**). Scanning electron Microscopy image (**D**) showing the imprint of a bacterial cell involved in calcium carbonate precipitation

geochemistry around their cells favoring mineralization. As a result, the type of mineral formed by BIM processes is strongly dependent on environmental conditions (Rivadeneyra et al. 1994; Ben Omar et al. 1997; Brennan et al. 2004). Different mechanisms/processes responsible for bacterial carbonatogenesis have been proposed, which result in extracellular mineral growth and contribute to the formation of calcium carbonate sediments and rocks (Chafetz et al. 1991; Folk 1993; Vasconcelos et al. 1995; Paerl et al. 2001; Zavarzin 2002; Sharma 2013). In addition to the bacterial metabolic pathways that shift the environmental pH to alkalinity and, in the presence of calcium ions, foster calcium carbonate precipitation, bacterial surfaces also play an important role in such a mineral formation by serving as heterogeneous nucleation sites (Fortin 1997; Rodriguez-Navarro et al. 2003). As a result of successive stratification of the carbonate precipitates on the bacterial external surface, bacterial cells can be embedded or entombed within the growing calcium carbonate crystals (Fig. 13.1).

BIM is the most commonly applied process in the field of stone protection and consolidation, an emerging interdisciplinary biotechnology that consists in the use of bacteria and their biological and metabolic pathways for the precipitation in the stone of different biominerals, depending on bacteria and substrate types (Rodriguez-Navarro et al. 2012). Nearly 64 varieties of minerals of phosphates, carbonates, silicates, iron and manganese oxides, sulfide minerals, and amorphous silica can be synthetized by different microorganisms using specific metabolic routes (Knoll 2003). Microbially induced carbonate mineral precipitation occurs via various heterotrophic pathways such as the nitrogen cycle, the sulfur cycle and iron reduction. In the nitrogen metabolism, calcium carbonate precipitation is thought to

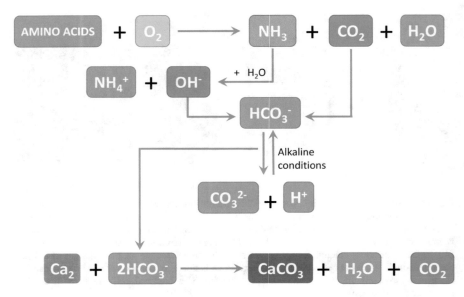

**Fig. 13.2** Bacterially induced mineralization occurring by oxidative deamination of amino acids via complex biochemical reactions

operate in the presence of oxygen (aerobiosis) by two different metabolic pathways, including amino acid and peptide catabolism (e.g.,: Rodriguez-Navarro et al. 2003, 2007; González-Muñoz et al. 2010; Jroundi et al. 2012), or degradation of urea or uric acid (Le Métayer-Levrel et al. 1999; Okwadha and Li 2010; De Muynck et al. 2010, 2013; Heidari Nonakaran et al. 2015; De Belie et al. 2018); both of which are important in increasing the local pH and inducing the production of carbonate ions.

Through the degradation of urea, ureolytic bacteria increase the concentration of dissolved inorganic carbon (DIC) and pH. During this process, carbamate is first produced by urease-catalyzed urea hydrolysis (Eq. 13.1). This carbamate intermediate spontaneously hydrolyzes to ammonia and carbonic acid (Eq. 13.2). Afterwards, these products equilibrate in the presence of water, forming bicarbonate, carbonate, and ammonium ions, as well as hydroxyl ions, thereby increasing the pH (Eqs. 13.3, 13.4). Finally, in the presence of calcium, these reactions induce the precipitation of calcium carbonate (Eq. 13.5) in an alkaline environment (De Muynck et al. 2010).

Urease

$$CO(NH_2)_2 + H_2O \rightarrow NH_2COOH + NH_3 \tag{13.1}$$

$$NH_2COOH + H_2O \rightarrow NH_3 + H_2CO_3 \tag{13.2}$$

$$H_2CO_3 \rightarrow 2H^+ + CO_3{}^{2-} \tag{13.3}$$

$$NH_3 + H_2O \rightarrow NH_4{}^+ + OH^- \tag{13.4}$$

$$Ca^{2+} + CO_3{}^{2-} \rightarrow CaCO_3 \tag{13.5}$$

Some Gram-negative aerobic microbial strains are able to use amino acids as their sole source of energy to initiate the biomineralization of calcium carbonate through amino acid and peptide catabolism in aerobiosis (Fig. 13.2). As shown in Fig. 13.2, carbon dioxide and ammonia are produced by oxidative deamination of amino acids. Ammonia (Eq. 13.4) increases the alkalinity in the microenvironment around the cell, whereas $CO_2$ dissolves and transforms into either $HCO_3^-$ or $CO_3^{2-}$ at higher pH. Here also, the presence of calcium and an alkaline environment favors the precipitation of calcium carbonate by heterogeneous nucleation on the bacterial cell walls and cell debris (Rodriguez-Navarro et al. 2003), and on the extracellular polymeric substances (EPS) forming a biofilm (Jroundi et al. 2017).

Incorporation of organics (bacterial cells or EPS) into the bacterial calcium carbonate crystals (vaterite and calcite) has been shown (Rodriguez-Navarro et al. 2007; Jroundi et al. 2017). EPS produced by bacteria have been demonstrated to be involved in biocalcification by entrapping calcium and carbonate ions and serving as a nucleation site (Braissant et al. 2003; Zamarreño et al. 2009). In *Pseudomonas fluorescens*, morphology and mineralogy of the carbonate crystals formed was demonstrated to be influenced by the composition and concentration of EPS (Braissant et al. 2003). Also, EPS appears to be involved in the aggregation of smaller crystals resulting in the growth of the final larger crystals (Buczynski and Chafetz 1991). This process is commonly extracellular, occurring in the area surrounding the cells, and often leading to the join mineralization of EPS and bacterial cells (Fig. 13.1D) that become the nucleus of the biominerals (Ehrlich 1998; Rodriguez-Navarro et al. 2007).

The pathways of calcium carbonate formation described above are ubiquitous in nature, accounting for the common occurrence of microbial carbonate precipitation (MCP) and confirms what Boquet et al. (1973) earlier reported, i.e., that most soil bacteria are able to induce the precipitation of carbonates under suitable conditions. This ability that bacteria have to promote carbonate mineral formation allowed the search and proposition, back in the 1990s, of a new kind of stone conservation treatment based on bacterial biomineralization. In the last few decades, the potential of bacterial biomineralization processes and their application to the conservation and consolidation of cultural heritage have drawn a widespread interest and since a while this is a matter of considerable attention and investment (e.g.,: Adolphe et al. 1990; Orial et al. 1993; Castanier et al. 2000; Rodriguez-Navarro et al. 2003; DaSilva 2004).

# 5 International Research Groups Involved in Bacterial Biomineralization

The biomineralization strategy for stone protection have had several propositions. Among the firsts to consider the use of bacterially induced calcium carbonate precipitation to stone conservation and protection were Adolphe et al. (1990)

through the CALCITE Bioconcept, a patented technique developed at the Université Pierre et Marie Curie in Paris, which involved the application to the stone of carbonatogenic bacteria. These authors demonstrated the ability of carbonatogenic bacteria, typically those inhabiting soils (Boquet et al. 1973), to precipitate calcite on damaged limestone surfaces, thus inducing their protection and consolidation. Further research was performed by Orial et al. (1993) from the University of Paris VI who explored the application of living cultures of selected carbonatogenic bacterial strains for calcite formation on stone and the replacement of the natural reinforced layer (CALCIN).

The technique was further optimized and industrialized by a collaboration between the French Company CALCITE, the University of Nantes, and the Historical Monuments Research Laboratory (LRMH). They carried out in situ tests and the application to historic monuments observing the reinforcement and consolidation of weathered stone. The first in situ application was carried out in 1993 in Thouars on the tower of the Saint Médard Church (Le Métayer-Levrel et al. 1999). The researchers also described the "successful" application of the CALCITE method on a sixteenth century limestone castle. Afterwards, a number of buildings in France were treated using this technique, including a castle at Châteaudun (Eure-et-Loir) and the Bordeaux cathedral (Castanier et al. 2000). There, it was reported the successful precipitation of the biocalcin layer, which resulted in no side effects on the esthetic appearance and highly reduced water absorption rate. Nevertheless, Orial (2000) concluded that every 10 years a new treatment would be needed to restore the protective effect of the biocalcin. In addition, upon investigating different bacteria, the authors selected *Bacillus cereus* cultures for in situ applications, which may pose different health risks (Perito et al. 1999). Furthermore, since only a few micrometers thick consolidated layer was observed, the ineffectiveness for in-depth consolidation was another main drawback of this technique (Le Métayer-Levrel et al. 1999; Rodriguez-Navarro et al. 2011).

Tiano (1995) proposed another approach for the conservation of calcareous stone based on the use of the "organic matrix" macromolecules extracted from shells of *Mytilus californianus* (a mollusc), which were applied on a sample of limestone to induce the mineralization process. With this technique, a good efficiency of the calcium carbonate precipitation process was showed leading to an increase in the strength of the surface of treated stone and a decrease in the water absorption rate. However, in situ tests were not successful and the application of organic matter posed a risk as it could promote the development of harmful heterotrophic microorganisms. This was one of the two research lines attempted by the European project BIOREINFORCE (BIOmediated calcite precipitation for monumental stones REINFORCEment, 2001–2004), which aimed at utilizing molecular biology and bacterial genetic engineering to elucidate the genetic background of crystal formation as an innovative alternative to the production of biomineralization-inducing macromolecules for the improvement of the biomediated calcite precipitation. At first, the consortium looked for molecules or structures from bacterial origin able to induce calcium carbonate precipitation. They demonstrated the ability of dead cells and cell fragments (BCF) from active carbonatogenic strains (*Bacillus cereus* and *Bacillus subtilis*) to induce higher and/or faster production of calcium carbonate crystals with a more complex shape than those produced by dead cells from less active strains (*Escherichia coli*). BCF application to stone resulted in a slight decrease in water absorption, with better effects on highly porous stones. Here again, this method revealed to be more useful for delicate small calcareous stone pieces than for entire buildings (Mastromei et al. 2008). The second research line consisted in identifying genes involved in the bacterial carbonatogenesis and the production at large-scale of macromolecules in bacterial cultures (Barabesi et al. 2007). The authors showed the presence in *Bacillus subtilis* of *lcfA* and *etfA* as operon and gene, respectively, which were involved in calcite precipitation and suggested a possible link between this precipitation and the fatty-acid metabolism (Barabesi et al. 2007; Perito and Mastromei 2011).

Tiano et al. (1999) evaluated the ability of *Micrococcus* sp. and *B. subtilis* in the consolidation and protection of a highly porous limestone (Pietra de Lecce). They reported the physical obstruction of the stone pores by the presence of a consistent layer of biological material (biofilm formation) and pointed out other potential drawbacks such as the reaction of stone and bacterial activity by-products or the modification of stone color.

Dick et al. (2006) proposed the application of several *Bacillus* species in a treatment that proceeded sequentially and in alternative cycles to biofilm formation and $CaCO_3$ precipitation on Euville limestone (from quarries of the "Département de la Meuse" in France). The aim of their study was to determine the microbial key factors contributing to the performance of such treatment. Overall, several parameters were examined such as deposition of dense crystal layers, urea degrading capacity, pH increase, calcite deposition on limestone, EPS-production, biofilm formation, the $\zeta$-potential. According to the authors, optimal deposition of a $CaCO_3$ layer in the treated limestone was based on a decrease in the water absorption rate. They finally concluded that the dense calcium carbonate crystals formed by *Lysinibacillus sphaericus* (=*Bacillus sphaericus*) were most suitable for coherent calcite production on degraded limestone and should be regarded as coating systems (Dick et al. 2006). Subsequently, the same bacterium was used for the bacterial deposition of a layer of calcite on the surface of concrete and mortars, which resulted in an improvement of the mortar resistance and the concrete durability with a more remarkable decrease in water uptake and a limited change of the chromatic aspect as compared to a treatment based on mixed ureolytic cultures (De Muynck et al. 2008a, b).

The BIOBRUSH (BIOremediation for Building Restoration of the Urban Stone Heritage) EU project was aimed at sequentially integrating salt removal and stone consolidation (May 2005). Within this project, Cappitelli et al. (2007) proposed a Carbogel as a delivery system for highly carbonatogenic bacteria such as *Desulfovibrio vulgaris*, which in addition to the elimination of black crusts also resulted in the conversion of gypsum to calcite. According to these authors, this delivery system could be employed for the control of potential harmful side effects of bacteria (Cappitelli et al. 2007). However, the practical applicability of such methodology was challenged by the complexity of the procedure and the necessity of a 2-weeks treatment for the confirmation of its effectiveness as consolidant. Furthermore, no information was provided neither on how the new calcite crystals were attached to the stone surface nor on the degree of stone reinforcement.

Daskalakis et al. (2013) proposed the use of *Pseudomonas*, *Pantoea* and *Cupriavidus* strains as candidate for bioconsolidation and protection of ornamental stones. They consolidated marble specimens and demonstrated a weight increase, due to the partial or even complete covering of the marble surfaces with (mainly) vaterite. Calcium carbonate precipitation induced by bacteria using only *Cupriavidus metallidurans* was then investigated to elaborate an environmentally friendly technique for the preservation and restoration of ornamental stones (Daskalakis et al. 2014). Further, these authors investigated the capacity of *Bacillus pumilus* for biomineralization on marble and reported that the rate of stone loss was reduced by the fine layer of bacterial calcium carbonate precipitated and *B. pumilus* proved to be useful as a candidate for in situ applications for stone conservation (Daskalakis et al. 2015).

Rodriguez-Navarro et al. (2003) proposed the use of *Myxococcus xanthus*, a non-pathogenic Gram-negative soil bacterium, and a nutritional solution that induced the precipitation of calcium carbonate within decayed stones. Depending on the culture media in which it grows, *M. xanthus* can induce the formation of a wide variety of biominerals, including phosphates (struvite: Gonzalez-Muñoz et al. 1993; Ben Omar et al. 1994, 1995, 1998), carbonates (calcite, Mg-calcite, vaterite: González-Muñoz et al. 2000; Rodriguez-Navarro et al. 2003; Chekroun et al. 2004; Rodriguez-Navarro et al. 2007) and sulfates (barita, taylorita: González-Muñoz et al. 2003). According to Rodriguez-Navarro et al. (2003), the treatment resulted in a deeper consolidation of calcarenite with a newly formed calcium carbonate cement in comparison to other known strategies for $CaCO_3$ precipitation in stone. Calcium carbonate precipitation typically occurred on the bacterial cell walls, where heterogenous nucleation was facilitated, and the calcite crystals grew as a coherent cement (often epitaxially) on the stone support without plugging or blocking the porous system. In this case, bacterial calcium carbonate cement was observed up to a depth of several hundred micrometers ($>500$ µm). These authors also showed that the newly produced bacterial cement was tougher and more resistant to dissolution than the original calcite in the stone, resulting from the incorporation of organic by-products of bacterial activity into the precipitated bacterial calcite. Interestingly, Chekroun et al. (2004) and Rodriguez-Navarro et al. (2007) showed that polymorph selection in carbonate precipitation was not strain specific, since in the case of bacteria such *M. xanthus* vaterite and calcite (or other minerals) can be precipitated just by modifying the composition of the culture medium (Fig. 13.3). These results suggested that an accurate selection of the growth media and the bacterial strains could yield to improved protection and consolidation of ornamental stones with different composition and textures. Similar results were also reported on non-porous marble, demonstrating that bacterial carbonatogenesis was not only effective in preserving porous stones but also in conserving non-porous ones (Rodriguez-Navarro et al. 2011).

**Fig. 13.3** SEM photomicrographs of different crystals induced by *M. xanthus*. (**A**) Vaterite crystals in M-3P culture medium. (**B**) Struvite crystals in CT medium (Ben Omar et al. 1998). Bar represent 10 and 100 μm in A and B, respectively

One of the most interesting aspects of this research has arisen when trying to apply this treatment on altered ornamental rocks, simulating real conditions of application in historic buildings (Jimenez-Lopez et al. 2007, 2008). In this case, it was observed that the degree of production of bacterial carbonate cement, and therefore the degree of consolidation achieved, was substantially higher than after the treatment of sterilized rocks. Such data pointed to the existence of a synergistic action between the bacteria existing in the altered stone and the biomineralization treatment with *M. xanthus* (Jimenez-Lopez et al. 2007). Nevertheless, exogenous bacteria or stone-isolated single bacterial cultures are likely at a competitive disadvantage with respect to the bacteria already present and/or better adapted to the local microenvironment in the stone. Furthermore, the application of an exogenous culture or a stone-isolated single bacterium may unpredictably alter the dynamics of the bacterial community in the stone, which is an important limitation of this strategy.

## 6  The University of Granada Stone Consolidation Patent

Subsequently, successful consolidation effects were achieved by the application of a sterile nutritional solution that activates the indigenous microbiota present on the stone and promotes the selective growth of carbonatogenic bacteria, as demonstrated by Gonzalez-Muñoz et al. (2008). This finding led to the development of a nutritional solution (M-3P) and a bacterial consolidation method patented in 2008. Here again, as with the first strategy, this new one, i.e., the application of sterile nutritional solution to the stone, resulted in the formation of calcium carbonate cement that protected and consolidated weathered stones.

This patented strategy was tested in different situations. In situ treatments were carried out in three historical monuments (Monastery of San Jeronimo, Royal Hospital, and Royal Chapel) placed in an urban environment in Granada, taking into account different locations, orientations, and degree of stone deterioration, so that more extensive information could be provided on the suitability of the patented methodology (Jroundi et al. 2010; Ettenauer et al. 2011; Rodriguez-Navarro et al. 2015). The results showed that in the first two historical buildings, a very similar degree of consolidation was achieved regardless of the type of treatment performed (with or without the additional foreign bacteria *M. xanthus*). Besides, the strengthening associated with the production of the newly formed calcium carbonate cement at both buildings allowed reinforcing the porous system and binding the loose grains of the stone with such a bacterial cement, which lasted (at least) for up to 4 years after application (Jroundi et al. 2010; Rodriguez-Navarro et al. 2015). Regarding the Royal Chapel, the in situ treatment was simultaneously performed during the application of ethyl silicate as a conventional treatment, to compare the degree of consolidation obtained by both types of treatments, in combination with a biocide application as a cleansing pre-treatment performed four-month prior to the consolidation treatments. The stone strengthening in this case was due to the new formation of bacterial calcite crystals without plugging the pores, thereby proving that the bacterial consolidation using a sterile nutrititional solution was as efficient as the traditional ethyl silicate consolidant (Jroundi et al. 2013). In addition, the evolution of the microbiota in all three historical buildings was assessed by culture-dependent and independent techniques and the results revealed that the majority of carbonatogenic bacteria were activated by the application of the patented M-3P nutritional solution and no negative effects associated with the possible activation of deleterious microbiota were observed (Ettenauer et al. 2011; Jroundi et al. 2010, 2015; Rodriguez-Navarro et al. 2015). Overall, the results of these studies showed that the in situ application of the bacterial consolidation method demonstrated no detrimental side effects on the stones with a highly proved medium- and long-term effectiveness.

# 7    Self-Inoculation with Indigenous Carbonatogenic Bacterial Community

When applied to heavily salt weathered stones, both previously described strategies (with or without additional bacterium) demonstrated to have a limited efficacy. Their effectiveness depended also on the stone characteristics (e.g., porosity, mineralogy, level of deterioration), the type of exogenous microorganism and bacterial load applied, and its interaction with indigenous microbiota. Salt weathering is known to be highly deleterious (Rodriguez-Navarro and Doehne 1999; Schiro et al. 2012; Flatt et al. 2014) and usually results in the partial filling of stone pores with soluble salts (Webster and May 2006), which may further inhibit the proliferation of indigenous or exogenous carbonatogenic bacteria. Therefore, a novel environmentally friendly bacterial self-inoculation approach was recently proposed to overcome the limitations associated to the existing two bacterial biomineralization strategies (i.e., (i) inoculation with a single carbonatogenic bacterial strain, and (ii) activation

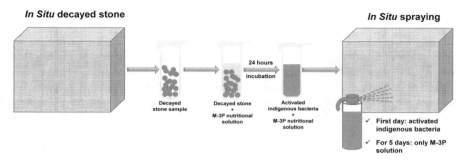

**Fig. 13.4** Workflow for the application of the environmentally friendly bacterial self-inoculation conservation treatment

of the indigenous carbonatogenic microbiota with a sterile nutrititional solution) currently applied in the field of stone conservation. This novel technique was based on the isolation from heavily salt damaged stone of an indigenous community of carbonatogenic bacteria followed by their activation by culturing (in the laboratory) and re-application back onto the same stone (Fig. 13.4). This methodology was in situ tested in a historic building (San Jeronimo Monastery) in Granada, Spain, where the indigenous bacterial community was collected from carbonate stone that had suffered considerable damage due to salt weathering. In the laboratory, the bacterial community was identified following activation with the patented M-3P nutritional solution and the whole carbonatogenic bacterial community was subsequently applied in situ, i.e., self-inoculation, along with M-3P, to the same stone from which it was isolated. The protection and consolidation effectiveness of this novel conservation method was evaluated and proved both in situ and in the laboratory. Results showed that this third bacterial conservation method provided a greater protection and consolidation to the salt weathered stones than the two previous methods above described. In this case, the effective stone protection and consolidation resulted from the formation of an abundant and exceptionally strong organic-inorganic hybrid cement consisting of a combination of nanostructured bacterial $CaCO_3$, bacterially-derived organics, and calcified bacterial EPS, all of which covered the calcitic substratum protecting the stone from chemical weathering (i.e., dissolution) (Jroundi et al. 2017).

The proposed self-inoculation approach for the conservation of stone has proven to have superior advantages over other techniques as it facilitates the formation of calcium carbonate cement by the activated indigenous bacteria and offers protection to the stone which guarantees the activity of the stone indigenous carbonatogenic community and the further consolidation of the heavily salt weathered stone. Remarkably, such a treatment helps to reduce the amount of deleterious microorganisms involved in biodeterioration, likely because they are in a competitive disadvantage with carbonatogenic bacteria selectively activated by the treatment. This is important, as it implies that the bacterial conservation treatment not only helps to protect and consolidate the treated substrate, but also reduces potential microbial biodeterioration processes.

# 8 Conclusions and Future Prospects

The sculptural and built stone heritage suffers several weathering processes that endangers its survival. To halt or mitigate the effects of such weathering phenomena, different protection and consolidation treatments have been developed and applied in the past, most of them with limited effectiveness. In recent years, the biomineralization of calcium carbonates by carbonatogenic bacteria has been proposed as a novel, highly effective, environmentally friendly method for the protection and consolidation of decayed stone. Three basic bacterial biomineralization-based conservation methods have been proposed: (i) Inoculation of a single bacterial strain (either exogenous or collected from the stone substrate to be treated) culture; (ii) Activation of the indigenous carbonatogenic bacteria present in the stone via the application of a sterile nutritional solution; and (iii) Self-inoculation of the stone with the indigenous carbonatogenic bacterial community extracted from the decayed stone, cultured in the laboratory and re-applied to the stone. Each of these methods has its advantages and shortcomings. Nonetheless, the high effectiveness and easiness of application of the second method (application of a sterile nutritional solution) offers a most versatile and efficient bacterial conservation method for stone protection and consolidation. However, in specific cases where extreme deterioration is observed, as could be the case of heavily damaged stone subjected to salt weathering, the third method, i.e., self-inoculation, seems to be the most effective conservation approach. For the successful application of the last two methods, a prerequisite is the existence in the stone to be treated of a population of indigenous carbonatogenic bacteria. Remarkably, numerous studies have shown that in nearly all stone substrates and environments on the Earth surface, carbonatogenic bacteria are abundant. This explains why so far all in situ applications of the second and third bacterial conservation methods have been successful.

Future studies should further explore if the proposed bacterial conservation methods are equally effective in extreme environments, such as hot and humid tropical sites, desert areas, or very cold environments. Studies should also be performed to identify and quantify carbonatogenic bacteria in stones located in such different environments. Also, the effectiveness of the bacterial protection and consolidation treatments discussed here should be further tested for the case of non-carbonate stone (e.g., sandstones, granites) and man-made materials, such as lime-based mortars, gypsum plasters, ceramics, and earthen constructions. Finally, it would be enlightening to evaluate in detail what is the impact of the above-mentioned bacterial conservation methods on the microbiota existing in the treated materials, and whether these treatments can help to reduce microbial biodeterioration.

**Acknowledgement** This work was carried out with the financial support of the European Regional Development Fund (ERDF) co-financed grant RTI2018-099565-B-I00 (MINECO Secretaría de Estado de Investigación, Desarrollo e Innovación, Spain), Proyecto de Excelencia RNM-3943, Research Groups RNM-179 and BIO-103 (Consejería de Economía, Innovación, Ciencia y Empleo, Junta de Andalucía), and the Unit of Research Excellence UCE- PP2016-05 of the University of Granada.

# References

Adolphe JP, Loubière JF, Paradas, J, Soleilhavoup F (1990) Procédé de traitement biologique d'une surface artificielle. European Patent 90400G97.0 (after French patent 8903517, 1989)

Baeuerlein E (2004) Biomineralization: progress in biology, molecular biology and application. John Wiley & Sons, Weinheim

Barabesi C, Galizzi A, Mastromei G, Rossi M, Tamburini E, Perito B (2007) *Bacillus subtilis* gene cluster involved in calcium carbonate biomineralization. J Bacteriol 189:228–235

Bäuerlein E (2003) Biomineralization of unicellular organisms: an unusual membrane biochemistry for the production of inorganic nano- and microstructures. Angew Chem Int Ed 42:614–641

Bazylinski DA, Moskowitz BM (1997) Microbial biomineralization of magnetic iron minerals: microbiology, magnetism and environmental significance. Rev Mineral 35:217–223

Bazylinski DA, Schübbe S (2007) Controlled biomineralization by and applications of magnetotactic bacteria. Adv Appl Microbiol 62:21–62

Ben Omar N, Arias JM, González-Muñoz MT (1997) Extracellular bacterial mineralization within the context of geomicrobiology. Microbiologia 13:161–172

Ben Omar N, Entrena M, González-Muñoz MT, Arias JM, Huertas F (1994) Effects of pH and phosphate on the production of struvite by *Myxococcus xanthus*. Geomicrobiol J 12:81–90

Ben Omar N, Gonzalez-Muñoz MT, Peñalver JMA (1998) Struvite crystallization on *Myxococcus* cells. Chemosphere 36:475–481

Ben Omar N, Martínez-Cañamero M, González-Muñoz MT, Arias JM, Huertas F (1995) *Myxococcus xanthus*' killed cells as inducers of struvite crystallization. Its possible role in the biomineralization processes. Chemosphere 30:2387–2396

Boquet E, Boronat A, Ramos-Cormenzana A (1973) Production of calcite (calcium carbonate) crystals by soil bacteria is a general phenomenon. Nature 246:527–529

Braissant O, Cailleau G, Dupraz C, Verrecchia EP (2003) Bacterially induced mineralization of calcium carbonate in terrestrial environments: the role of exopolysaccharides and amino acids. J Sediment Res 73:485–490

Brennan ST, Lowenstein TK, Horita J (2004) Seawater chemistry and the advent of biocalcification. Geology 32:473–476

Buczynski C, Chafetz HS (1991) Habit of bacterially induced precipitates of calcium carbonate and the influence of medium viscosity on mineralogy. J Sediment Res 61:226–233

Cappitelli F, Toniolo L, Sansonetti A, Gulotta D, Ranalli G, Zanardini E, Sorlini C (2007) Advantages of using microbial technology over traditional chemical technology in removal of black crusts from stone surfaces of historical monuments. Appl Environ Microbiol 73:5671–5675. https://doi.org/10.1128/AEM.00394-07

Castanier S, Le Métayer-Levrel G, Orial G, Loubiere JF, Perthuisot JP (2000) Bacterial carbonatogenesis and applications to preservation and restoration of historic property. In: Ciferri O, Tiano P, Mastromei G (eds) Of microbes and art: the role of microbial communities in the degradation and protection of cultural heritage. Plenum, New York, pp 201–216

Chafetz H, Rush PF, Utech NM (1991) Microenvironmental controls on mineralogy and habit of $CaCO_3$ precipitates: an example from an active travertine system. Sedimentology 38:107–126

Chekroun KB, Rodríguez-Navarro C, González-Muñoz MT, Arias JM, Cultrone G, Rodríguez-Gallego M (2004) Precipitation and growth morphology of calcium carbonate induced by *Myxococcus xanthus*: implications for recognition of bacterial carbonates. J Sediment Res 74:868–876

Crispim CA, Gaylarde PM, Gaylarde CC (2003) Algal and cyanobacterial biofilms on calcareous historic buildings. Curr Microbiol 46:79–82

DaSilva EJ (2004) Art, biotechnology and the culture of peace. Electron J Biotechnol 7:130–166

Daskalakis MI, Magoulas A, Kotoulas G, Katsikis I, Bakolas A, Karageorgis AP, Mavridou A, Doulia D, Rigas F (2013) *Pseudomonas, Pantoea and Cupriavidus* isolates induce calcium carbonate precipitation for biorestoration of ornamental stone. J Appl Microbiol 115:409–423

Daskalakis MI, Magoulas A, Kotoulas G, Katsikis I, Bakolas A, Karageorgis AP, Mavridou A, Doulia D, Rigas F (2014) *Cupriavidus metallidurans* biomineralization ability and its application as a bioconsolidation enhancer for ornamental marble stone. Appl Microbiol Biotechnol 98:6871–6883

Daskalakis MI, Rigas F, Bakolas A, Magoulas A, Kotoulas G, Katsikis I, Karageorgis AP, Mavridou A (2015) Vaterite bio-precipitation induced by *Bacillus pumilus* isolated from a solutional cave in Paiania, Athens, Greece. Int Biodeterior Biodegrad 99:73–84

De Belie N, Wang J, Bundur ZB, Paine K (2018) Bacteria-based concrete. In: Pacheco-Torgal F, Melchers RE, Shi X, Belie ND, Tittelboom KV, Sáez A (eds) Eco-efficient repair and rehabilitation of concrete infrastructures, Woodhead Publishing Series in Civil and Structural Engineering. Woodhead Publishing, Duxford, pp 531–567

De Muynck W, Cox K, De Belie N, Verstraete W (2008a) Bacterial carbonate precipitation as an alternative surface treatment for concrete. Constr Build Mater 22:875–885

De Muynck W, Debrouwer D, De Belie N, Verstraete W (2008b) Bacterial carbonate precipitation improves the durability of cementitious materials. Cem Concr Res 38:1005–1014

De Muynck W, De Belie N, Verstraete W (2010) Microbial carbonate precipitation in construction materials: a review. Ecol Eng 36:118–136

De Muynck W, Verbeken K, De Belie N, Verstraete W (2013) Influence of temperature on the effectiveness of a biogenic carbonate surface treatment for limestone conservation. Appl Microbiol Biotechnol 97:1335–1347

Dhami NK, Reddy MS, Mukherjee A (2014) Application of calcifying bacteria for remediation of stones and cultural heritages. Front Microbiol 5:209

Dick J, De Windt W, De Graef B, Saveyn H, Van der Meeren P, De Belie N, Verstraete W (2006) Bio-deposition of a calcium carbonate layer on degraded limestone by *Bacillus* species. Biodegradation 17:357–367

Doehne E, Price CA (2010) Stone conservation: an overview of current research. The Getty Conservation Institute, Los Angeles

Ehrlich HL (1998) Geomicrobiology: its significance for geology. Earth Sci Rev 45:45–60

Ettenauer J, Piñar G, Sterflinger K, Gonzalez-Muñoz MT, Jroundi F (2011) Molecular monitoring of the microbial dynamics occurring on historical limestone buildings during and after the in situ application of different bio-consolidation treatments. Sci Total Environ 409:5337–5352

Flatt RJ, Caruso F, Sanchez AMA, Scherer GW (2014) Chemo-mechanics of salt damage in stone. Nat Commun 5:4823

Folk RL (1993) SEM imaging of bacteria and nannobacteria in carbonate sediments and rocks. J Sediment Res 63:990–999

Fortin D (1997) Surface-mediated mineral development. Rev Mineral 35:162–180

Gonzalez-Muñoz MT, Arias JM, Montoya E, Rodriguez-Gallego M (1993) Struvite production by *Myxococcus coralloides* D. Chemosphere 26:1881–1887

González-Muñoz MT, Chekroun KB, Aboud AB, Arias JM, Rodriguez-Gallego M (2000) Bacterially induced Mg-calcite formation: role of $Mg^{2+}$ in development of crystal morphology. J Sediment Res 70:559–564

González-Muñoz MT, Fernández-Luque B, Martínez-Ruiz F, Ben Chekroun K, Arias JM, Rodríguez-Gallego M, Martínez-Cañamero M, de Linares C, Paytan A (2003) Precipitation of barite by *Myxococcus xanthus*: possible implications for the biogeochemical cycle of barium. Appl Environ Microbiol 69:5722–5725

Gonzalez-Muñoz MT, Rodriguez-Navarro C, Jimenez-Lopez C, Rodriguez-Gallego M (2008) Method and product for protecting and reinforcing construction and ornamental materials. Spanish patent WO 2008/009771 A1

González-Muñoz MT, Rodriguez-Navarro C, Martínez-Ruiz F, Arias JM, Merroun ML, Rodriguez-Gallego M (2010) Bacterial biomineralization: new insights from *Myxococcus*-induced mineral precipitation. Geol Soc Lond Spec Publ 336:31–50

Hansen EF, Doehne E, Fidler JM, Larson JD, Martin BL, Matteini M, Rodriguez-Navarro C, Pardo EMS, Price CS, de Tagle A, Teutonico JM, Weiss N (2003) A review of selected inorganic consolidants and protective treatments for porous calcareous materials. Stud Conserv (sup 1) 49:13–25

Heidari Nonakaran S, Pazhouhandeh M, Keyvani A, Abdollahipour FZ, Shirzad A (2015) Isolation and identification of *Pseudomonas azotoformans* for induced calcite precipitation. World J Microbiol Biotechnol 31:1993–2001

Jimenez-Lopez C, Jroundi F, Pascolini C, Rodriguez-Navarro C, Piñar-Larrubia G, Rodriguez-Gallego M, González-Muñoz MT (2008) Consolidation of quarry calcarenite by calcium carbonate precipitation induced by bacteria activated among the microbiota inhabiting the stone. Int Biodeterior Biodegradation 62:352–363

Jimenez-Lopez C, Rodriguez-Navarro C, Piñar G, Carrillo-Rosúa FJ, Rodriguez-Gallego M, Gonzalez-Muñoz MT (2007) Consolidation of degraded ornamental porous limestone stone by calcium carbonate precipitation induced by the microbiota inhabiting the stone. Chemosphere 68:1929–1936

Jroundi F, Fernández-Vivas A, Rodriguez-Navarro C, Bedmar EJ, González-Muñoz MT (2010) Bioconservation of deteriorated monumental calcarenite stone and identification of bacteria with carbonatogenic activity. Microb Ecol 60:39–54

Jroundi F, Gómez-Suaga P, Jimenez-Lopez C, González-Muñoz MT, Fernandez-Vivas MA (2012) Stone-isolated carbonatogenic bacteria as inoculants in bioconsolidation treatments for historical limestone. Sci Total Environ 425:89–98

Jroundi F, Gonzalez-Muñoz MT, Rodriguez-Navarro C, Martin-Peinado B, Martin-Peinado J (2013) Conservation of carbonate stone by means of bacterial carbonatogenesis: evaluation of *in situ* treatments. In: Koui M, Koutsoukos P, Zezza F (eds) Proceedings of the 8th international symposium on the conservation of monuments in the Mediterranean basin. Patras. Technical Chamber of Greece, Patras, pp 159–171

Jroundi F, Gonzalez-Muñoz MT, Sterflinger K, Piñar G (2015) Molecular tools for monitoring the ecological sustainability of a stone bio-consolidation treatment at the Royal Chapel, Granada. PLoS One 10:e0132465

Jroundi F, Schiro M, Ruiz-Agudo E, Elert K, Martín-Sánchez I, González-Muñoz MT, Rodriguez-Navarro C (2017) Protection and consolidation of stone heritage by self-inoculation with indigenous carbonatogenic bacterial communities. Nat Commun 8:279

Knoll AH (2003) Biomineralization and evolutionary history. Rev Mineral Geochem 54:329–356

Lazzarini L, Laurenzi Tabasso M (2010) Il Restauro della Pietra. Utet Scienze Techniche, Torino

Le Métayer-Levrel G, Castanier S, Orial G, Loubière J-F, Perthuisot J-P (1999) Applications of bacterial carbonatogenesis to the protection and regeneration of limestones in buildings and historic patrimony. Sediment Geol 126:25–34

Lowenstan HA, Weiner S (1989) On biomineralization. Oxford University Press, New York

Mann S (1995) Biomineralization and biomimetic materials chemistry. J Mater Chem 5:935–946

Mastromei G, Marvasi M, Perito B (2008) Studies on bacterial carbonate precipitation for stone conservation. In: Proc. of 1st BioGeoCivil engineering conference, Delft, Netherlands. Delft University of Technology, Delft, pp 104–106

May E (2005) Biobrush research monograph: novel approaches to conserve our European heritage. EVK4-CT-2001-00055

Mihajlovski A, Gabarre A, Seyer D, Bousta F, Di Martino P (2017) Bacterial diversity on rock surface of the ruined part of a French historic monument: the Chaalis abbey. Int Biodeterior Biodegradation 120:161–169

Negi A, Sarethy IP (2019) Microbial biodeterioration of cultural heritage: events, colonization, and analyses. Microb Ecol 78:1014–1029

Okwadha GDO, Li J (2010) Optimum conditions for microbial carbonate precipitation. Chemosphere 81:1143–1148

Orial G (2000) La biominéralisation appliquée à la conservation du patrimoine: bilan de dix ans d'experimentation. In: XXI Siglo SA (ed) Restaurar la memoria, Actas del Congreso, Internacional AR&PA 2000, Fundación del Patrimonio Histórico de Castilla y León. Arganda del Rey, Madrid

Orial G, Castanier S, Metayer GL, Loubière JF (1993) The biomineralization: a new process to protect calcareous stone; applied to historic monuments. In: Ktoishi H, Arai T, Yamano K (eds) Proceedings of the 2nd international conference on biodeterioration of cultural property. International Communications Specialists, Tokyo, Japan, pp 98–116

Paerl HW, Steppe TF, Reid RP (2001) Bacterially mediated precipitation in marine stromatolites. Environ Microbiol 3:123–130

Perito B, Biagiotti L, Daly S, Galizzi A, Tiano R, Mastromei G (1999) Bacterial genes involved in calcite crystal precipitation. In: Ciferri O, Tiano P, Mastromei G (eds) Of microbes and art: the role of microbial communities in the degradation and protection of cultural heritage, international conference on microbiology and conservation. Kluwer, New York, pp 219–230

Perito B, Mastromei G (2011) Molecular basis of bacterial calcium carbonate precipitation. Prog Mol Subcell Biol 52:113–139

Price C, Ross K, White G (1988) A further appraisal of the "lime technique" for limestone consolidation, using a radioactive tracer. Stud Conserv 33:178–186

Rivadeneyra MA, Delgado R, del Moral A, Ferrer MR, Ramos-Cormenzana A (1994) Precipatation of calcium carbonate by *Vibrio* spp. from an inland saltern. FEMS Microbiol Ecol 13:197–204

Rodriguez-Navarro C, Doehne E (1999) Salt weathering: influence of evaporation rate, supersaturation and crystallization pattern. Earth Surf Process Landf 24:191–209

Rodriguez-Navarro C, González-Muñoz MT, Jimenez-Lopez C, Rodriguez-Gallego M (2011) Bioprotection. In: Reitner J, Thiel V (eds) Encyclopedia of geobiology. Springer, Netherlands, Dordrecht, pp 185–189

Rodriguez-Navarro C, Jimenez-Lopez C, Rodriguez-Navarro A, Gonzalez-Muñoz MT, Rodriguez-Gallego M (2007) Bacterially mediated mineralization of vaterite. Geochim Cosmochim Acta 71:1197–1213

Rodriguez-Navarro C, Jroundi F, Gonzalez-Muñoz MT (2015) Stone consolidation by bacterial carbonatogenesis: evaluation of in situ applications. Restor Build Monum 21:9–20

Rodriguez-Navarro C, Jroundi F, Schiro M, Ruiz-Agudo E, González-Muñoz MT (2012) Influence of substrate mineralogy on bacterial mineralization of calcium carbonate: implications for stone conservation. Appl Environ Microbiol 78:4017–4029

Rodriguez-Navarro C, Rodriguez-Gallego M, Chekroun KB, Gonzalez-Muñoz MT (2003) Conservation of ornamental stone by *Myxococcus xanthus*-induced carbonate biomineralization. Appl Environ Microbiol 69:2182–2193

Rodriguez-Navarro C, Ruiz-Agudo E (2018) Nanolimes: from synthesis to application. Pure Appl Chem 90:523–550

Rodriguez-Navarro C, Sebastian E (1996) Role of particulate matter from vehicle exhaust on porous building stones (limestone) sulfation. Sci Total Environ 187:79–91

Rodriguez-Navarro C, Suzuki A, Ruiz-Agudo E (2013) Alcohol dispersions of calcium hydroxide nanoparticles for stone conservation. Langmuir 29:11457–11470

Saiz-Jimenez C (1997) Biodeterioration *vs* biodegradation: the role of microorganisms in the removal of pollutants deposited on historic buildings. Int Biodeterior Biodegradation 40:225–232

Sassoni E, Naidu S, Scherer GW (2011) The use of hydroxyapatite as a new inorganic consolidant for damaged carbonate stones. J Cult Herit 12:346–355

Schiro M, Ruiz-Agudo E, Rodriguez-Navarro C (2012) Damage mechanisms of porous materials due to in-pore salt crystallization. Phys Rev Lett 109:265503

Sebastián E, Rodríguez-Navarro C (1995) Alteración y conservación de materiales pétreos ornamentales: antecedentes y estado actual de conocimientos. Ingeniería Civil 96:167–178

Seifan M, Berenjian A (2019) Microbially induced calcium carbonate precipitation: a widespread phenomenon in the biological world. Appl Microbiol Biotechnol 103:4693–4708

Sharma O (2013) Characterization of novel carbonic anhydrase from halophilic bacterial isolates and it's role in microbially induced calcium carbonate precipitation (Ms Thesis). Thapar University, India

Strzelczyk AB (1981) Stone. In: Rose AH (ed) Microbial biodeterioration. Academic Press, London, pp 61–80

Tiano P (1995) Stone reinforcement by calcite crystal precipitation induced by organic matrix macromolecules. Stud Conserv 40:171–176

Tiano P, Biagiotti L, Mastromei G (1999) Bacterial bio-mediated calcite precipitation for monumental stones conservation: methods of evaluation. J Microbiol Methods 36:139–145

Urosevic M, Yebra-Rodríguez A, Sebastián-Pardo E, Cardell C (2012) Black soiling of an architectural limestone during two-year term exposure to urban air in the city of Granada (S Spain). Sci Total Environ 414:564–575

Vasconcelos C, McKenzie JA, Bernasconi S, Grujic D, Tiens AJ (1995) Microbial mediation as a possible mechanism for natural dolomite formation at low temperatures. Nature 377:220–222

Warscheid T, Braams J (2000) Biodeterioration of stone: a review. Int Biodeterior Biodegrad 46:343–368

Webster A, May E (2006) Bioremediation of weathered-building stone surfaces. Trends Biotechnol 24:255–260

Wheeler G (2005) Alkoxysilanes and the consolidation of stone. The Getty Conservation Institute, Los Angeles

Zamarreño DV, Inkpen R, May E (2009) Carbonate crystals precipitated by freshwater bacteria and their use as a limestone consolidant. Appl Environ Microbiol 75:5981–5990

Zavarzin GA (2002) Microbial geochemical calcium cycle. Microbiology 71:1–17

# Chapter 14
# Siderophores and their Applications in Wood, Textile, and Paper Conservation

Stavroula Rapti, Stamatis C. Boyatzis, Shayne Rivers, and Anastasia Pournou

**Abstract** Since the 1950s, siderophores have been acknowledged as nature's chelating powerhouse and have been given considerable attention concerning their crucial roles in microorganisms and plants for capturing non-bioavailable iron from aquatic and terrestrial environments, as well as for their applications in agriculture, health, and materials science and environmental research. In recent years, the exceptional affinity and complexing efficacy, as well as the high selectivity of these potent chelators towards iron(III), have led to investigations by researchers aiming at understanding their capacity for removing potentially harmful and aesthetically unacceptable iron stains from organic substrates in cultural heritage objects. In the context of the conservation of cultural heritage objects, potent chelators have been proposed to remove iron from surfaces by transferring it to the more soluble complexed phase. In this review, the origins and the types of bio-environments of siderophores as well as their structure and chemistry are investigated and related to the requirements of conservation. It is evident that, given the enormous potential that these chelators have, the research for their application in cultural heritage is at a preliminary level, and has to date been within the rather narrow context of cellulosic materials such as paper and wood. The results of research conducted to date are presented in this review and questions regarding the optimal use of siderophores as iron-removing agents are posed.

**Keywords** Siderophores · Green chelators · EDTA (ethylene-diamino tetra-acetic acid) · DTPA (diethylene-triaminopenta-acetic or pentetic acid) · Cultural heritage · Conservation

S. Rapti · S. C. Boyatzis (✉) · A. Pournou
Department of Conservation of Antiquities and Works of Art, University of West Attica,
Athens, Greece
e-mail: srapti@uniwa.gr; sboyatzis@uniwa.gr; pournoua@uniwa.gr

S. Rivers
West Dean College of Arts and Conservation, Chichester, West Sussex, UK

E. Joseph (ed.), *Microorganisms in the Deterioration and Preservation of Cultural Heritage*, https://doi.org/10.1007/978-3-030-69411-1_14

301

# 1   Introduction

Siderophores are among nature's successful solutions to overcome the challenges of bioavailable iron in aqueous systems (Barton and Hemming 1993; Neilands 1995; Boukhalfa and Crumbliss 2002; Crichton 2019).

The origin of the word siderophore is Greek and means an "iron carrier" (Neilands 1995; Saha et al. 2013); siderophores are natural metal-chelating agents, employed by a wide range of microorganisms and plants to mobilize iron from its least available sources (Höfte 1993; Neilands 1995; Kraemer 2004; Ali and Vidhale 2013).

They are secondary metabolites of low molecular masses (200–2000 Da) and high affinity for iron ($K_f > 10^{30}$) (Hider and Kong 2010; Saha et al. 2013; Wang et al. 2014; Ahmed and Holmström 2014; Khan et al. 2018).

Siderophores have been classified depending on the oxygen ligands for Fe(III) coordination, as catecholate (sensu stricto, catecholates, and phenolates; better termed as aryl caps), hydroxamate, carboxylate, and mixed types (Miethke and Marahiel 2007; Ahmed and Holmström 2014; Saha et al. 2016; Khan et al. 2018). More than 500 different types of siderophores are known, of which 270 have been structurally characterized (Hider and Kong 2010; Ali and Vidhale 2013; Wang et al. 2014; Ahmed and Holmström 2014).

They have a high biotechnological potential and a wide range of applications not only in ecology, agriculture, and medicine but also in the field of conservation of cultural heritage, mostly for removing iron corrosion products from artifacts (Albelda-Berenguer et al. 2019).

The question of removing iron from organic substrates of cultural heritage has been previously considered through the use of chelators such as citric acid, EDTA (ethylene-diaminetetraacetic acid), DTPA (diethylene-triaminepenta-acetic, or pentetic acid), etc. (Almkvist et al. 2005; Burgess 1991; Margariti 2003; Rivers and Ummey 2003; Richards et al. 2012). Water-soluble iron(III) ions are highly reactive, as they participate in Fenton-type reactions (see Sect. 14.4.1) (Koppenol 1993; Kolar and Strlic 2001; Bulska and Wagner 2002; Dunford 2002; Burkitt 2003; Jablonský et al. 2010; Corregidor et al. 2019) causing severe oxidative stress to organic substrates.

The removal of iron stains from wood, textiles, and paper substrates emerges as an important issue, which although it has been typically dealt by the above-mentioned chelators, these were shown to not fully encompass iron(III) in their complexes, and significantly, not to prevent the harmful oxidative stress on cellulosic materials (Kolar 1997; Kolar and Strlič 2004). In addition, these complexing media do not offer the best possible solution due to a lack of specificity. EDTA, for instance, complexes similarly iron, copper, manganese, cobalt, and other cultural heritage-significant ions and unwanted side-reactions (see below) which in the case of siderophores are avoided. Therefore the need for increased specificity from chelators has been recognized in the field of cultural heritage and efforts employing siderophores have been reported among some research groups (Wagner and Bulska

2003; Rapti et al. 2017; Albelda-Berenguer et al. 2019). In this review, these efforts and their results will be presented along with the questions that still need to be answered.

The enormous biodiversity among bacteria, fungi, and particular types of plants has led to optimized strategies for sequestering abundant, but minimally bioavailable iron from the environment by employing a vast palette of chelating agents, commonly termed as *siderophores*. This highly successful strategy can only be seen as a triumph of nature which produces custom-made iron chelators according to the types of organisms, the character of the environment itself, and the severity of the iron nutrient demand. Learning from nature can teach us the right approach under the right condition for our applications, including conservation. Within this review, this leads to the necessity for presenting siderophores, their biological origins, the structures, and their chemistry as a necessary background (Sects. 14.1–14.3) for effectively considering their applications in conservation regarding the removal of unwanted iron from certain types of materials (Sect. 14.4).

## 2 Biosynthesis, Roles, and Applications of Siderophores

Almost all microorganisms and living members of the animal and plant kingdom are dependent on iron (Renshaw et al. 2002; Hider and Kong 2010; Saha et al. 2013; Ahmed and Holmström 2014; Khan et al. 2018).

Iron's unique chemical properties, such as its ability to coordinate and activate oxygen and its ideal redox chemistry ($Fe^{2+}$, $Fe^{3+}$, $Fe^{5+}$), turn this element into an essential and vital nutrient of the growth and developmental processes of every living organism (Saha et al. 2013, 2016; Ahmed and Holmström 2014; Khan et al. 2018), because it is acting as a catalyst in enzymatic processes, oxygen metabolism, electron transfer, and DNA and RNA syntheses (Hider and Kong 2010; Ahmed and Holmström 2014; Saha et al. 2016).

Although iron is abundant in nature, it is not easily bioavailable because at neutral pH and in the presence of atmospheric oxygen it undergoes rapid oxidation from ferrous ($Fe^{2+}$) to ferric ($Fe^{3+}$) iron and finally forms insoluble ferric hydroxides and oxy-hydroxide, which is almost unavailable for acquisition by organisms (Renshaw et al. 2002; Saha et al. 2013; Wang et al. 2014). More specifically, the solubility and consequently the bioavailability of iron as ferric oxy-hydroxide are less than $10^{-18}$ M, while microbes require $10^{-5}$–$10^{-8}$ M for carrying out vital physiological and metabolic processes and have optimal growth (Saha et al. 2013; Wang et al. 2014; Khan et al. 2018).

To solve this problem, a vast number of bacteria, fungi, plants, and even higher eukaryotes under iron-deficient conditions produce siderophores, which is an essential metabolic feature that allows them to survive (Renshaw et al. 2002; Saha et al. 2013, 2016; Ahmed and Holmström 2014; Khan et al. 2018).

Hydroxamate types of siderophores comprise the most common group of siderophores found in nature and are produced by both bacteria and fungi (Ali and

Vidhale 2013; Saha et al. 2016; Khan et al. 2018). Catecholate siderophores are found only in certain bacteria (Ali and Vidhale 2013; Saha et al. 2016; Khan et al. 2018), while the carboxylate type is produced by few bacteria such as *Rhizobium* and *Staphylococcus* and fungi such as *Mucorales* belonging to the phylum *Zygomycota* (Saha et al. 2016; Khan et al. 2018).


## 2.1    Biosynthesis

Siderophore biosynthesis is typically regulated by iron levels of the environment where the organism lives (Hider and Kong 2010). Although iron deficiency is the key factor, other external factors such as pH, temperature, carbon source, and the presence of other metals play an important role (Saha et al. 2013). Thus some hydroxamate siderophores are prevalent in lower or acidic pH, whereas other catecholate siderophores are produced in neutral to alkaline pH (Saha et al. 2013). The distinction of the various siderophore types and their significance regarding applications in conservation can be found in Sect. 14.3.2.

There are two biosynthesis pathways involved in siderophores' synthesis (a) the dependent on non-ribosomal peptide synthetases (NRPSs) and (b) the independent of NRPS (Miethke and Marahiel 2007; Saha et al. 2013; Khan et al. 2018). The non-ribosomal peptide synthetases are large multi-enzyme complexes responsible for the synthesis of several biologically important peptidic products without an RNA template (Miethke and Marahiel 2007; Saha et al. 2013). In general, NRPSs are essentially assembly lines of specialized domains that link amino acids via thioester intermediates. They function in a similar manner to that of fatty acid synthetase (Hider and Kong 2010). They consist of three domains required for peptide bond formation (a) adenylation domain, (b) peptidyl carrier protein domain (PCP or thiolation), and (c) condensation domain, responsible for the assembly of a wide array of amino, carboxy, and hydroxy acids in various combinations to produce macrocyclic peptidic with high structural variability (Miethke and Marahiel 2007; Hider and Kong 2010; Saha et al. 2013).

NRPSs are responsible mainly for the synthesis of aryl-capped siderophores (Miethke and Marahiel 2007). Hydroxamate and carboxylate siderophores are assembled by NRPS-independent mechanisms in the majority of cases, even though in some cases, NRPSs are partially involved in the synthesis of hydroxamate and carboxylate siderophores to build a peptidic backbone to which the iron-coordinating residues are attached (Miethke and Marahiel 2007).

Nonetheless, organisms such as bacteria, fungi, and plants have different ways to synthesize their siderophores (Khan et al. 2018; Albelda-Berenguer et al. 2019) and therefore the siderophores can be categorized based on the organism that produces them: fungal, bacterial, plant (phytosiderophores), cyanobacterial, or mammalian siderophores (Khan et al. 2018).

## 2.2  Roles

Fungal siderophores are mainly of hydroxamate type, belonging to the ferrichrome family (i.e., ferrichrome A), coprogen and triacetylfusarinine families, and carboxylate type of siderophores (Renshaw et al. 2002; Hider and Kong 2010; Ahmed and Holmström 2014; Khan et al. 2018). *Aspergillus fumigatus* and *Aspergillus nidulans* have been thoroughly studied for their siderophore production (Khan et al. 2018).

Bacterial siderophores comprise extracellular forms of siderophore (Khan et al. 2018). Most of the bacterial siderophores are catecholates (i.e., enterobactin), while there are also some carboxylates (i.e., rhizobactin), hydroxamates (i.e., ferrioxamine B), and mixed types (i.e., pyoverdine) (Ahmed and Holmström 2014). The Gram-negative, facultative anaerobe *Escherichia coli*, found normally in the intestine, is the most widely studied bacterium for siderophore production. It principally produces enterobactin, with the highest affinity towards iron(III) ion, of any known siderophore (Khan et al. 2018). Members of *Actinobacteria* such as *Streptomyces* are also well-recognized for their ability to produce multiple siderophores such as the characteristic desferrioxamines G, B, and E (Wang et al. 2014).

Mammalian siderophores may have structural and functional similarity to bacterial siderophores (Khan et al. 2018). Studies have confirmed the existence of a mammalian siderophore and provided insight into its structure, biosynthesis, and function (Devireddy et al. 2010). Siderocalin is the only mammalian siderophore-binding protein currently known however, compounds that serve endogenously as siderophore equivalents have been identified and characterized through associations with siderocalin (Correnti and Strong 2012).

Marine organisms such as phytoplankton and cyanobacteria can also produce siderophores (Ahmed and Holmström 2014). Cyanobacterial siderophores are mainly of dihydroxamate-type (such as schizokinen and anachelin H) (Khan et al. 2018). Moreover, the coastal marine cyanobacterium *Synechococcus* sp. produces three amphiphilic siderophores, synechobactins A, B, and C (Ito and Butler 2005).

Finally, plant siderophores (phytosiderophores) are hexadentate ligands that coordinate Fe(III) with their amino and carboxyl groups (Ahmed and Holmström 2014). These are prevalent in members of *Poaceae* and belong to the mugineic acid family that form a hexadentate Fe-phytosiderophore complex (Kumar et al. 2016; Khan et al. 2018). Mugineic acid has been found to be a much better complexing agent for the ferrous ion than ligands which contain hydroxamate or catecholate (Sugiura and Nomoto 1984).

Under low iron availability, which is a critical growth-limiting factor for virtually all aerobic microorganisms, siderophores become important in several ecological niches. These are included in four major habitats: soil and surface water, marine water, plant tissue (pathogens), and animal tissue (pathogens) (Hider and Kong 2010). Therefore, mixtures of siderophores control the supply of iron in critical environments such as rivers, marine surface water, forest soils and agricultural land and pathogenic bacteria and fungi employ siderophores in a chemical competition with their plant and animal hosts (Hider and Kong 2010).

The primary role of siderophores is to chelate the ferric iron from different terrestrial and aquatic habitats and make it available for microbial and plant cells (Ahmed and Holmström 2014). Thus, they exhibit growth factor activity for nearly all known fungi and plants (Renshaw et al. 2002; Ahmed and Holmström 2014). Additionally, many siderophores have several significant roles, such as virulence in pathogens, oxidative stress tolerance, and antimicrobial properties (Renshaw et al. 2002; Khan et al. 2018).

Several studies have demonstrated the role of siderophores in mediating pathogen multiplication and development of virulence, because iron is critical for many pathogenic species (Neilands 1995; Saha et al. 2013). Moreover, siderophores have been extensively reported to reduce oxidative stress in microorganisms producing them (Khan et al. 2018). Another important role of siderophores is their contribution in biofilm formation: studies have shown that intracellular iron concentration plays an important role in the development of the complex community of microorganisms growing in various substrates in an aqueous environment (Saha et al. 2013).

Nonetheless, apart from their primary role in iron chelation and mobilization, siderophores have the ability to bind a variety of other heavy metals such as $Al^{3+}$, $Zn^{2+}$, $Cu^{2+}$, $Pb^{2+}$, and $Cd^{2+}$ (Saha et al. 2013; Ahmed and Holmström 2014; Kumar et al. 2016; Złoch et al. 2016; Khan et al. 2018). Thus, siderophores are also seen serving in bioremediation for cadmium and lead toxicity (Khan et al. 2018) and can affect homeostasis and heavy metal tolerance of microorganisms (Złoch et al. 2016). This is the reason why siderophore-producing bacteria have been used to assist in phytoremediation of heavy metals from contaminated environments or in the scarcity of nutrients (Saha et al. 2013; Thiem et al. 2018).

It becomes thus apparent why siderophores have wide applications in various fields such as agriculture, medicine, ecology, and environmental applications (Ali and Vidhale 2013; Saha et al. 2016).

## 2.3   Applications

In agriculture, siderophores can be considered to be an eco-friendly alternative to hazardous chemical pesticides as they can play a significant role in the biological control mechanism against certain phytopathogens. Moreover, they can promote plant growth as iron is required for chlorophyll biosynthesis, redox reactions, and some important physiological activities in plants (Saha et al. 2016).

Medical siderophores have been exploited to deliver drugs inside the cells, a concept adopted from the naturally occurring sideromycins (daunomycins, albomycins, microsins) (Saha et al. 2013). Sideromycins are conjugates of siderophores and antibiotics, in which the antibiotic utilizes siderophores as mediators to enter the cells via the iron-uptake machinery. This process is known as the Trojan horse strategy (Ali and Vidhale 2013; Saha et al. 2013, 2016; Khan et al. 2018). They can be also used to treat diseases caused by pathogenic microorganisms by exploiting their antimicrobial property (Khan et al. 2018). Moreover, medical siderophores have been used to treat iron overload conditions such as ß-thalassemia and aluminum overload (Renshaw et al. 2002; Ali and Vidhale 2013; Saha et al. 2016; Khan et al. 2018). Desferrioxamine and other hydroxamate siderophores have also been studied for use in the treatment of other medical conditions, including cancer, infectious diseases such as malaria, due to their antimicrobial properties and for decontamination by actinides such as neptunium and plutonium (Renshaw et al. 2002; Saha et al. 2016; Khan et al. 2018).

The ability of siderophores to complex actinides, which result mainly by the production and testing of weapons and by nuclear power stations and reprocessing plants, indicates their use for reprocessing of nuclear fuel, bioremediation of metal-contaminated sites, and the treatment of industrial waste, including radioactive waste (Renshaw et al. 2002; Saha et al. 2016; Khan et al. 2018).

Another application of siderophores is in the field of microbial ecology as it has been demonstrated that a siderophore-based approach has markedly facilitated the growth and cultivation in the pure culture of many unculturable organisms (Saha et al. 2016).

Finally, some siderophores have been characterized to function as sensitive, robust, and specific $Fe^{3+}$ biosensors. Nonetheless, most of them have not yet been characterized and thus it could be just hypothesized that some of them may also turn out to be novel and potential biosensors (Saha et al. 2016).

Besides significant ecological agricultural, environmental, medical, and biotechnological applications, siderophores have recently found a new field for applications in the conservation of cultural heritage, and in particular, paper and wood (Albelda-Berenguer et al. 2019), which will be thoroughly discussed in a following section.

# 3   The Structure and Chemistry of Siderophores

## 3.1   Chelation of Iron

Exceptionally stable complexes can be formed between metal ions such as iron(III) and multidentate organic Lewis bases, called chelators (from the Greek "χηλή," meaning *hoof* or *claw*, suggesting two-sided, or bidentate gripping ability). Chelators are capable for two-, four-, or six-sided coordination, accordingly termed as bi-, tetra-, or hexadentate ligands, respectively (Cotton et al. 1995; Crichton 2019). This way, highly stable multidentate coordination complexes are formed.

Two forms of iron are the most common in the environment: (a) ferric or iron(III) (in its ionic form it is depicted as $Fe^{3+}$) which is the most abundant and (b) ferrous or iron(II) which is the least stable, often oxidizable to ferric compounds. The former is mostly the case in the discussion within this paper as it is the main species in iron corrosion products and therefore a usual target for conservators. The two forms of iron are in a dynamic relationship between each other depending on factors such as the presence of oxygen, the pH, complexing agents, the solubility of their compounds, etc. Excellent books and reviews are available for further reading (Cornell and Schwertmann 2003; Crichton 2019). The reader also needs to have an adequate background on the acid and base character and their inter-relations, as well as the significance of pH with respect to the acidity constant of an acid; these are key to understand the siderophore action, and more generally, chelation (see for instance, Harris 2015).

Ferric ion ($Fe^{3+}$) is a Lewis acid capable of accepting an electron pair from functional groups in chelators, which on their part act as Lewis bases; the latter can be electron pair-rich groups such as hydroxyl (:OH), hydroxide (:OH$^-$), and amino (:NH$_2$) groups (Cotton et al. 1995; Marusak et al. 2006; Harris 2015; Crichton 2019). The coordination of metals by organic or inorganic molecules is a ubiquitous phenomenon in nature. For instance, metal ions, such as $Fe^{3+}_{(aq)}$, can be coordinated in aqueous solutions by six water molecules arranged according to octahedral geometry; in these, the electron-rich oxygen atom in each water molecule behaves as the Lewis acid. The coordination process occurs stepwise involving ligand molecules (such as $H_2O$) in a sequential manner, so that most $Fe^{3+}$ ions in their aqueous solutions exist as the hexa-aquo complex. Equation (14.1) shows the overall reaction for the complexation of $Fe^{3+}$ by water in the formation of the iron hexa-aquo complex. This complex behaves as an acid as it further interacts with water and reduces the pH of the solutions; with pKa value 2.2, a 0.1 M aqueous iron(III) solution, assumed to be entirely in is hexa-aquo complex form, is expected to show a pH as low as 1.6.

$$\underset{\text{ferric ion}}{Fe^{3+}} + 6H_2O \rightleftharpoons \underset{\text{hexa-aquo complex}}{Fe(H_2O)_6^{3+}} \quad \beta = \underset{\text{stability constant}}{K_f = \frac{\left[Fe(H_2O)_6\right]^{3+}}{\left[Fe^{3+}\right]}} \tag{14.1}$$

A chelate complex forms as a ligand that approaches the ferric ion ($Fe^{3+}$) and arranges in its first coordination sphere; a stable hexacoordinate complex is thus formed, in which properties such as solubility, hydrolysis, and reduction are utterly controlled (Dhungana and Crumbliss 2005; Pepper and Gentry 2015; Albelda-Berenguer et al. 2019).

The overall process is governed by the combination of enthalpy and entropy factors and characterized by an equilibrium constant, which is called the *stability constant* often symbolized as log$\beta$; the higher the value of log$\beta$, the more stable is the coordination complex (Nakamoto 2009; Harris 2015; Crichton 2019).

The chelate complexes of ferrous ions ($Fe^{2+}$) are much less stable with stabilities more than ten orders of magnitude lower than those of iron(III). As a result, the reduction of iron(III) to iron(II) has enormous implications on the dechelation (or release) mechanisms of uptaken iron inside the living systems and is also a significant factor that needs to be taken into account in the various applications of siderophores (Boukhalfa et al. 2006; Hider and Kong 2010; Albelda-Berenguer et al. 2019).

The hard-soft acid base (HSAB) concept proposed by Pearson further helps to evaluate and understand the high tendency for reaction between Lewis acids and bases of similar hardness (Cotton et al. 1995; Marusak et al. 2006; Harris 2015; Crichton 2019). As a result, chelators having a hard base character can complex iron (III) (a hard acid) so efficiently that the latter can be removed from its insoluble and minimally bioavailable iron oxides, hydroxy-oxides, and hydroxide. Siderophores possess such ability, which has been developed as one of nature's successful strategies to sequester iron (Bauer and Exner 1974; Agrawal 1979; Boukhalfa and Crumbliss 2002; Kraemer 2004; Tseng et al. 2006; Reichard et al. 2007; Hider and Kong 2010; Sharpe et al. 2011; Ahmed and Holmström 2014; Albelda-Berenguer et al. 2019). In conservation, a similar issue needs to be addressed: potent chelators can be employed to remove iron from surfaces of cultural heritage objects by transferring it to the more soluble complexed phase; this way, a cleaning effect can be achieved.

An addition, most iron(III) chelate complexes are colored due to specific electron transitions facilitated by charge transfer between the iron(III) and the ligand. This offers an extra asset for visual observation, and additionally, a widely followed methodology for titrating of such complexes (Boukhalfa et al. 2006; Harris 2015).

## 3.2 The Tool: Siderophores

Siderophores have been developed by living systems based on organic functional groups (phenolic hydroxyls, carboxyl groups, hydroxamic and sulfhydryl groups) capable of efficient complexation of metals. Their development is also based on the number and the stereochemical arrangement that these groups possess for more efficient complexation in order to overcome the limited bioavailability of iron as a micronutrient (Kraemer 2004; Kalinowski and Richardson 2005; Bertrand et al. 2009; Hider and Kong 2010).

Siderophores, as most chelators, form thermodynamically stable hexacoordinate complexes with iron(III) accommodated in the most favorable molecular geometry (see, for instance, Fig. 14.1b) (Crumbliss 1990; Kraemer 2004); this along with the high affinity of siderophores for iron(III) produce strong driving forces towards complexation. The stability constants for iron(III) are generally higher than iron (II) which means that ferric complexes are generally much easier to form as compared to ferrous. This preference is more significant in the case of siderophores, as compared to other chelators such as EDTA. This accordingly plays a major role in stability of ferric complexes (see Sect. 14.3.3.3); in the context of conservation applications, it has the significance that in the presence of the wrong chelator, or even more, in the absence of any chelator, ferric ions can be reduced to ferrous with detrimental effects on many organic substrates (see Sect. 14.4.1).

The action of chelation by siderophores implies multiple acidic functional groups (such as OH, COOH) capable of chelating metals which are involved in successive deprotonation schemes (Edwards et al. 2005; Boukhalfa et al. 2006; Reichard et al.

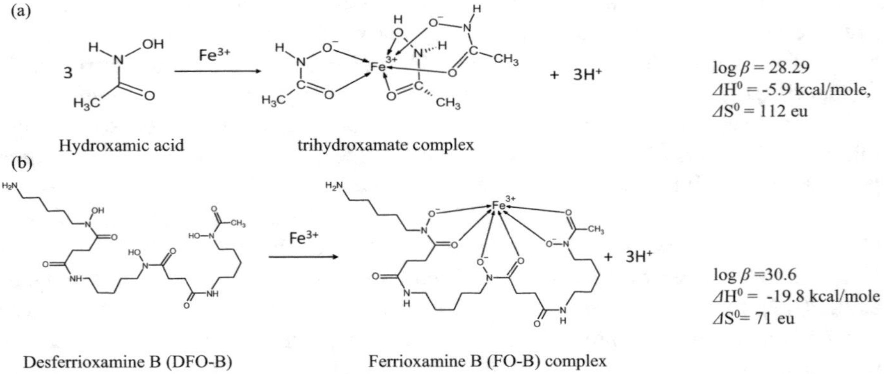

(a)

$$3 \quad \xrightarrow{Fe^{3+}} \quad + \ 3H^+$$

$\log \beta = 28.29$
$\Delta H^0 = -5.9$ kcal/mole,
$\Delta S^0 = 112$ eu

Hydroxamic acid      trihydroxamate complex

(b)

$$\xrightarrow{Fe^{3+}} \quad + \ 3H^+$$

$\log \beta = 30.6$
$\Delta H^0 = -19.8$ kcal/mole
$\Delta S^0 = 71$ eu

Desferrioxamine B (DFO-B)      Ferrioxamine B (FO-B) complex

**Fig. 14.1** Chelation schemes: (**a**) model chelate complex: iron(III) trihydroxamate; (**b**) ferrioxamine (or desferrioxamine B complex with iron); and (**c**) ferrioxamine E (or desferrioxamine E complex with iron); modified from Crumbliss (1990)

2007). As acids, they tend to dissociate in their aqueous solutions releasing free hydronium ions ($H_3O^+$), thus decreasing the pH of the solution. Their tendency to dissociate is characterized by their pKa values; the lower the pKa value, the stronger the acid. As the deprotonated (mostly anionic) forms are generally capable of metal complexation, their $pK_a$ values are important as they control the range for their most efficient action.

Siderophores can be categorized according to their functional groups as carbox-ylates, catecholates, hydroxamates, and mixed type (Neilands 1995; Boukhalfa and Crumbliss 2002; Kraemer 2004; Crichton 2019). A siderophore database can be found on the web (Bertrand 2010); chemical structures of selected siderophores are shown in Fig. 14.2. A brief presentation of the main types of siderophores follows.

### 3.2.1 Carboxylate Siderophores

The carboxylate siderophores contain $\alpha$-hydroxycarboxylic acid moieties, which renders them as acids with a broad range of acidities as seen through their pKa values. Citric acid has been considered as a model compound (Fig. 14.2a) with pKa values 3.13, 4.76, and 6.40 (Harris 2015) the lowest value assigned to the a-OH carboxyl group (Silva et al. 2009; Heller et al. 2012). The pKa values of carboxyl groups in the siderophore molecules range between 3.5 and 5 (Miethke and Marahiel 2007) and are therefore active for complexation in relatively low pH values.

These siderophores can coordinate with the metal ion through their electron-rich $\alpha$-OH carboxylate oxygens; as a result, they can form chelates at low pH values, which offer an asset for applications that require an acidic environment. Nature has taken advantage of this fact by utilizing these agents in acidic microbial

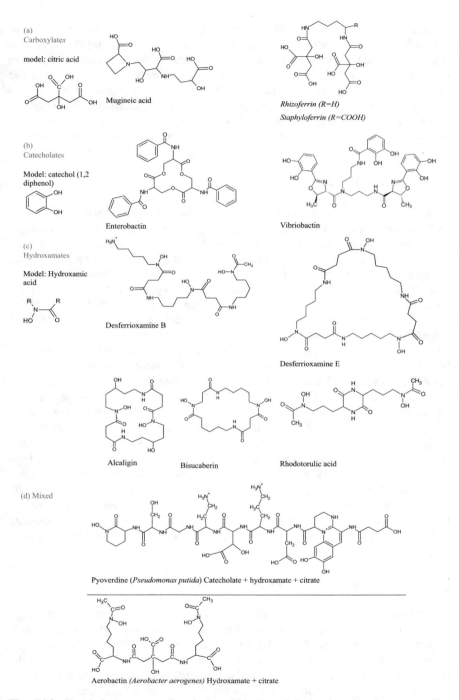

**Fig. 14.2** Chemical structures of typical multidentate chelator classes and corresponding siderophores: (**a**) carboxylates model: citric acid; (**b**) catecholates model: catechol (1,2 diphenol); (**c**) hydroxamates model: hydroxamic acid; and (**d**) mixed

environments (Harris et al. 1979). Their $\log\beta$ values are among the lowest in siderophores, typically around 20–30, suggesting that they are among the least stable siderophore complexes; their relatively low chelate stabilities are compensated in nature by achieving concentrations in the rhizosphere as high as 1 mM, while the more potent desferrioxamine chelators (see below) can be successful at three orders of magnitude lower concentration (approximately, 1 μM) (Kraemer 2004; Hider and Kong 2010; Ahmed and Holmström 2014). These concentration ranges are also expected to play a similar role in other applications of siderophores in conservation.

### 3.2.2 Catecholate Siderophores

This siderophore type is based on catechols, or 1,2-dihydroxybenzenes (Avdeef et al. 1978; Romero et al. 2018) (Fig. 14.2b), which due to the geometrically favorable acidic phenolic groups can efficiently coordinate to iron. Catechols have been extensively studied due to their antioxidant, as well as pro-oxidant properties (i.e., the indirect oxidative action, by enabling the formation of reactive oxygen species, also suggesting potential antimicrobial action), depending on the conditions (Iwahashi et al. 1989; Moran et al. 1997; Nakamura 2000; Salgado et al. 2013). Especially the pro-oxidant activities occur by increasing the activity of Fenton reagent for reducing Fe(III) to Fe(II) and $O_2$ to peroxides ($\bullet$OOH). These highly energetic and reactive oxygen species play significant roles in wood biodegradation by white and brown rot fungi (Goodell et al. 2006; Romero et al. 2018). These activities are related to their ability for chelation through their catechol phenolate ions, which are formed over pH 9–10 (Avdeef et al. 1978; Hynes and Ó Coinceanainn 2001). This issue will be further discussed in the following sections.

Catechol-based siderophores show among the highest stability constants for iron chelation; specifically, enterobactin, which is produced by *Escherichia coli* and other bacteria (Harris et al. 1979; Kraemer 2004) shows the highest known $\log\beta$ value, 39.5 (Table 14.1), meaning that it is the most stable siderophore complex known; in other words, ferric species in contact with this siderophore will prefer the chelation route over any other, leading to the enterobactin–Fe(III) complex. Enterobactin forms hexadentate chelates by coordinating to iron in a "salicylate mode," i.e., using a carbonyl adjacent to a phenolic OH (Correnti and Strong 2012).

### 3.2.3 Hydroxamate Siderophores

Hydroxamate siderophores have drawn a lot of attention as they compose one of the largest and most efficient categories (Ahmed and Holmström 2014). They contain hydroxamic acid moieties (Fig. 14.2c) which bear relatively high acidities (comparable to phenols), thus allowing the formation of hydroxamate ion, which is the actual chelator. Hydroxamic acids have been systematically studied since the 1970s, based on their "unexpectedly" not-so-low acidities (Agrawal 1979; Crumbliss 1990; Yang et al. 2006). Aceto-hydroxamic acid has been used as a model compound for

**Table 14.1** Selected iron(II)chelators with some of their most important properties

| Chelator | Origin | log $(\beta Fe^{III})$ | log $(\beta Fe^{II})$ | $E^0_{(aq)} - 0.059\log[\beta(Fe^{3+})/\beta(Fe^{2+})]^a$ | $E_{\frac{1}{2}}$,mV (vs NHE) | pFe | log $(Ksol)^a$ | Denticity of chelate |
|---|---|---|---|---|---|---|---|---|
| *Siderophores* | | | | | | | | |
| Enterobactin | *Escherichia coli* | 49 | 23.91 | −0.71 | −750 | 35.5 | 14.62 | Hexadentate |
| Pyoverdine | *Pseudomonas* spp. | 30.8 | 9.78 | −0.47 | −510 | 27 | 6.97 | Hexadentate |
| Ferrichrome | *Aspergillus* spp. *Penicillium* spp. *Ustilago sphaerogena* | 32.0 | 9.91 | −0.53 | −440 | 25.2 | 5.45 | Hexadentate |
| Desferrioxamine B | *Streptomyces pilosus* | 30.6 | 10.29 | −0.43 | −468 | 26.6 | 6.83 | Hexadentate |
| Desferrioxamine E | *Streptomyces* spp. | 32.5 | 11.16 | −0.49 | −477 | 27.7 | 7.87 | Hexadentate |
| Aerobactin | *Escherichia coli* | 22.5 | 4.86 | −0.27 | −336 | 23.3 | 3.67 | Tetradentate |
| Alcaligin | *Rhodotorula Pilmanae* | 32.3 | 12.3 | −0.41 | – | 23.0 | 8.28 | Binuclear tetradentate |
| Rhodotorulic acid | *Rhodotorula Pilmanae* | 31.2 | 10.6 | −0.45 | – | 21.8 | 7.6 | Binuclear tetradentate |
| *Model hydroxamate chelators*[b] | | | | | | | | |
| Aceto-hydroxamic acid | | 28.29 | 11.2 | −0.24 | −293 | 12.5 | 3.68 | Hexadentate |
| N-methylaceto-hydroxamic acid | | 29.4 | 11.2 | −0.3 | −348 | 16.2 | 6.87 | Hexadentate |
| Hydroxamic acid- iron | | 46.3 | 26.3 | −0.41 | <−450 | 20.94 | | Hexadentate |

Compiled data from (Schwertmann 1991; Boukhalfa and Crumbliss 2002; Kraemer 2004; Hider and Kong 2010; Cézard et al. 2014)
[a]Values calculated from Eq. (14.6)
[b]Refer to text, in *"The stability of siderophore–iron complexes"*

the study of related siderophores; its iron(III) complex shows significant stability with a log$\beta$ value of 28.3 (Fig. 14.1a) (Boukhalfa and Crumbliss 2002).

The *desferrioxamines* (often known as deferoxamines) A, B, and E are among the most well-studied hydroxamate siderophores, carrying three such groups which form 1:1 hexadentate complexes with iron(III), called *ferrioxamines* (Fig. 14.1b); these show very high log$\beta$ values 32.0, 30.6, and 32.5, respectively (Konetschny-Rapp et al. 1992; Edwards et al. 2005; Sharpe et al. 2011).

Acid dissociation plays here, too, a significant role. The typical p$K_a$ values of hydroxamate siderophores are generally at 8.5–11 (Singh et al. 2015). In particular, the p$K_a$ values of desferrioxamine B, one of the most studied, are 8.32, 9.16, 9.94, and 11.44, meaning that their anionic forms (hydroxamate ions) are gradually produced above the corresponding pH values, and therefore effective complexation may occur in this region.

Denticity is another ruling factor. Rhodotulic acid, alcaligin, and bisucaberin form tetradentate chelates; the former has relatively lower stability ($\beta$ value 31.2), while alcaligin and bisucaberin possess higher stabilities ($\beta$ values 32.3 and 32.3, respectively) due to their closed ring molecular geometries which are pre-arranged for efficient coordination to iron (Crumbliss 1990; Konetschny-Rapp et al. 1992; Hider and Kong 2010). This closed ring effect is more pronounced for the hexadentate desferrioxamine E, (log$\beta$ value 32.5). Possibly, the most studied hydroxamate siderophore is desferrioxamine B, an open-ring, hexadentate chelator with log$\beta$ value of 30.6, which has also been considered for conservation-related applications (Rapti et al. 2017; Albelda-Berenguer et al. 2019). All above structures are listed in Fig. 14.2.

### 3.2.4 Mixed Type Siderophores

A large number of siderophores carry more than one type of functional group. For instance, aerobactin (see structure in Fig. 14.2d) contains three carboxylate and two hydroxamate groups, which allow complexation with log$\beta$ of 22.5 (Neilands 1995; Miethke and Marahiel 2007), which is considered as a relatively low stability constant (cf. log$\beta_{Fe(III)}$ values in Table 14.1). The participation of the groups in complexation varies significantly over pH, with three protonation steps below 4.5 (carboxylates) and two steps at over 9 (hydroxamates) It is significant to note that complexation starts from pH value as low as 1, where half of the total iron is complexed (Harris et al. 1979).

Quite often, mixed type siderophores contain amino acids. For instance, pyoverdine incorporates glycine, serine, aspartic acid, ornithine, and diaminobutyric acid, as well as succinic acid and catechol moieties; in these, $\alpha$-hydroxy-carboxylate, catecholate, and hydroxamate groups are located, all participating in complexation (Boukhalfa and Crumbliss 2002; Boukhalfa et al. 2006; Albelda-Berenguer et al. 2019). Pyoverdine has pKa values 2.72, 3.48, 4.97, 6.40, 8.12, 9.56, 10.24, and 12.88. As a result, complexation may start at low pH values, while at neutral pH almost all iron is complexed (Boukhalfa et al. 2006).

## 3.3 The Action: Dissolution of Iron Oxides and Hydroxy-Oxides

### 3.3.1 Acidic Dissolution of Iron Species

Dissolution of iron oxides and hydroxy-oxides or their transformation into a water-soluble form is necessary for the cleaning of iron-stained objects. The following discussion attempts to present the scientific basis on which the action of siderophores is critical for the dissolution of iron and allowing its availability for nature's nutrition systems as well as for our applications.

In the absence of a chelator, the solubilities of iron oxides, hydroxy-oxides, and hydroxides are characterized by their ion dissociation in water and depend significantly on the pH. The soluble form is the iron(III) aquo complex, $Fe^{3+}(H_2O)_6$, sometimes described as $Fe^{3+}_{(aq)}$. At neutral pH, the ionic solubility of iron oxides, hydroxy-oxides, and hydroxide can be expressed by using goethite ($\alpha$-FeOOH) as example, according to Eq. (14.2) (Cornell and Schwertmann 2003).

$$\text{Neutral environment}\ \alpha - FeOOH_{(s)} + H_2O \rightleftharpoons Fe^{3+}_{(aq)} + 3\ OH^-\ K_{sp}$$
$$= \left[Fe^{3+}\right][OH^-]^3 \tag{14.2}$$

From Eq. (14.2), the solubility of iron (expressed as $[Fe^{3+}]$) can be calculated; for this, selected values can be found in Table 14.2, which can be as low as $10^{-13}$ mol/L (Crumbliss 1990; Cornell and Schwertmann 2003; Kraemer 2004). The lowest the $[Fe^{3+}]$ concentration, the least soluble and bioavailable is the oxide or hydroxy-oxide. The low solubilities of iron oxides and hydroxy-oxides at neutral pH consequently demand strong dissolution strategies in applications aiming at iron uptake. Specifically, goethite and hematite are the least soluble compounds and despite their

**Table 14.2** Selected properties of some iron compounds considered targets for siderophores

| Name | Chemical structure | Iron oxidation state | Crystal structure | Solubility, [Fe] at pH 7[a] |
|---|---|---|---|---|
| Hematite | $\alpha$-Fe$_2$O$_3$ | III | Rhombohedral hexagonal | $10^{-8.8}$–$10^{-5.8}$ |
| Maghemite | $\gamma$-Fe$_2$O$_3$ | III | Cubic/tetragonal | $10^{-5.0}$ |
| Wüstite | FeO | II | Cubic | – |
| Magnetite | Fe$_3$O$_4$ | Mixed II+III | Cubic | – |
| Bernalite | Fe(OH)3 | III | Orthorhombic | – |
| Ferrihydrite | Fe$_5$O$_7$(OH)$_4$.H$_2$0 | III | Hexagonal | $10^{-3.1}$ |
| Goethite | $\alpha$-FeOOH | III | Orthorhombic | $10^{-10.9}$–$10^{-5.9}$ |
| Akaganeite | $\beta$-FeOOH | III | Monoclinic | – |
| Lepidocrocite | $\gamma$-FeOOH | III | Orthorhombic | $10^{-5.2}$ |
| Ferroxyhyte | $\delta$-FeOOH | III | Hexagonal | – |

[a]Values from Schwertmann (1991)

high abundance in the environment, their bioavailability is extremely low because of their low solubilities; as a consequence, their mobilization (or bioavailability through dissolution) becomes a real problem. Also, the same compounds, either as corrosion products or as unwanted iron stain factors, are among the most common iron species, and drastic tools for their removal through dissolution, such as citrate and cystine have been proposed (Kolar et al. 2006; Henniges et al. 2008; Sharpe et al. 2011; Liu et al. 2017).

At acidic pH, the solubility for goethite can be expressed according to Eq. (14.3). The acidic dissolution constant ($K^H_s$) values are relatively high as compared to neutral dissolution ones (expressed by $K_{sp}$ values); this means that in order to effectively solubilize such iron species and formation of the water-soluble $Fe^{3+}$ ion, the pH needs to be significantly lowered (pH < 5). However, this can be quite prohibitive for substrates in many cultural heritage objects for which pH values of 6–8 are generally acceptable; on the other hand, this may be a workable region in the case of materials, such as East Asian lacquer, which are stable in acidic pH. The aquo complex of iron $[Fe(HO_2)_6{}^{3+}]$ is also very acidic (Siddall and Vosburgh 1951; Milburn and Vosburgh 1955; Sapieszko et al. 1977; Wilkins 1991; Cotton et al. 1995), and therefore, its presence may lower the solution pH. Moreover, iron(III) can initiate Fenton reactions with detrimental effects on organic substrates (see below).

To complete the picture, the kinetics of dissolution (studying *how fast* the phenomenon occurs) is equally important as the thermodynamics (studying *how stable* the chemical species are). Kinetics is directly linked to the mechanism of iron dissolution, which has been studied and has been shown that it is heavily proton dependent. This happens because, as $H^+$ ions are adsorbed on the mineral surface, the surface $Fe^{3+}$-OH bonds weaken and the detachment of iron ions is facilitated (Stumm and Furrer 1987; Cornell and Schwertmann 2003). In particular, the initial dissolution of iron(III) is very fast and is subsequently followed by a slower step (Samson and Eggleston 1998; Reichard et al. 2007). This affects the way siderophore molecules work which is able to capture a "burst" of liberated $Fe^{3+}$ ions during this first step. This proton-mediated process has been found to control all types of dissolution, including the action of siderophores (see below).

### 3.3.2 Dissolution of Iron Species in the Presence of Siderophores

*The stability of siderophore–iron complexes*: In the presence of a siderophore, the complexation of iron occurs, which is expressed in Eq. (14.4); this involves an acidic environment as protons ($H^+$) are produced during complexation; however, summing these two equations lead to the combined dissolution reaction, shown in Eq. (14.5) (Kraemer 2004).

Acidic environment $\alpha - \text{FeOOH}_{(s)} + 3\text{H}^+ \rightleftharpoons \text{Fe}^{3+}_{(aq)} + 2\text{H}_2\text{O}$

$$K^H_s = \frac{[\text{Fe}^{3+}]}{[\text{H}^+]^3} \qquad (14.3)$$

Chelation by siderophore $\text{Fe}^{3+}_{(aq)} + \text{H}_3\text{Sid} \rightleftharpoons \text{Fe} - \text{Sid}_{(aq)} + 3\text{H}^+$

$$K_{\text{Sid}} = \frac{[\text{Fe} - \text{Sid}][\text{H}^+]^3}{[\text{H}_3\text{Sid}][\text{Fe}^{3+}]} \qquad (14.4)$$

Sum of (2) and (3) $\alpha - \text{FeOOH}_{(s)} + \text{H}_3\text{Sid} \rightleftharpoons \text{Fe}^{3+}_{(aq)} + 2\text{H}_2\text{O}$

$$K^H_{s,\text{Sid}} = \frac{[\text{Fe} - \text{Sid}]}{[\text{H}_3\text{Sid}]} \qquad (14.5)$$

where Sid is siderophore, $K^H_s$ the acidic dissolution constant, $K_{\text{Sid}}$ the complex stability constant (often symbolized as log$\beta$), and $K^H_{s,\text{Sid}}$ the combined dissolution constant. It appears that the overall process in Eq. (14.5) for siderophores, such as desferrioxamine B is thermodynamically stable over a wide range of pH (Kraemer 2004; Kraemer et al. 2005); this has obvious consequences for the usage of such siderophores in their applications.

Focusing on the hydroxamate siderophores, smaller hydroxamic acids, often called "model molecules," are typically used in chemical methodology to better understand the action of more complex, but similar molecules. In this context, chelate complexation by model hydroxamic acids, as compared to that of desferrioxamine B siderophores, can be shown in Fig. 14.1.

For complexation to occur successfully, all energy (or *enthalpy*) of molecules and ions involved in the destruction of goethite Eq. (14.2) and the formation of the siderophore complex Eq. (14.3) need to decrease. On the other hand, the very strict arrangement of the complex formed in Eq. (14.3) involves increase of molecular ordering (or decrease in *entropy*); keeping in mind that entropy reflects the disordering of all molecules and ions involved, its decrease imposes a geometric difficulty to complexation. In other words, the decrease in enthalpy is favorable for the reaction, while the decrease in entropy is disfavoring; however, it appears that the contribution of enthalpy is much bigger, and therefore, the complexation occurs successfully.

However, the contribution of entropy is not negligible, and this can explain the more favorable action of cyclic or corona-type siderophores, such as DFO-E, which is more stable (higher log$\beta$ value) than the non-cyclic DFO-B (Crumbliss 1990). Figure 14.1 shows the chelation reactions of a model hydroxamic acid (Fig. 14.1a) in comparison to DFO-B (Fig. 14.1b) and DFO-E (Fig. 14.1c); in the latter case, the entropy factor is comparatively more favorable. Of the three, desferrioxamine E (a corona-type, or cyclical molecule) is "pre-arranged" for a favorable chelation; therefore, the final complex has similar cyclic geometry to the precursor and therefore, entropy only *slightly* decreases. The stability constants of chelation are mathematically linked to both enthalpy and entropy (Crichton 2019), the combined consideration of enthalpy and entropy finally explains why DFO-E ($\log\beta = 32.5$) is better complexing agent toward iron(III) than DFO-B ($\log\beta = 30.6$), by almost two orders of magnitude, or 100 times.

*The role of pH*: As has already been mentioned above, carboxylate siderophores (Sect. 14.3.2.1) can form complexes in relatively low pH values, while hydroxamates (Sect. 14.3.2.3) at generally are higher than 8–9. Finally, mixed siderophores, such as pyoverdine (Sect. 14.3.2.4) may form their anionic forms at a much larger span, ranging from 2.7 to 12.9, meaning that complexes can be produced in significantly lower pH. Therefore, optimal complexation is clearly depending on the acidities in each case. As a result, the simultaneous consideration of the $\log\beta$ values and the deprotonation constants may not allow a straightforward evaluation of the chelation effectiveness of siderophores towards iron(III).

As a more practical approach, the concentration of dissolved iron (expressed as pFe) has been proposed to reflect more objectively the effectiveness of complexation (Harris et al. 1979; Albelda-Berenguer et al. 2019). The pFe, defined as $-\log[\text{Fe}_{\text{free}}]$, has been introduced for iron (and accordingly, pM for metals in general), at certain pH, metal and ligand concentrations. High pFe values mean low free iron concentration and therefore effective complexation. For instance, the high $\log\beta$ value of desferrioxamine B (30.6) corresponds to reasonably high pFe value (26.6); on the other hand, aerobactin, although with significantly lower $\log\beta$ (22.5), has relatively high pFe (23.3) suggesting that it can still adequately complex iron(III). On the other hand, aceto-hydroxamic acid with $\log\beta = 28.3$ has low pFe (12.5). Values of pFe for various siderophores are shown in Table 14.1.

*Selectivity of siderophores towards iron (III)*: The selectivity of complexation regarding metal ions is an obvious demand for several fields, including that of conservation. Cultural heritage objects often contain metal ions as pigments, corrosion products, depositions, etc., where the selective removal of iron(III) calls for drastic chelating strategies; these generally involve considerable differences for stability constants of chelators towards ferric compounds as compared to other metal ions. In other words, selective cleaning action is needed where the conservator aims to remove a specific metal ion/stain/corrosion product without affecting others.

The drive for choosing iron(III) chelators better than EDTA and DTPA emanates from the fact that these two widely used reagents complex effectively a wide range of metals, such as aluminum, cobalt, nickel, copper, and manganese, which all are present in cultural heritage objects. On the other hand, siderophores show a comparative preference towards iron, which makes these chelators a suitable choice for selective complexation of iron(III) towards other metals. Ferric ion ($Fe^{3+}$) is a hard Lewis acid (see Sect. 14.3.2 for a brief introduction of the hard-soft acid base concept) and therefore shows a high affinity for hard Lewis bases such as most siderophores (Pearson 1973; Crumbliss 1990; Jensen 1991).

In addition, it has been shown that siderophores are particularly sensitive to the electronegativity ($\chi_M$) of metal ions as well as their charge ($Z$); the value of $\chi_M \cdot Z$ for $Fe^{3+}$ is 5.9, while for $Al^{3+}$ and $Cu^{2+}$ is 4.9 and 4.0, respectively. As a result of the above, the stability constants of DFO-B towards iron(III) is by far higher than any other metal ion ($\log\beta = 30.7$), followed by aluminum ($\log\beta = 23.1$), $Cu^2$

**Table 14.3** Comparison of overall stability constants ($\beta$) between representative chelators towards common metal ions

| Metal | Stability constants ($\beta$) | | |
|---|---|---|---|
| | DFO[a] | EDTA[b] | DTPA[c] |
| $Fe^{3+}$ | 30.7 | 25.1 | 27.5 |
| $Fe^{2+}$ | 7.2 | 14.3 | 16.0 |
| $Ca^{2+}$ | 9.0[d] | 10.6 | 10.9 |
| $Ni^{2+}$ | 10.9 | 18.4 | 20.2 |
| $Cu^{2+}$ | 14.1 | 18.8 | 21.1 |
| $Zn^{2+}$ | 10.1 | 16.5 | 18.4 |
| $Cd^{2+}$ | 7.9 | 16.5 | 19.9 |
| $Al^{3+}$ | 23.1 | 16.4 | – |

[a]Adapted from (Shenker et al. 1996; Hernlem et al. 1996; Harris 2015)
[b]Harris (2015)
[c]Values from Rivers and Ummey (2003)
[d]Value from Keberle (2006)

[+](log$\beta$ = 14.1), $Zn^{2+}$ (log$\beta$ = 10.1) (Shenker et al. 1996; Hernlem et al. 1996). For comparison reasons, values of stability constants for other chelators towards common metal ions are given in Table 14.3. These values can be used to predict the effectiveness of cleaning solutions to selectively chelate metals.

### 3.3.3  Effects of Siderophores on the Reduction of Iron(III)

A significant issue underlying why we mostly care about the presence of iron(III) in cultural heritage artifacts is the fact that it is readily reduced generally to iron (II) through complex chemical cycles involving deleterious reactive intermediates; because of this, iron(III) is often termed as "active iron" (Bulska and Wagner 2005), responsible for significant damage in cellulosic materials (see Sect. 14.4.1). Therefore, from the conservator's point, stabilizing iron(III) may prevent deleterious cycles as the Fenton reaction on this type of substrates.

Iron(III) species may readily be reduced to iron(II); these two forms represent a couple involving the oxidizing agent and the corresponding reducing agent, respectively; these are termed the *redox couple* iron(III) /iron(II). As a measure of the tendency for the iron(III) $\rightarrow$ iron(II) reaction, the reduction potential ($E_0$) measured in Volts (25°C, 1.0 M aqueous solution) is a critical factor. In aqueous solution, the reduction potential of $Fe^{3+} \rightarrow Fe^{2+}$ is $E_0 = +0.77$ V, which means that this reduction is a spontaneous process in solution (Cotton et al. 1995, Cornell and Schwertmann 2003, Crichton 2019). Generally, the higher the $E_0$ value, the higher the tendency for reduction of iron(III) to iron(II). On the other hand, lower $E_0$ values mean lower tendency for reduction; even more, when these values are negative, the tendency is minimal and the iron(III) form is stabilized (Cotton et al. 1995; Pepper and Gentry 2015).

*Controlling iron(III) reduction through siderophore chelation*: In the light of the above, the tendency for reduction of iron(III) may change for other forms such as iron oxides and chelate complexes. In particular, siderophore chelates of iron(III) show negative reduction potentials, meaning that iron(III) is practically not reduced to iron(II); for instance, values for DFO-B and enterobactin are $-478$ and $-750$ mV, respectively, meaning that enterobactin is a superior stabilizing agent and that reduction of iron(III) is no longer spontaneous, unless a very strong reducing agent is available (Dhungana and Crumbliss 2005).

Microorganisms and plants after complexing non-bioavailable iron(III) in its oxides and hydroxy-oxides, they need to utilize it as nutrient, often by reducing it to a more available form (see Sect. 14.3.3.4). For this, exceptionally strong reducing agents are needed, with negative reduction potentials (for instance, NADH, or hydrogen-nicotinamide dinucleotide, with $E_0 = -320$ mV) to override the complex stability and finally rendering it in the more bioavailable iron(II). This strategy succeeds in many cases, but fails for the more stable complexes, such as the enterobactin–iron(III) complex (Boukhalfa and Crumbliss 2002). This is a desired situation for applications such as conservation, because mild reducing agents (such as polyphenols) are typically present in cellulosic materials, which consequently are not strong enough to reduce iron from its siderophore complexes.

Very negative $E_0$ values, such as those mentioned, do not allow unwanted redox reactions, such as the Haber–Weiss cycle and more particular the Fenton reaction (see Sect. 14.4.1) which produces hydroxyl radical ($\bullet$OH) and the superoxide ion ($\bullet O^{2-}$), which are highly reactive oxygen species, harmful to organic substrates (Hider and Kong 2010). For the benefit of conservation practice, a conservator needs to be aware of processes such as the Fenton reaction, which is responsible for most of damage in cellulosic materials; because of these, strategies for the decrease for unwanted redox reactions are direly needed (Dhungana and Crumbliss 2005; Albelda-Berenguer et al. 2019). The importance of this process in conservation will be discussed in the following section.

Careful choice of complexing agents towards iron involves the selection of chelators with suitable $\beta$ values for their iron(III) and (II) chelates. It has been proposed that the higher the ratio of the $\beta$ value of siderophore complex of iron (III) ($\beta^{Fe(III)-Sid}$) with respect to that of iron(II) ($\beta^{Fe(II)-Sid}$), the better the suppression of the unwanted redox cycles (Boukhalfa et al. 2006; Albelda-Berenguer et al. 2019).

$$E_{Fe(III)-Sid} = E_{aq}^0 - 59 \log \frac{\beta^{Fe(III)-Sid}}{\beta^{Fe(II)-Sid}} \tag{14.6}$$

Selected log$\beta$ values towards iron(III) and iron(II) are shown in Table 14.3. Other strong iron chelators, such as EDTA notoriously fail to prevent Fenton reactions, as the $\beta$ values between the two iron species are not very different. This is a strong motif towards a better practice concerning the removal of iron from organic substrates.

### 3.3.4  The Kinetics and Mechanisms of Siderophore Action

The thermodynamic stability of complexes (often called thermodynamic stability) and the rates of reactions such as dissolution and complexation have been already encountered in Sect. 14.3.3.1 The thermodynamics of siderophore-mediated iron(III) and the corresponding chemical equilibria of complexation do not offer a complete picture of the process and the conditions that favor it unless the kinetics are equally considered (Römheld 1991). As already mentioned in Sect. 14.3.3.1, understanding goethite removal from substrates is critical to comprehend the necessity for potent chelators. This has been recognized in conservation, as it has been recognized in living systems.

The rates of goethite dissolution in the absence of chelator are depending on acidity. It has been shown that the detachment of iron from its mineral surface is triggered by an $H^+$-depending mechanism (Samson et al. 2000). It appears that pH plays a significant role in dissolution rates, which are related to the mechanisms of iron removal from its crystal lattice sites on goethite and other minerals (Monzyk and Crumbliss 1979; Brink and Crumbliss 1984).

A mechanism has been proposed, generally involving three steps. It has been found that during the first step, which is the faster, siderophore molecules of the solution are consumed, thus decreasing their local concentration around iron spots; this may negatively influence the overall effectiveness of complexation in an application, unless high concentrations of siderophores are employed (Hider and Kong 2010). In the subsequent steps, gradual complexation occurs, involving bidentate and hexadentate complexes of iron(II) in a progressive mode (Furrer and Stumm 1986; Kraemer 2004; Reichard et al. 2007).

Specifically in the first step, the complexation process involves competition between the chelator and protons provided by the solution towards iron(III). A similar competition has been observed when a second chelator (or a co-chelator) is added. This is an interesting outcome at the application level: as the rates of complexation by siderophores are favored by acidity, as well as, synergistically to other agents as co-chelators, such as oxalic acid (Brink and Crumbliss 1984; Crumbliss 1990; Cheah et al. 2003; Kraemer 2004).

Regarding applications in cultural heritage, the above mechanistic considerations, in combination with what has been presented in the context of thermodynamic stability of siderophore–iron complexes may impose extra difficulties regarding the effect of acidity, and they consequently demand for careful design of siderophore-based strategies.

*Dechelation of captured iron(III)*: The stability of siderophore complexes does not allow for their easy destruction (often called *dechelation*). The iron-uptake process by microorganisms and plant roots from their immediate geo-environment eventually involves a later stage for releasing chelated iron. This involves specific mechanisms, as the exceptionally high stabilities of the chelate complexes do not allow for easy complex dissociation, often called *dechelation* (Boukhalfa and Crumbliss 2002; Dhungana and Crumbliss 2005).

Although this is a desired property regarding conservation—related applications, living systems need (and finally succeed) to achieve dechelation, because complexed iron, although soluble, is still non-bioavailable. The reduction strategy by bacteria involving potent reducing agents (such as NADH) has been presented above; in conservation, although reducing agents in substrates such as wood are generally milder, this needs to be taken into account for targeted cleaning strategies.

This is achieved by proton-assisted mechanisms and/or by reducing iron(III) to iron(II), where the corresponding complex of the latter is less stable by many orders of magnitude (Table 14.3). The proton-assisted dechelation mechanism has been studied extensively and showed that dechelation rates significantly increase at lower pH values. Although the siderophore complexes are extremely stable, in low pH can be labile, this factor has been found to favorably affect dechelation pathways for the benefit of living systems. In the case of iron stain cleaning, pH needs to be seriously considered regarding the balance between iron capture by chelation and release by dechelation (Boukhalfa and Crumbliss 2002).

### 3.3.5 Side Reactions, Synergistic Effects, and Light Sensitivity

The Fenton-type reactions that were mentioned in the Sect. 14.1 are of significant concern for the various applications as they act detrimentally to most substrates. While chelators such as EDTA do not prevent Fenton reactivity of iron(III) (Albelda-Berenguer et al. 2019), it has been reported that siderophores, such as desferrioxamine B and enterobactin do not favor this type of unwanted side-reactions.

A synergistic effect between dicarboxylic acids and siderophores has been observed; increase of iron(III) dissolution rates through complexation by desferrioxamine B in the presence of oxalate was observed (Reichard et al. 2007; Loring et al. 2008; Albelda-Berenguer et al. 2019). The similar accelerating effect was observed with other dicarboxylic acids, such as fumaric, malonic, and succinic; on the other hand, citric acid was found to inhibit the rates, possibly due to adsorption competition effects (Reichard et al. 2007).

Light also seems to play a significant role in the final dechelation stages and the bioavailability of iron in marine systems. In particular, some iron(III) complexes with marine siderophores have been found to dechelate the metal through a photolysis mechanism which leads to the final release of ferrous ions (Hider and Kong 2010). Studies using aquachelin (a mixed type siderophore contains hydroxamate and $\alpha$-hydroxy-carboxylate groups) have been shown to react photochemically towards releasing an iron(III) photoproduct complex, which undergoes a ligand-to-metal charge-transfer process, resulting in the reduction of iron(III) to iron (II) (Barbeau et al. 2001, 2003). The latter is easily released and therefore, available to the benefit of iron-starving organisms. On the reasonable basis that photolytic processes may affect the effectiveness of applications employing iron(III) chelation by siderophores, this issue merits further investigation for other siderophore types as well.

# 4    Application of Siderophores in Conservation

## 4.1    The Detrimental Effect of Iron

Museum artifacts such as furniture, toys, musical instruments, tools, household utensils, manuscripts, and religious items are often consisted of organic materials, such as wood, textile, paper, leather, and elements made of iron. When these composite objects are become wet or are exposed to humid environments, their iron elements corrode, and the formed corrosion products may accumulate on the surface of the organic substrate or they may impregnate its matrix.

The most common iron corrosion products formed are iron(III) (ferric) and iron (II) (ferrous) oxides, such as $Fe_2O_3$, $Fe_3O_4$, and $FeO$, hydroxy-oxides such as $FeOOH$, occurring in various polymorphs, often also called oxy-hydroxides and iron hydroxide $Fe(OH)_3$. Iron corrosion products generally exhibit very low solubilities, which depend on their grain size, since the smaller the size the higher the solubility. Therefore, for some compounds such as hematite and goethite, solubility increases, as their grain size can be extremely small ranging between 10 and 150 nm. In addition, other hydroxy-oxides of iron(III), such as ferrihydrite ($Fe_5O_7(OH).4H_2O$), may also exist as corrosion products and are considered significant participants in the overall iron corrosion process (Kraemer 2004). Hematite (red), magnetite (black), and goethite (pale brown-yellow, beige) have also been historically used as pigments. In Table 14.2, a list of the most abundant forms of iron compounds found on cultural heritage objects is presented (Cornell and Schwertmann 2003; Scott and Eggert 2009; Richards et al. 2012).

The destructive effect of iron on organic substrates, which can be either cellulosic or proteinaceous, has been the subject of several studies. Metal ions, when found in proximity to organic substrates, take part in redox reactions and act as catalysts to oxidation processes of these substrates. These processes are a wide variety of chemical reactions leading to the formation of various deterioration products and to depolymerization of main chemical components of organic substrates and consequently to the reduction of their mechanical strength (Baker 1980; Timár-Balázsy and Eastop 1998; Strlič et al. 2001; Scott and Eggert 2009; Badillo-Sanchez et al. 2019).

The oxidation of organic materials is probably caused by the free hydroxyl radical formation and hydrogen peroxide, which allow the Fenton-type reactions to occur. Thus, during the oxidation of ferrous iron to its ferric state, the free radical formation might be enhanced if organic substrates are present and this in turn leads to the formation of organic radicals, as seen in Eqs. (14.7a)–(14.7d), or to the formation of hydrogen peroxide, as seen in Eqs. (14.7e)–(14.7f) (Neevel 1995). The hydrogen peroxide produced during oxidation of organic substrates is decomposed by reacting further with ferrous ions and forms additional hydroxyl radicals, according to Eqs. (14.7e) and (14.7f).

$$Fe^{2+} + O_2 \rightarrow Fe^{3+} + O - O \bullet^- \tag{14.7a}$$

$$Fe^{3+} + O - O \bullet^- + RH \rightarrow R \bullet + HOO \bullet + Fe^{2+} \tag{14.7b}$$

$$R \bullet + O_2 \rightarrow ROO \bullet \tag{14.7c}$$

$$ROO \bullet + RH \rightarrow ROOH + R \bullet \tag{14.7d}$$

$$Fe^{2+} + HOO \bullet + H^+ \rightarrow Fe^{3+} + H_2O_2 \tag{14.7e}$$

$$Fe^{2+} + H_2O_2 \rightarrow Fe^{3+} + HO \bullet + OH^- \tag{14.7f}$$

The latter equation is often referred to as *"Fenton reaction"* (Gupta et al. 2016), although there is still considerable controversy among chemists in the reaction mechanism and the products obtained from that reaction. Some researchers claim that the *Fenton mechanism* involves hydroxyl radicals (OH•) formation, whereas others report the generation of other oxidizing iron species, such as ferryl ion ($FeO^{2+}$) (Wardman and Candeias 1996; Strlič et al. 2001; Barbusiński 2009). Nevertheless, the mixture of hydrogen peroxide and the ferrous ion is called the *Fenton reagent* and it is a strong oxidant of organic materials, such as wood, textile, and paper. Its oxidative properties depend on the hydrogen peroxide concentration, the ratio of ferrous ion to hydrogen peroxide, the pH, and the reaction time (Barbusiński 2009). The severe oxidation mechanism of organic substrates has been intensely studied and the catastrophic role of hydroperoxides and the hydroxyl free radicals (OH•) has been also demonstrated (Emery and Schroeder 1974; Zeronian and Inglesby 1995; Kolar 1997).

Based on the above it becomes clear that the removal of iron corrosion products from museum artifact combining organic substrates, such as wood, textile, paper is an urgent issue during the conservation processes.

## 4.2 The Problem of Conventional Chelators' Use in Conservation

There are many published studies on the removal of iron corrosion products from substrates, such as paper, textile, stone, metal, paintings, waterlogged wood by using chelating agents such as citric acid, di-ammonium and tri-ammonium citrate, EDTA (ethylene-diaminetetraacetic acid), DTPA (diethylene-triaminepenta-acetic, or pentetic acid) (Hart 1981; Hofenk de Graaff 1982; Banik and Ponahlo 1983; Slavin 1990; Burgess 1991; Häkäri 1992; Phenix and Burnstock 1992; Chapman 1997; Margariti 2003; Rivers and Ummey 2003; Almkvist et al. 2005; Fors 2008). However, those substances are not always giving consistent results concerning their efficacy in removing iron corrosion products and as mentioned previously they lack specificity for iron and promote unwanted side reactions due to their tendency for the production of hydroxyl radicals. Furthermore, there are several studies giving considerable attention to the environmental impact and the biodegradability of conventional chelators, such as EDTA and DTPA, which are extensively used, in the conservation field, especially on waterlogged objects (Sillanpää 1997; Nörtemann 1999; Grčman et al. 2001; Nörtemann 2005; Kołodyńska 2011). EDTA, however, has been found to be more persistent than DTPA in degradation by conventional biological and physicochemical methods (Nörtemann 1999; Grčman et al. 2001; Nörtemann 2005). Although DTPA is more biodegradable, it has been reported that EDTA can be one of its decomposition products (Sillanpää 1997). Nonetheless, both EDTA and DTPA may result in several risks when released to the aquatic environment, as they may contribute to toxicity at significantly low concentrations (Sillanpää 1997; Grčman et al. 2001). Additionally, their degradation products can be used as a nitrogen source for algae growth and thus EDTA and DTPA have the potential to contribute to eutrophication (Sillanpää 1997; Kołodyńska 2011).

## 4.3   Applications of Siderophores in Conservation of Cultural Heritage Objects

The tendency nowadays is to find safer alternative materials and methods for conservation purposes. This tendency though, becomes a necessity, considering the established national legislations and international environmental restrictions (Balliana et al. 2016) and therefore the replacement of conventional chelators with new ones with improved biodegradability, is critical.

There are a few studies regarding the removal of iron corrosion products from cultural heritage objects, with *"green"* compounds such as siderophores (Wagner and Bulska 2003; Rapti et al. 2015, 2017; Albelda-Berenguer et al. 2019). These studies have investigated the application of siderophores, mainly desferrioxamine B (DFO-B), on different organic substrates and preservation states. The application methodologies followed by both siderophores and conventional chelators were not similar and therefore comparative analysis of the obtained results cannot be done.

### 4.3.1   Extraction of Iron from Historical Manuscripts With Iron Gall Ink

Bulska and Wagner (2005) dealt with historic manuscripts written with iron gall inks[1] that contained a substantial amount of "active" iron(II) ions, which can act as

---

[1]Historical recipes for gall ink production involve the use of aqueous solutions of iron(II) sulfate with extracts of gall nuts to form complexes of oxidized iron(III) with gallic acid. However, most of historical manuscripts contain an excess of iron(II) ions that were not complexed and thus are active iron(II) ions that can act as catalysts to oxidative degradation processes of paper.

catalysts and are responsible for promoting the degradation processes of paper cellulose via Fenton reactions. In order to prevent this paper degradation process, which is often accompanied by acid hydrolysis among other deteriorative mechanisms, Bulska and Wagner (2005) tried to find a cleaning treatment to remove the excess of active iron ions. The treatment was expected, without altering the iron gall ink color, to impede both Fenton reaction and acid hydrolysis leaving the paper of the manuscripts unaffected by the cleaning procedure. Therefore, they comparatively evaluated conventional chelators versus siderophores for their efficiency of iron extraction on mock-ups mimicking historical manuscripts with iron gall inks and on mock-ups simulating only the amount of active iron ions. Chelators were required to demonstrate low cleaning efficacy on the first mock-up type and concurrently high on the second type of mock-ups. Moreover, they attempted to simultaneously neutralize the acidity of papers (deacidification) during the cleaning treatment.

Several physicochemical analyses were implemented in order to design the cleaning methodology and to evaluate the efficacy of the cleaning treatments, including Scanning Electron Microscopy (SEM), X-ray Fluorescence Spectrometry (XRF), Graphite Furnace Atomic Absorption Spectrometry (GFAAS), Inductively Coupled Plasma Spectrometry (ICM-MS), Molecular spectroscopy (UV/VIS), X-ray Absorption Near-Edge Structure (XANES) (Bulska et al. 2001; Wagner et al. 2001a, 2001b; Wagner and Bulska 2003; Bulska and Wagner 2004; Wagner and Bulska 2004).

The chelators used for iron extraction and iron complex formation included bidentate ligand acetylacetone, dibenzoylmethane, diethylene-triaminepenta-acetic acid (DTPA), the potassium-magnesium salt of phytic acid and the mesylate salt of DFO-B. All compounds were selected due to their inhibition effect of further formation of hydroxyl radicals and to their high stability constants of complexes with iron as it was expected to be more effective and phytate, a known compound for its beneficial iron deactivation effect, as it reduces iron ability to undergo further reactions and occupy all coordination sites of iron (Neevel 1995). The solutions of the examined chelators were prepared at various concentrations (0.001, 0.005, and 0.01 M) and pH values in the range of 7–9, in addition to that obtained after the chelators' dissolution in water (low pH < 3). Ethanol/water solutions (10, 25, and 50%) of DTPA, phytic acid, and DFO-B were also examined at 0.005 M. The mock-ups were immersed in the various chelators' solutions for 20 min (Bulska et al. 2001; Wagner et al. 2001b; Wagner and Bulska 2003).

Results obtained by physicochemical analyses showed that iron–DFO-B complex formation at both types of mock-ups was above 80%. Moreover, on mock-ups mimicking the excess of active iron ions, DFO-B revealed an increase of iron extraction, which depended mainly on the pH of solution. The concentration of all prepared solutions of chelators was pronouncedly higher than the amount of iron ions need to be extracted. In particular, at pH 8, a value typically used in paper conservation to enhance deacidification, and solution concentration of 0.005 M, iron extraction reached ca. 91%. Among all the tested pH, the lowest extraction of 67–72% was observed at the highest concentration 0.01 M. Concerning applications on mock-ups simulating manuscripts with iron gall ink, DFO-B with a concentration of 0.005 M also showed high extraction capacity which reached 83 and 85%, at pH 8 and 9, respectively (Wagner and Bulska 2003). However, this was considered a drawback because iron extraction was not selective, as the color of iron gall ink was severely affected. Consequently, the most effective solution for removing excess iron was proved unsuitable for iron gall ink mock-ups.

The influence of pH on siderophores' efficacy has been also the subject of several environmental studies unrelated to conservation. Borer et al. (2009) among others (Cheah et al. 2003; Borer et al. 2005; Reichard et al. 2005; Loring et al. 2008) investigated DFO-B and aerobactin in order to improve the understanding of these siderophores biogeochemistry and to gain information about siderophores reactivity towards the dissolution and mobilization of iron hydroxy-oxides, such as goethite and lepidocrocite, which are often found in terrestrial and aquatic environments. Their study revealed that pH influences the iron adsorption in ferric compounds, for instance, in lepidocrocite via DFO-B and that was also associated with siderophores' functional group and their deprotonation stage. More specifically, they showed that the maximum absorption of lepidocrocite by DFO-B occurred at pH 8.6, which is close to its first pKa (8.3), whereas at lower or higher pH values, the adsorption was decreased.

### 4.3.2   Extraction of Iron Compounds from Composite Museum Objects

Rapti et al. (2015) investigated in a preliminary study the removal of iron ions from dry wooden substrates, which cannot tolerate waterborne treatments, due to the hygroscopic and anisotropic nature of wood. Desferrioxamine, in the form of a commercial injectable drug, Desferal® manufactured by Novartis, was compared with the conventional chelators EDTA and DTPA to treat wooden mock-ups artificially stained with iron corrosion products. To avoid immersion in aqueous solutions and minimize the amount of water that would penetrate into the wood and at the same time maintain the cleaning potency of the solution applied, the chelators were applied by cotton swab and carboxymethyl cellulose gel (CMC) 4.5% w/v. The effectiveness of the two conventional chelators was also examined with the addition of a reducing agent, sodium dithionite (SDT) (Selwyn and Tse 2009). The solutions of EDTA and DTPA were prepared at 2.5% w/v with or without the addition of SDT (5% w/v) at pH 6.2, and 6.8, respectively, whereas DFO-B concentration was $5 \times 10^{-5}$ M, at pH 8.5. Colorimetry and elemental analysis by Energy Dispersive X-ray (EDX) spectroscopy were implemented before and after cleaning for evaluating the efficacy of the chelators. Results showed that DFO-B was an effective chelator, as efficient as EDTA when SDT was added to it. However, residues of Na and S ions were detected, owed to SDT, which have the potential to promote future deterioration of substrates and therefore, this suggest the need to avoid SDT as part of the chelators solutions on this type of substrate; thus, siderophore was considered more appropriate. Conventional chelators when applied without the reducing agent were shown clearly less efficient.

Further comparative investigation of Desferal®, EDTA, and DTPA was undertaken by Rapti et al. (2017). In this study, the cleaning potential of chelators combined with gel formulations when applied on dry composite objects comprising of wood, textiles, and iron was examined. Gel formulations were selected as they could be easily applied, clean the substrate in a more controlled and selective way and at the same time minimize the diffusion of the aqueous solution within the wood without though reducing its cleaning efficiency. Moreover, the use of gels has the advantage of avoiding dismantling the composite object, as cleaning can be achieved selectively at the stained substrate. For that purpose, wood and textile mock-ups were artificially stained with iron corrosion products. Chelator solutions (Desferal®, EDTA, and DTPA) were prepared ($3 \times 10^{-2}$M), and accordingly applied with several gel formulations (Klucel G 10% w/v, agarose A0701 4% w/v, Carbopol 940 2.5 w/v neutralized by Ethomeen C25 15 v/v and xanthan gum 2.2 w/v). The pH of Desferal® was adjusted to 8.45 and 6.8, of EDTA to 6.2 and DTPA 6.8, according to their acid dissociation constants and to the substrate compatibility in order to avoid deterioration processes caused by the cleaning procedure. The time of application was not set, since it varied depending on the color change caused by complexation. The gel was removed after colored complexes were formed (EDTA ferric complex is pale yellow, DTPA complex is yellow, while DFO-B is orange-red) (Almkvist et al. 2005; Dominguez-Vera 2004); the process was repeated with fresh identical gels until complexation was no longer observed, followed by a rinse with swab wetted in 50% v/v ethanol aqueous solution. The effectiveness of all chelators was assessed by colorimetry and energy dispersive X-ray (EDX) spectroscopy before and after the cleaning procedure.

The colorimetry measurements (L*, a*, b*) showed that the L* and b* coordinates increased in all samples after cleaning, indicating that samples became brighter and yellower. In contrast, a* factor was reduced due to the partial removal of the red-brown iron corrosion products. The total color difference ($\Delta E^*$) indicates the cleaning efficacy; the higher the difference the more effective cleaning of mock-ups. Regarding the textile mock-ups, DFO-B at pH 6.8 showed the greatest change when gelled with Klucel G, followed by xanthan gum (Fig. 14.3a). In contrast, DFO-B at pH 8.4 exhibited the second highest $\Delta E^*$ of all the cleaning methods employed, when applied with xanthan gum instead of Klucel G. Moreover, the DTPA application provided similar $\Delta E^*$ to DFO at pH 8.4, whereas EDTA was the least effective, with the exception of the agarose gel. Although the results of $\Delta E^*$ values of both substrates suggested that the gelling agents were actually involved in the effectiveness in cleaning efficacy, it was not possible to draw any conclusions, as the results obtained were not consistent for all chelators. Analogous results on the chelators' efficacy were obtained for wooden mock-ups.

Furthermore, elemental analysis by Energy Dispersive X-ray spectroscopy (EDX) demonstrated the weight percent of the detected iron before and after the cleaning of wood and textile mock-ups (Fig. 14.3b). More specifically, DFO-B at both pHs (6.8 and 8.45) proved superior to EDTA and DTPA in its performance for removing iron corrosion products, and this is consistent with its higher stability constant. In contrast to colorimetry measurements, EDX results were more

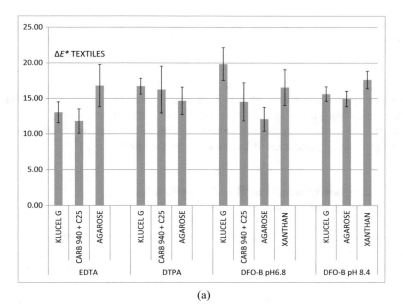

(a)

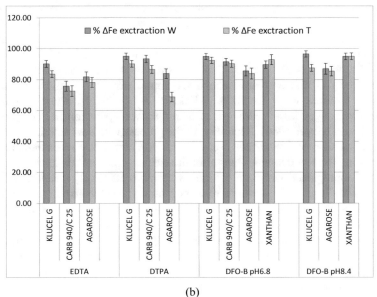

(b)

**Fig. 14.3** (**a**) ΔE* colorimetry values obtained after cleaning textile mock-ups with different chelators and gel formulations; (**b**) EDX results of iron extraction demonstrated by the weight percentages of iron difference (ΔFe) before and after the cleaning of wooden and textile mock-ups with different chelators and gel formulations, based on Rapti et al. (2017). *W* wood; *T* textile

consistent and showed that the efficacy of the chelators in iron extraction was influenced by the gelling agents. In the case of DFO-B at both pH values, EDTA and DTPA, gels prepared with Klucel G seemed to be more effective than other gelling agents.

Finally, as observed by EDX and confirmed by colorimetry results, wood was cleaned more effectively than the textile and this is possibly due to the fact that iron oxides have been impregnated deeper within the textile fibers (Glenn et al. 2015). Although iron corrosion products were not completely removed from either substrate, the overall appearance of both mock-ups was improved.

### 4.3.3 Extraction of Iron Oxides and Iron Sulfides From Waterlogged Wood

Albelda-Berenguer et al. (2019) under the framework of the MICMAC project (microbes for archeological wood conservation) investigated the production and subsequent application of siderophores in extraction and stabilization of sulfur and iron compounds found in waterlogged archeological wood. Pyoverdine (produced from the fluorescent *Pseudomonas putida*) was employed as the selected siderophore, due to the fact that its chromophore (see Fig. 14.2d) offers an additional asset for its spectrophotometric detection. This was studied along with Desferal® and EDTA on mock-ups made of balsa wood that were artificially impregnated either with iron oxides or iron sulfides through immersion for 24 h in each chelator solution (20 mM, pH 7). The formation of iron complexes was visually evaluated, based on the fact that each iron–chelator complex acquires a unique color upon formation (Dominguez-Vera 2004; Almkvist et al. 2005). More specifically, pyoverdine changes from yellow-green to red-brown upon iron(III) complexation (Hohnadel and Meyer 1986). Repetition of the process with new solutions is needed to be applied to promote further iron extraction. Colorimetry measurements were conducted on the wooden mock-ups after the cleaning process. Results showed that mock-ups impregnated with iron sulfides were more efficiently cleaned than those prepared with iron oxides, which remained almost unchanged. Pyoverdine was showed to be the most effective for the iron sulfides removal during the cleaning application.

Albelda-Berenguer et al. (2019) ongoing study is focusing on the development of an application protocol employing siderophores for extracting iron oxides or iron sulfides found in waterlogged archeological wood of shipwreck timbers. Moreover, the extraction of iron from known iron minerals by the use of siderophores is also a target of this ongoing work in order to demonstrate their ability in complexing iron. In this regard, iron sulfides present in three mineral phases, mackinawite (FeS), pyrite ($FeS_2$), and mineral sulfur $\alpha$-S8 along with iron oxides (hematite) and oxy-hydroxides (goethite) are being tested with pyoverdine, Desferal®, and EDTA at different pH. Preliminary results revealed that pyoverdine could extract iron from mackinawite, pyrite, and goethite, but is not capable to extract it from hematite (Monachon et al. 2019).

### 4.3.4 Other Studies with Siderophores

Besides the above-mentioned studies, other disciplines employing siderophores as possible preventive antioxidants of cellulose deterioration can be considered as indirect approaches to conservation issues (Strlič et al. 2001). The use of siderophores as antioxidants has been investigated in the food industry (Todokoro et al. 2016; Albelda-Berenguer et al. 2019). Strlič et al. (2001) studied whether DFO-B, phytate, EDTA and DTPA chelators were possible preventive antioxidants of cellulose degradation, as it typically constitutes several museum objects. Their antioxidant role was investigated concerning the formation of stable complexes and the occupation of all the available coordination sites of iron. The inhibition of hydroxyl radical production during degradation of cellulose was comparatively investigated in aqueous systems containing the above-mentioned chelators by calculating their rate constants. Their efficiency concerning their antioxidant behavior at 20°C (range tested: 20–80°C) and in the pH range from 6.5 to 8 was found to be higher for DFO-B followed by phytate and DTPA.

## 5 Conclusions

The research investigations on siderophores were initially conducted mostly due to their significant roles in making iron available for living systems by chelating ferric compounds from terrestrial and aquatic habitats. These roles have provided an impressive body of biological, chemical, and biochemical data which have greatly helped in introducing those agents for several applications in fields such as medicine, agriculture, environmental research, and materials science. However, despite their high potential, these materials have only recently been introduced in fields such as the conservation of cultural heritage. Nonetheless, some promising research results in this field show that the siderophores are highly useful materials due to their chelating abilities which allow for efficient and safe practices in conservation.

Siderophores, as naturally occurring chelators, have been considered as *"green"* chemical substances with low environmental risk. Their high affinity to iron and the potential for minimum harm to museum objects may rank them among the most promising novel materials with potentially significant uses in the conservation field. However, factors such as their complex stability constants, the working conditions (such as the pH range, interfering metal ions and light), along with kinetic considerations regarding chelation and dechelation need to be examined and balanced before these compounds are recommended for a specific use. So far, only desferrioxamine B and pyoverdine have been investigated for cleaning cultural heritage objects. Nonetheless their application although limited to few organic substrates seems to provide promising results for the effective removal of iron corrosion products. In order to develop widely accepted strategies employing these powerful agents for conservation treatments further investigation efforts are needed, mainly concerning the understanding of their properties and optimizing their application. The following issues are among those that need to be considered:

- The formation of highly stable complexes requires a certain pH range, depending on the nature of siderophores; therefore, pH needs to be adjusted to avoid both iron dechelation and the harmful side-reactions to the object. These demands are typically ad hoc defined, based on the specific properties of each cultural heritage substrate, so that the risk for further damage due to a siderophore-based application is minimized.
- Selectivity issues towards interfering metals need also to be investigated. Balancing acidity and redox conditions can potentially increase selectivity. Moreover selectivity of one type of iron compound over others based on their solubilities, redox behavior, etc., need also to be addressed.
- Application on complex or composite materials, in particular, those incorporating inorganic components need some attention as siderophores can be beneficial for one substrate and damaging for another.
- The significant synergistic effect of similarly chelating dicarboxylic acids present can be further exploited for enhanced and optimized action. The parallel functioning of some of these acids (such as malonic and succinic) as pH buffering agents may also offer an additional asset.
- The incorporation of siderophores in gel systems needs also to be addressed and their specific action towards various substrates needs to be further investigated.
- Sensitivity to UV and visible light is a clearly underestimated issue. In addition, the fluorescence chromophores that are included in the structures of some siderophores may point toward additional applications, as well as for some extra caution.

Taking into account these issues, the optimization and fine-tuning of siderophore-based applications on wood, textile, paper, etc. are possible. The information about siderophores' efficacy, in combination with different application methodologies, should be a starting point for their future broader application in conservation.

# References

Agrawal YK (1979) Hydroxamic acids and their metal complexes. Russ Chem Rev 48:948–963

Ahmed E, Holmström SJM (2014) Siderophores in environmental research: roles and applications. Microb Biotechnol 7:196–208

Albelda-Berenguer M, Monachon M, Joseph E (2019) Siderophores: from natural roles to potential applications, 1st edn. Elsevier, Amsterdam, The Netherlands

Ali SS, Vidhale NN (2013) Bacterial siderophore and their application: a review. Int J Curr Microbiol App Sci 2:303–312

Almkvist G, Dal L, Persson I (2005) Extraction of iron compounds from VASA wood. In: Hoffmann P, Spriggs JA, Strætkvern K, Gregory D (eds) Proceedings of the 9th ICOM-CC group on wet organic archaeological materials conference, Copenhagen, 7–11 June, 2004, ICOM, Bremerhaven, pp 202–210

Avdeef A, Sofen SR, Bregante TL, Raymond KN (1978) Coordination chemistry of microbial iron transport compounds. 9: stability constants for catechol models of enterobactin. J Am Chem Soc 100:5362–5370

Badillo-Sanchez D, Chelazzi D, Giorgi R, Cincinelli A, Baglioni P (2019) Understanding the structural degradation of south American historical silk: a Focal Plane Array (FPA) FTIR and multivariate analysis. Sci Rep 9:1–10

Baker AJ (1980) Corrosion of metal in wood products. In: Sereda PJ, Litvan GG (eds) Durability of building materials and components, American Society for Testing and Materials (ASTM) STP 691, pp 981–993

Balliana E, Ricci G, Pesce C, Zendri E (2016) Assessing the value of green conservation for cultural heritage: positive and critical aspects of already available methodologies. Int J Conserv Sci 7 (1):185–202

Banik G, Ponahlo J (1983) Some aspects of degradation phenomena of paper caused by green copper-containing pigments. The Paper Conservator 7(1):3–7

Barbeau K, Rue EL, Bruland KW, Butler A (2001) Photochemical cycling of iron in the surface ocean mediated by microbial iron(III)-binding ligands. Nature 413:409–413

Barbeau K, Rue EL, Trick CG, Bruland KW, Butler A (2003) Photochemical reactivity of siderophores produced by marine heterotrophic bacteria and cyanobacteria based on characteristic Fe(III) binding groups. Limnol Oceanogr 48:1069–1078

Barbusiński K (2009) Fenton reaction—controversy concerning the chemistry. Ecol Chem Eng 16 (3):347–358

Barton LL, Hemming BC (eds) (1993) Iron chelation in plants and soil microorganisms. Academic Press, New York

Bauer L, Exner O (1974) The chemistry of hydroxamic acids and n-hydroxyimides. Angew Chem Int Ed Engl 13:376–384

Bertrand S (2010) Siderophore base. The web data base of microbial siderophores. http://bertrandsamuel.free.fr/siderophore_base/siderophores.php. Accessed 2 Jan 2020

Bertrand S, Larcher G, Landreau A, Richomme P, Duval O, Bouchara J-P (2009) Hydroxamate siderophores of Scedosporium apiospermum. BioMetals 22:1019–1029

Borer P, Hug SJ, Sulzberger B, Kraemer SM, Kretzschmar R (2009) ATR-FTIR spectroscopic study of the adsorption of desferrioxamine B and aerobactin to the surface of lepidocrocite (γ-FeOOH). Geochim Cosmochim Acta 73(16):4661–4672

Borer PM, Sulzberger B, Reichard P, Kraemer SM (2005) Effect of siderophores on the light-induced dissolution of colloidal iron(III) (hydr)oxides. Mar Chem 93(2–4):179–193

Boukhalfa H, Crumbliss AL (2002) Chemical aspects of siderophore mediated iron transport. BioMetals 15:325–339

Boukhalfa H, Reilly SD, Michalczyk R, Iyer S, Neu MP (2006) Iron(III) coordination properties of a pyoverdin siderophore produced by Pseudomonas putida ATCC 33015. Inorg Chem 45:5607–5616

Brink CP, Crumbliss AL (1984) Kinetics, mechanism, and thermodynamics of aqueous iron(III) chelation and dissociation: influence of carbon and nitrogen substituents in hydroxamic acid ligands. Inorg Chem 23:4708–4718

Bulska E, Wagner B (2002) Investigation of iron-gall ink corrosion of ancient manuscript by non-destructive and microanalytical methods. In: van Grieken R, Janssens K, Van't dack L, Meersman G (eds) Art 2002: 7th international conference on non-destructive testing and microanalysis for the diagnostics and conservation of the cultural and environmental heritage, 2–6 June 2002, Congress Centre Elzenveld, Antwerp, Belgium: Proceedings, University of Antwerp

Bulska E, Wagner B (2004) A study of ancient manuscripts exposed to iron-gall ink corrosion. In: Janssens K, Van Grieken R (eds) Comprehensive analytical chemistry. Elsevier, Amsterdam, pp 755–788

Bulska E, Wagner B (2005) Investigation of a novel conservation procedure for historical documents. In: Van Grieken R, Janssens K (eds) Cultural heritage conservation and environmental impact assessment by non-destructive testing and micro-analysis. A.A Balkema Publishers, London, pp 101–116

Bulska E, Wagner B, Sawicki MG (2001) Investigation of complexation and solid-liquid extraction of iron from paper by UV/VIS and atomic absorption spectrometry. Microchim Acta 136 (1–2):61–66

Burgess H (1991) The use of chelating agents in conservation treatments. The Paper Conservator 15 (1):36–44

Burkitt MJ (2003) Chemical, biological and medical controversies surrounding the Fenton reaction. Prog React Kinet Mech 28:75–104

Cézard C, Farvacques N, Sonnet P (2014) Chemistry and biology of pyoverdines, pseudomonas primary siderophores. Curr Med Chem 22:165–186

Chapman V (1997) The conservation of a painted silk tambourine—and tri-ammonium citrate. In: Lockhead V (ed) Painted textiles, UKIC textile section forum postprints, 21 April 1997, UKIC, London

Cheah S-F, Kraemer SM, Cervini-Silva J, Sposito G (2003) Steady-state dissolution kinetics of goethite in the presence of desferrioxamine B and oxalate ligands: implications for the microbial acquisition of iron. Chem Geol 198:63–75

Cornell RM, Schwertmann U (2003) The iron oxides: structure, properties, reactions. In: Occurrences and uses, 2nd edn. VCH Verlag, Weinheim, Germany

Corregidor V, Viegas R, Ferreira LM, Alves LC (2019) Study of iron gall inks, ingredients and paper composition using non-destructive techniques. Heritage 2:2691–2703

Correnti C, Strong RK (2012) Mammalian siderophores, siderophore-binding lipocalins, and the labile iron pool. J Biol Chem 287:13524–13531

Cotton FA, Wilkinson G, Gaus PL (1995) Basic inorganic chemistry. John Wiley & Sons, New York

Crichton R (2019) Biological inorganic chemistry, 3rd edn. Academic Press, London

Crumbliss A (1990) Iron bioavailability and the coordination chemistry of hydroxamic acids. Coord Chem Rev 105:155–179

Devireddy LR, Hart DO, Goetz DH, Green MR (2010) A mammalian siderophore synthesized by an enzyme with a bacterial homolog involved in enterobactin production. Cell 141:1006–1017

Dhungana S, Crumbliss AL (2005) Coordination chemistry and redox processes in siderophore-mediated iron transport. Geomicrobiol J 22:87–98

Dominguez-Vera JM (2004) Iron(III) complexation of Desferrioxamine B encapsulated in apoferritin. J Inorg Biochem 98(3):469–472

Dunford H (2002) Oxidations of iron(II)/(III) by hydrogen peroxide: from aquo to enzyme. Coord Chem Rev 233–234:311–318

Edwards DC, Nielsen SB, Jarzecki AA, Spiro TG, Myneni SCB (2005) Experimental and theoretical vibrational spectroscopy studies of acetohydroxamic acid and desferrioxamine B in aqueous solution: effects of pH and iron complexation. Geochim Cosmochim Acta 69:3237–3248

Emery JA, Schroeder HA (1974) Iron-catalyzed oxidation of wood carbohydrates. Wood Sci Technol 8(2):123–137

Fors Y (2008) Sulfur-related conservation concerns for marine archaeological wood: the origin, speciation and distribution of accumulated sulfur with some remedies for the VASA. PhD Thesis, Department of Physical, Inorganic and Structural Chemistry, Stockholm University

Furrer G, Stumm W (1986) The coordination chemistry of weathering: I. dissolution kinetics of δ-Al 2 O 3 and BeO. Geochim Cosmochim Acta 50:1847–1860

Glenn S, Haldane E-A, Hackett J and Sung I (2015) Borrowing from the neighbours: using the technology of other disciplines to treat difficult textile conservation problems. In: Fairhurst A (ed) Learning curve: education, experience and reflexion icon textile spring forum postprints, 13 April 2015, Icon, London

Goodell B, Daniel G, Jellison J, Qian Y (2006) Iron-reducing capacity of low-molecular-weight compounds produced in wood by fungi. Holzforschung 60:630–636

Grčman H, Velikonja-Bolta Š, Vodnik D, Kos B, Leštan D (2001) EDTA enhanced heavy metal phytoextraction: metal accumulation, leaching and toxicity. Plant Soil 235(1):105–114

Gupta P, Lakes A, Dziubla T (2016) A free radical primer. In: Dziubla T, Butterfield DA (eds) Oxidative stress and biomaterials. Academic Press, Cambridge, pp 1–33

Häkäri AE (1992) The removal of rust stains from historic cellulosic textile material. Unpublished Diploma Dissertation, Textile Conservation Centre, London

Harris DC (2015) Quantitative chemical analysis. 8th edn

Harris WR, Carrano CJ, Raymond KN (1979) Coordination chemistry of microbial iron transport compounds. 16: isolation, characterization, and formation constants of ferric aerobactin. J Am Chem Soc 101:2722–2727

Hart JR (1981) Chelating agents in the pulp and paper industry. Tappi 64(3):43–44

Heller A, Barkleit A, Foerstendorf H, Tsushima S, Heim K, Bernhard G (2012) Curium(III) citrate speciation in biological systems: a europium(III) assisted spectroscopic and quantum chemical study. Dalton Trans 41:13969

Henniges U, Reibke R, Banik G, Huhsmann E, Hähner U, Prohaska T, Potthast A (2008) Iron gall ink-induced corrosion of cellulose: aging, degradation and stabilization. Part 2: application on historic sample material. Cellulose 15:861–870

Hernlem BJ, Vane LM, Sayles GD (1996) Stability constants for complexes of the siderophore desferrioxamine B with selected heavy metal cations. Inorg Chim Acta 244:179–184

Hider RC, Kong X (2010) Chemistry and biology of siderophores. Nat Prod Rep 27:637

Hofenk de Graaff JH (1982) Some recent developments in the cleaning of ancient textiles. In: Bromelle NS, Thomson G (eds) Science and technology in the service of conservation. IIC, London, pp 93–95

Höfte M (1993) Classes of microbial siderophores. In: BL L, HB C (eds) Iron chelation in plants and soil microorganisms. Academic Press, New York, pp 3–27

Hohnadel D, Meyer JM (1986) Pyoverdine-facilitated iron uptake among fluorescent Pseudomonads. In: Swinburne TR (ed) Iron, siderophores, and plant diseases. NATO ASI series (series a: life sciences), vol 117. Springer, Boston, pp 119–129

Hynes MJ, Ó Coinceanainn M (2001) The kinetics and mechanisms of the reaction of iron(III) with gallic acid, gallic acid methyl ester and catechin. J Inorg Biochem 85:131–142

Ito Y, Butler A (2005) Structure of synechobactins, new siderophores of the marine cyanobacterium Synechococcus sp. PCC 7002. Limnol Oceanogr 50:1918–1923

Iwahashi H, Morishita H, Ishii T, Sugata R, Kido R (1989) Enhancement by catechols of hydroxyl-radical formation in the presence of ferric ions and hydrogen peroxide. J Biochem 105:429–434

Jablonský M, Kazíková J, Holúbková S (2010) The effect of the iron-gall ink on permanence in paper by breaking length, degree of polymerisation and thermogravimetric stability of paper during accelerated ageing. Acta Chim Slov 3:63–73

Jensen WB (1991) Overview lecture: the Lewis acid-base concepts: recent results and prospects for the future. J Adhes Sci Technol 5:1–21

Kalinowski DS, Richardson DR (2005) The evolution of iron chelators for the treatment of iron overload disease and cancer. Pharmacol Rev 57:547–583

Keberle H (2006) The biochemistry of desferrioxamine and its relation to iron metabolism. Ann N Y Acad Sci 119:758–768

Khan A, Singh P, Srivastava A (2018) Synthesis, nature and utility of universal iron chelator—siderophore: a review. Microbiol Res 212–213:103–111

Kolar J (1997) Mechanism of autoxidative degradation of cellulose. Restaurator 18:163–176

Kolar J, Štolfa A, Strlič M, Pompe M, Pihlar B, Budnar M, Simčič J, Reissland B (2006) Historical iron gall ink containing documents—properties affecting their condition. Anal Chim Acta 555:167–174

Kolar J, Strlic M (2001) Stabilisation of ink corrosion. In: Jean E, Brown A (eds) The Iron gall ink meeting. University of Northumbria, Newcastle-upon-Tyne, UK, pp 135–139

Kolar J, Strlič M (2004) Ageing and stabilisation of paper. National and University Library, Turjaška 1, 1000 Ljubljana, Slovenia, Ljubljana

Kołodyńska D (2011) Chelating agents of a new generation as an alternative to conventional chelators for heavy metal ions removal from different waste waters. In: Ning RY (ed) Expanding issues in desalination. InTech, Rijeka, pp 339–370

Konetschny-Rapp S, Jung G, Raymond KN, Meiwes J, Zähner H (1992) Solution thermodynamics of the ferric complexes of new desferrioxamine siderophores obtained by directed fermentation. J Am Chem Soc 114:2224–2230

Koppenol WH (1993) The centennial of the Fenton reaction. Free Radic Biol Med 15:645–651

Kraemer SM (2004) Iron oxide dissolution and solubility in the presence of siderophores. Aquat Sci—Res Across Boundaries 66:3–18

Kraemer SM, Butler A, Borer P, Cervini-Silva J (2005) Siderophores and the dissolution of iron-bearing minerals in marine systems. In: Banfield JF, Cervini-Silva J, Nealson K (eds) Molecular geomicrobiology. De Gruyter, Berlin, Boston, pp 53–84

Kumar L, Meena NL, Singh U (2016) Role of phytosiderophores in acquisition of iron and other micronutrients in food legumes. In: Singh U, Praharaj C, Singh S, Singh N (eds) Biofortification of food crops. Springer, New Delh, pp 291–302

Liu Y, Kralj-cigić I, Strlič M (2017) Kinetics of accelerated degradation of historic iron gall ink-containing paper. Polym Degrad Stab. https://doi.org/10.1016/j.polymdegradstab.2017.07.010

Loring JS, Simanova AA, Persson P (2008) Highly mobile iron pool from a dissolution—readsorption process. Langmuir 24:7054–7057

Margariti C (2003) The use of chelating agents in textile conservation: an investigation of the efficiency and effects on the use of three chelating agents for the removal of copper and iron staining from cotton textiles. In: Dawson L, Berkouwer M (eds) Dust, sweat and tears: recent advances in cleaning techniques, UKIC Textile Section, AGM Spring Forum, 7 April 2003, UKIC, London, pp 28–38

Marusak RA, Doan K, Cummings SD (2006) Integrated approach to coordination chemistry: an inorganic laboratory guide, John Wiley & Sons, New Jersey

Miethke M, Marahiel MA (2007) Siderophore-based iron acquisition and pathogen control. Microbiol Mol Biol Rev 71:413–451

Milburn RM, Vosburgh WC (1955) A spectrophotometric study of the hydrolysis of iron(III) ion. II: polynuclear species. J Am Chem Soc 77:1352–1355

Monachon M, Albelda-Berenguer M, Joseph E (2019) Bio-based treatment for the extraction of problematic iron sulphides from waterlogged archaeological wood. In: Book of abstracts of the 14th ICOM-CC group on wet organic archaeological materials conference, Portsmouth, UK, 20–24 May 2019, ICOM, Portsmouth, pp 46–47

Monzyk B, Crumbliss AL (1979) Mechanism of ligand substitution on high-spin iron(III) by hydroxamic acid chelators: thermodynamic and kinetic studies on the formation and dissociation of a series of monohydroxamatoiron(III) complexes. J Am Chem Soc 101:6203–6213

Moran JF, Klucas RV, Grayer RJ, Abian J, Becana M (1997) Complexes of iron with phenolic compounds from soybean nodules and other legume tissues: prooxidant and antioxidant properties. Free Radic Biol Med 22:861–870

Nakamoto K (2009) Infrared and Raman spectra of inorganic and coordination compounds. In: Theory and applications to inorganic chemistry, 6th edn. John Wiley & Sons, New York

Nakamura Y (2000) A simple phenolic antioxidant protocatechuic acid enhances tumor promotion and oxidative stress in female ICR mouse skin: dose- and timing-dependent enhancement and involvement of bioactivation by tyrosinase. Carcinogenesis 21:1899–1907

Neevel JG (1995) Phytate: a potential conservation agent for the treatment of ink corrosion caused by iron gall inks. Restaurator 16(3):143–160

Neilands JB (1995) Siderophores: structure and function of microbial iron transport compounds. J Biol Chem 270:26723–26726

Nörtemann B (1999) Biodegradation of EDTA. Appl Microbiol Biotechnol 51(6):751–759

Nörtemann B (2005) Biodegradation of chelating agents: EDTA, DTPA, PDTA, NTA, and EDDS. Biogeochemistry of chelating agents. ACS symposium series, American Society, Washington, DC, pp 150–170

Pearson R (1973) Hard and soft acids and bases. Dowden, Hutchinson and Ross, Stroudsburg

Pepper IL, Gentry TJ (2015) Earth environments. In: Pepper IL, Gerba CP, Gentry TJ (eds) Environmental microbiology. Elsevier, Amsterdam, pp 59–88

Phenix A, Burnstock A (1992) The removal of surface dirt on paintings with chelating agents. The Conservator 16(1):28–38

Rapti S, Boyatzis S, Rivers S, Velios A, Pournou A (2017) Removing iron stains from wood and textile objects: assessing gelled siderophores as novel green chelators. In: Angelova LV, Ormsby B, Townsend JH, Wolbers R (eds) Gels in the conservation of art. Archetype Books, London, pp 343–348

Rapti S, Rivers S, Pournou A (2015) Removing iron corrosion products from museum artefacts: Investigating the effectiveness of innovative green chelators. In: International conference Science in Technology, SCinTE, Athens

Reichard PU, Kraemer SM, Frazier SW, Kretzschmar R (2005) Goethite dissolution in the presence of phytosiderophores: rates, mechanisms, and the synergistic effect of oxalate. Plant Soil 276 (1–2):115–132

Reichard PU, Kretzschmar R, Kraemer SM (2007) Dissolution mechanisms of goethite in the presence of siderophores and organic acids. Geochim Cosmochim Acta 71:5635–5650

Renshaw JC, Robson GD, Trinci APJ, Wiebe MG, Livens FR, Collison D, Taylor RJ (2002) Fungal siderophores: structures, functions and applications. Mycol Res 106:1123–1142

Richards V, Kasi K, Godfrey I (2012) Iron removal from waterlogged wood and the effects on wood chemistry. In: Straetkvern K, Williams E, (eds) Proceedings of the 11th ICOM-CC group on wet organic archaeological materials conference, ICOM, Greenville, 2010. Lulu, Greenville, pp 383–400

Rivers S, Ummey N (2003) Conservation of furniture. Butterworth-Heinemann, Oxford

Romero R, Salgado PR, Soto C, Contreras D, Melin V (2018) An experimental validated computational–method for pKa determination of substituted 1,2-dihydroxybenzenes. Front Chem 6:1–11

Römheld V (1991) The role of phytosiderophores in acquisition of iron and other micronutrients in graminaceous species: an ecological approach. Plant Soil 130:127–134

Saha M, Sarkar S, Sarkar B, Sharma BK, Bhattacharjee S, Tribedi P (2016) Microbial siderophores and their potential applications: a review. Environ Sci Pollut Res 23:3984–3999

Saha R, Saha N, Donofrio RS, Bestervelt LL (2013) Microbial siderophores: a mini review. J Basic Microbiol 53:303–317

Salgado P, Melin V, Contreras D, Moreno Y, Mansilla HD (2013) Fenton reaction driven by iron ligands. J Chil Chem Soc 58:2096–2101

Samson SD, Eggleston CM (1998) Active sites and the non-steady-state dissolution of hematite. Environ Sci Technol 32:2871–2875

Samson SD, Stillings LL, Eggleston CM (2000) The depletion and regeneration of dissolution-active sites at the mineral-water interface: I. Fe, Al, and in sesquioxides. Geochim Cosmochim Acta 64:3471–3484

Sapieszko RS, Patel RC, Matijevic E (1977) Ferric hydrous oxide sols. 2: thermodynamics of aqueous hydroxo and sulfato ferric complexes. J Phys Chem 81:1061–1068

Schwertmann U (1991) Solubility and dissolution of iron oxides. In: Plant and soil; selected papers from the fifth international symposium on iron nutrition and interactions in plants, pp 1–25

Scott DA, Eggert G (2009) Iron and steel in art: corrosion, colorants, conservation. Archetype Publications, London

Selwyn L, Tse S (2009) The chemistry of sodium dithionite and its use in conservation. Stud Conserv 54:61–73

Sharpe PC, Richardson DR, Kalinowski DS, Bernhardt PV (2011) Synthetic and natural products as iron chelators. Curr Top Med Chem 11:591–607

Shenker M, Chen Y, Hadar Y (1996) Stability constants of the fungal siderophore rhizoferrin with various microelements and calcium. Soil Sci Soc Am J 60:1140

Siddall TH, Vosburgh WC (1951) A spectrophotometric study of the hydrolysis of iron(III) ion. J Am Chem Soc 73:4270–4272

Sillanpää M (1997) Environmental fate of EDTA and DTPA. Reviews of environmental contamination and toxicology, vol 152. Springer, New York, pp 85–111

Silva AMN, Kong X, Hider RC (2009) Determination of the pKa value of the hydroxyl group in the α-hydroxycarboxylates citrate, malate and lactate by 13C NMR: implications for metal coordination in biological systems. BioMetals 22:771–778

Singh N, Karpichev Y, Sharma R, Gupta B, Sahu AK, Satnami ML, Ghosh KK (2015) From α-nucleophiles to functionalized aggregates: exploring the reactivity of hydroxamate ion towards esterolytic reactions in micelles. Org Biomol Chem 13:1–8

Slavin J (1990) The removal of salt deposits from decorative paintings on paper. In: Hackney S, Townsend JH, Eastaugh N (eds) Dirt and pictures separated, UKIC and tate gallery conference, UKIC, London, pp 49–50

Strlič M, Kolar J, Pihlar B (2001) Some preventive cellulose antioxidants studied by an aromatic hydroxylation assay. Polym Degrad Stab 73(3):535–539

Stumm W, Furrer G (1987) The dissolution of oxides and aluminum silicates: examples of surface-coordination-controlled-kinetics. In: Stumm W (ed) Aquatic surface chemistry. John Wiley and Sons, Somerset, NJ, pp 97–219

Sugiura Y, Nomoto K (1984) Phytosiderophores structures and properties of mugineic acids and their metal complexes. In: Siderophores from microorganisms and plants: structure and bonding, vol 58. Springer, Heidelberg; Berlin, pp 107–135

Thiem D, Złoch M, Gadzała-Kopciuch R, Szymańska S, Baum C, Hrynkiewicz K (2018) Cadmium-induced changes in the production of siderophores by a plant growth promoting strain of Pseudomonas fulva. J Basic Microbiol 58:623–632

Timár-Balázsy AT, Eastop D (1998) Chemical principles in textile conservation. Butterworths-Heinemann, London

Todokoro T, Fukuda K, Matsumura K, Irie M, Hata Y (2016) Production of the natural iron chelator deferriferrichrysin from Aspergillus oryzae and evaluation as a novel food-grade antioxidant. J Sci Food Agric 96(9):2998–3006

Tseng CF, Burger A, Mislin GLA, Schalk IJ, Yu SSF, Chan SI, Abdallah MA (2006) Bacterial siderophores: the solution stoichiometry and coordination of the Fe(III) complexes of pyochelin and related compounds. J Biol Inorg Chem 11:419–432

Wagner B, Bulska E (2003) Towards a new conservation method for ancient manuscripts by inactivation of iron via complexation and extraction. Anal Bioanal Chem 375:1148–1153

Wagner B, Bulska E (2004) On the use of laser ablation inductively coupled plasma mass spectrometry for the investigation of the written heritage. J Anal At Spectrom 19 (10):1325–1329

Wagner B, Bulska E, Hulanicki A, Heck M, Ortner HM (2001a) Topochemical investigation of ancient manuscripts. Fresenius J Anal Chem 369(7-8):674–679

Wagner B, Bulska E, Meisel T, Wegscheider W (2001b) Use of atomic spectrometry for the investigation of ancient manuscripts. J Anal At Spectrom 16(4):417–420

Wang W, Qiu Z, Tan H, Cao L (2014) Siderophore production by actinobacteria. BioMetals 27:623–631

Wardman P, Candeias LP (1996) Fenton chemistry: an introduction. Radiat Res 145(5):523–531

Wilkins RG (1991) Kinetics and mechanism of reactions of transition metal complexes, 2nd edn. VCH, Weinheim

Yang J, Bremer PJ, Lamont IL, McQuillan AJ (2006) Infrared spectroscopic studies of siderophore-related hydroxamic acid ligands adsorbed on titanium dioxide. Langmuir 22:10109–10117

Zeronian SH, Inglesby MK (1995) Bleaching of cellulose by hydrogen peroxide. Cellulose 2 (4):265–272

Złoch M, Thiem D, Gadzała-Kopciuch R, Hrynkiewicz K (2016) Synthesis of siderophores by plant-associated metallotolerant bacteria under exposure to Cd2+. Chemosphere 156:312–325

# Chapter 15
# Organic Green Corrosion Inhibitors Derived from Natural and/or Biological Sources for Conservation of Metals Cultural Heritage

**Vasilike Argyropoulos, Stamatis C. Boyatzis, Maria Giannoulaki, Elodie Guilminot, and Aggeliki Zacharopoulou**

**Abstract** In the last decade, there has been an increase in research related to green corrosion inhibitors for conservation of metals cultural heritage to help promote sustainable practices in the field that are safe, environmentally friendly, and ecologically acceptable. The most common are organic substances derived either from natural and/or biological sources: plant extracts and oils, amino acids, microorganisms, and biopolymers. The chapter will provide a review of these substances as corrosion inhibitors for metals conservation, by discussing the state-of-the-art research to date, with a special focus on cysteine. Most of the research has focused on the examination of such inhibitors on metal coupons with or without corrosion products using electrochemical techniques or weight-loss measurements to determine their effectiveness. Some of these studies have also considered the conservation principles for practice, i.e., reversibility of the treatment and the visual aspect of the modification of the treated metal surface. However, before such green inhibitors can be routinely applied by conservators, more research is required on their application to real artefacts/monuments using in situ corrosion measurements. Furthermore, given that the composition of a green inhibitor is highly dependent on its extraction process, research must also involve identifying the specific adsorption models and involved mechanisms to ensure reproducibility of results.

V. Argyropoulos (✉) · S. C. Boyatzis · M. Giannoulaki
Department of Conservation of Antiquities and Works of Art, University of West Attica, Athens, Greece
e-mail: bessie@uniwa.gr; sboyatzis@uniwa.gr; mgiann@uniwa.gr

E. Guilminot
GPLA Arc'Antique, Nantes, France
e-mail: Elodie.GUILMINOT@loire-atlantique.fr

A. Zacharopoulou
School of Chemical Engineering Materials Science and Engineering Department, National Technical University of Athens, Athens, Greece
e-mail: azachar@chemeng.ntua.gr

© The Author(s) 2021
E. Joseph (ed.), *Microorganisms in the Deterioration and Preservation of Cultural Heritage*, https://doi.org/10.1007/978-3-030-69411-1_15

**Keywords** Green corrosion inhibitors · Conservation · Metals · Cultural heritage · Cysteine · Natural or biological sources · sustainable practice

# 1    Introduction

Corrosion inhibitors are used in conservation of metals cultural heritage to slow down or even prevent their corrosion during pretreatment, stabilization treatment, or storage/exhibition. Protection systems are used for maintenance of metal cultural heritage objects and monuments often with corrosion inhibitors, applied many times with or even in coatings (e.g., varnishes, waxes, and paints), or found in the packing materials used to store the objects. These systems must be effective at protecting long-term important monuments, such as the Eiffel Tower in Paris or rare objects such as the Antikythera mechanism at the National Archaeological Museum in Athens each with millions of visitors each year. There has been much discussion, research, and review into the application of corrosion inhibitors in metals conservation (Keene 1985; Rocca and Mirambet 2007; Cano and Lafuente 2013). In the last decade there has been a trend in conservation research as for industry into the application of green corrosion inhibitors or eco-friendly substances that have biocompatibility in nature. Sustainable conservation practices are growing concern especially given the effects of climate change to the environment (De Silva and Henderson 2011), and for health and safety concerns for conservators, even though there is a lack of exposure studies due to their work-related practices (Hawks et al. 2010; Schrager et al. 2017). As such, conservation research in metals should focus on the use of green materials, and the choices for corrosion inhibitors must conform to applicable standards and regulations concerning toxicity and environmental protection (Sharma and Sharma 2012; Goni and Mazumder 2019). This chapter will discuss and review organic green corrosion inhibitors (OGCIs) derived from natural and/or biological sources for conservation of metals with the aim in understanding their effectiveness, limitations, and way forward in research for conservation.

Specifically, when reviewing and discussing the literature on corrosion inhibitor applications for metals cultural heritage, most often metal surfaces have corrosion products that need to be retained since they preserve their original surface; also, their application occurs in solutions at alkaline or neutral (near-neutral) pHs either to preserve the metal during stabilization treatments (i.e., removal of corrosive species) or as a modification of the surface as a final conservation step for added protection. For industrial metals, corrosion inhibitors are applied to clean metal surfaces (without corrosion products), and most often in solutions at acidic pHs with the aim to strip the metal of those corrosion products, while at the same time protecting the metal substrate from corrosion.

For metals conservation, corrosion inhibitors can be applied anywhere during the conservation process: before, during, and after conservation. As such, corrosion inhibitors are applied in the following situations:

**Fig. 15.1** The application of tannic acid as a final conservation step in coating cast iron cannon

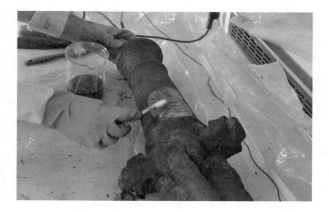

1. In aqueous solutions, where the objects are immersed during their storage prior to treatment and/or during stabilization treatments (e.g., archaeological objects from terrestrial or marine sites);
2. In surface treatments applied with a solvent either by immersion or brush to the objects/monuments as the final step in conservation with or without an additional coating;
3. Included in paints and coating systems to provide added protection to the coated objects/monuments;
4. In vapour phase, as capsules or in packing materials for long-term storage of metal objects/items;
5. As reactive particles found in packaging materials to help produce a microenvironment leading to the preservation of the metal object either before or after conservation.

The conservation field usually follows industry trends in testing the application of new types of corrosion inhibitors. The two best known corrosion inhibitors applied from industry and used in conservation of objects made from copper and iron alloys for over 50 years are benzotriazole (BTA) (Madsen 1967) and tannins (Pelikan 1966), respectively (see Fig. 15.1). These inhibitors are usually applied as surface treatment prior to coating, or even found in coatings systems, such as BTA in Incralac (used to absorb ultraviolet light preventing breakdown of the protective film) (Erhardt et al. 1984; Boyatzis et al. 2017). Today, BTA is classified by the European Chemicals Agency as toxic (ECHA 2019a), but conservators still favour its use, until a suitable replacement is found (Cano and Lafuente 2013).

Industry over the past 10 years has shown a great interest in the use of green corrosion inhibitors for industrial applications, which is apparent by the dramatic increase in the publications on the topic when searching the Web of Knowledge, due to the hazards caused by commonly used toxic corrosion inhibitors (Shehata et al. 2018; Marzorati et al. 2019). Although there has also been conservation research during this period on the topic (see discussion in Sect. 15.3), metal conservators do not yet routinely apply these new green materials for conservation since they want

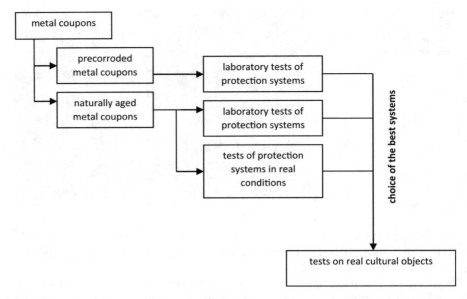

**Fig. 15.2** Methodology followed during PROMET project for testing traditional and innovative protection systems for metals cultural heritage

more in-depth research into the long-term effects of these new products on cultural heritage metal objects. As a result, even today many traditional materials (e.g., chemicals including solvents) used by conservators may be unsafe both to their health and to the environment, since their primary concern is the effectiveness at stabilizing or protecting their objects long-term.

Research into green corrosion inhibitors for conservation of metals cultural heritage was greatly influenced by a collaborative project established over 10 years ago under the acronym PROMET entitled: 'Innovative Conservation Approaches for Monitoring and Protecting Ancient and Historic Metals Collections from the Mediterranean Region', financially supported by European project FP6-INCO (2004–2008) (Argyropoulos 2008; De Silva and Henderson 2011). The research results of this project set a milestone in sustainable practices for conservation of metals by:

1. Establishing a protocol for the testing of coatings and corrosion inhibitors on material culture made of metals involving a common methodology for coupon preparation and coating application as well as the performance of accelerated and natural aged corrosion tests (see Fig. 15.2) (Degrigny 2008; Argyropoulos et al. 2013).
2. Supporting and carrying out research into new and safe materials for protecting metal objects, the results of which inspired further research projects into this area (Cano and Lafuente 2013).

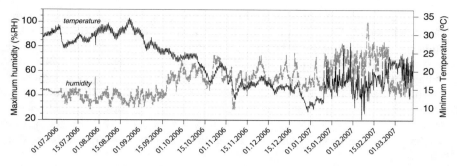

**Fig. 15.3** Relative humidity and temperature measurements at the Palace Armoury collection in Malta between May–November 2006

**Fig. 15.4** Coating appearance at the end of the accelerated ageing process (ageing cycles defined by the PROMET project) on artificially aged steel PROMET coupons with Paraloid B72® (using a stereo microscope x4) showing filiform corrosion

The PROMET project published a survey in 2007 on the use of corrosion inhibitors and coatings by conservators for metals museum collections in the Mediterranean region, which found that BTA in ethanol as a corrosion inhibitor and Paraloid B72 as a coating were the most popular in those countries (Argyropoulos et al. 2007b). However, these materials/practices were often developed and tested for metal objects in northern European and North American museums and were then brought into use in Mediterranean countries by foreigners working on missions in those countries or conservators who received their education abroad. In some cases, some materials and methods were found to be unsuitable for objects stored in a museum environment with high and fluctuating temperature and relative humidity including the presence of salt aerosols from close proximity to the sea (see Figs. 15.3 and 15.4) (Degrigny et al. 2007; Argyropoulos et al. 2007a). Also, the protection of metal collections from active corrosion is often the main concern both for archaeological and historical metal indoor and outdoor collections making the use of corrosion inhibitors and/or coatings essential in the Mediterranean region along with the necessary stabilization treatments.

In reviewing the literature on the topic presented in Sect. 15.3 of this chapter, countries in the Mediterranean region are the leaders in research of green corrosion inhibitors to cultural heritage metals. This chapter will define the types of green organic corrosion inhibitors commonly used by industry, how they work, and highlight the state-of-the-art research for conservation into these types of inhibitors used in different types of applications. Finally the chapter will provide a summary of the authors' research into cysteine as a corrosion inhibitor for copper alloy metal artefacts and provide a statement for the way forward into research for metals conservation using green organic corrosion inhibitors from natural and biological sources.

## 2 Definition of Green Corrosion Inhibitors

Green corrosion inhibitors are biodegradable and do not contain heavy metals or other toxic compounds (Rani and Basu 2012). Traditionally industry used corrosion inhibitors that were non-biodegradable synthetic organic corrosion inhibitors and traditional inorganic corrosion inhibitors often contained heavy metals, such as chromates, that became restrictive with environmental regulations due to their hazardous effects, causing both problems to human health and/or ecological systems (Popoola 2019). These environmental issues have led corrosion scientists and engineers to move towards the use of green corrosion inhibitors from natural organic compounds and sources that are inexpensive, readily available, environmentally friendly and ecologically acceptable, and renewable (Goni and Mazumder 2019).

## 2.1 The Different Types or Classes of Corrosion Inhibitors

The different types or classification of corrosion inhibitors are either scavengers or interface inhibitors, where the former scavenges aggressive substances from the corrosive medium, and the latter inhibits corrosion through film formation at the metal–environment interface (Popoola 2019).

The best known scavenger type inhibitors used in metals conservation either prevent oxygen induced corrosion or trap volatile gaseous compounds by either trapping or reacting with those compounds. For conservation, examples of products commonly used in packaging for storage/exhibition are Ageless®(Grattan and Gilberg 1994) or RP/ESCAL system (Mathias et al. 2004) for $O_2$ scavengers and activated charcoal or silver particles impregnated in fabrics, e.g., Pacific Silvercloth® for reacting with sulphureous compounds (see Fig. 15.5). There are also many types of oxygen-scavenging compounds that are directly incorporated into packaged materials, such as flexible films, plastics, etc.: one such product is Corrosion Intercept® used for packaging, where the polymer matrix is composed of highly reactive copper particles, making the matrix film reactive and neutralizing

**Fig. 15.5** An iron artefact stored in a sealed microenvironment containing RP/ESCAL to control the relative humidity and the $O_2$ levels with scavenger inhibitor

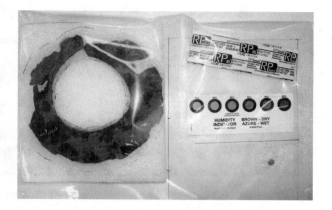

corrosive gases they come in contact with. There are also examples of such films, plastics that are bio-based containing, for example, gallic acid as an oxygen scavenger (Pant et al. 2017), but to date there is no application in conservation to the authors' knowledge.

Interface inhibitors can be classified as either vapour phase or liquid phase, with many known applications in conservation of metals. The vapour-phase inhibitors or vapour corrosion inhibitors (VCI) provide temporary atmospheric corrosion protection especially in closed environments by loosely impregnating wrapping paper inside a closed container. Some popular type paper rolls used for wrapping different types metals from museum collections contain VCI, such as Zerust® and Cortec VpCI films.

However, the most prominent corrosion inhibitors used both by industry and for conservation of metals are the liquid-phase inhibitors, which can be further subdivided into cathodic, anodic, or mixed based on the reaction-type inhibition, which can be any of the cathodic, anodic, or both electrochemical reactions. Cano and Lafuente (2013) provide a good review for these types of corrosion inhibitors used in metals conservation according to the different types of metal alloys. They confirm through their review that the most commonly applied inhibitors to the surface of a copper and iron alloyed objects are BTA and tannins, respectively. They also state that as opposed to industrial applications, for the metal conservation field, interface inhibitors are used to produce surface modifications or films through adsorption onto the metal surface using most often immersion followed by drying and coating. They briefly touch upon research in conservation using green inhibitors, but state that their application for metals conservation is limited.

Like regular corrosion inhibitors, green inhibitors have the same classifications but are derived from organic natural and/or biological sources as opposed to synthetic and for inorganic do not contain heavy metals. For the latter, some rare earth compounds (e.g., $CeCl_3$ and $LaCl_3$) are considered as inorganic green corrosion inhibitors and for the former natural and/or biological sources for organic green corrosion inhibitors are plants, drugs, amino acids, surfactants, biopolymers, and ionic liquids (El Ibrahimi et al. 2020). Others classify organic green inhibitors

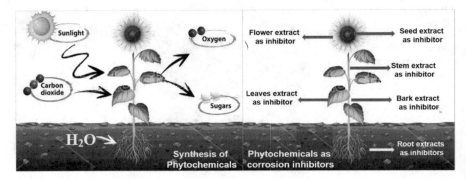

**Fig. 15.6** Plant extracts as corrosion inhibitors can be taken from a diverse part of the plants (Reproduced from (Verma et al. 2018) with permission from Elsevier)

according to their natural compounds or products they contain, such as amino acids, alkaloids, phenols and polyphenols, fatty acids (Marzorati et al. 2019), or as biological (chitosan, amino acids, bacteria, and fungi), vegetable (plant extracts, shells, tannins), and pharmaceutical drugs (Montemor 2016). Regardless of their classification, approximately 80% of organic corrosion green inhibitors are categorized as mixed inhibitors that protect the metal from corrosion by chemical and physical adsorption and film formation (Popoola 2019).

Lastly, it is important to remember that even green inhibitors may not always be inexpensive or even environmentally friendly due to the cost and time needed to extract and purify naturally occurring substances, like from some plants (see Fig. 15.6 (Verma et al. 2018)), drugs or ionic liquids (El Ibrahimi et al. 2020). Also later authors state that organic solvents may be needed for the extraction process, which can also damage the environment. Thus, abundance, renewal, and bio disposal are the key for applying such green inhibitors in real situations (Montemor 2016).

The rest of this chapter will only focus on organic green corrosion inhibitors derived from natural and/or biological sources, which are the most commonly researched by industry and conservation. It does not include discussions of organic ligands or synthetic surfactants as corrosion inhibitors, such as salicylaldoxime and their derivatives (Abu-Baker et al. 2013; Abu-baker 2019) and Hostacor IT (Argyropoulos et al. 1999), since although eco-friendly and biodegradable, they are not obtained from natural and/or biological sources.

## 2.2 How Do They Work?

Green inhibitors have properties that are similar to the 'non-green' inhibitors (Kesavan et al. 2012). The most commonly used in conservation as for industry are corrosion interface inhibitors, which act by: (i) forming a film that is adsorbed on the metal surface; (ii) producing corrosion products that acts as a passivator; and (iii) yielding precipitates that can eliminate or inactivate aggressive constituents (Goni and Mazumder 2019). Their action on the metal surface takes place via physical or chemical adsorption which removes water or corrosive species from the surface by providing a barrier film formation (Goni and Mazumder 2019). Organic green corrosion inhibitors are effective due to the availability of organic compounds with polar functional groups with N, O, and S heteroatoms, which have a shielding effect and by repelling aqueous corrosive species away from the metal surface (Popoola 2019).

The most common organic green inhibitors to contain these heteroatoms, which act to inhibit metals from corrosion are: derived from plant extracts adsorbing on the metal surface and forming a compact barrier film (Sharma and Sharma 2012); amino acids which contain molecules with functional groups like carboxyl (-COOH) group, amino (-NH$_2$), and S thiol (Goni and Mazumder 2019); and drugs because they are synthesized from natural products that contain the presence of heteroatoms, benzene ring, and heterocycles, such as thiophenes, pyridine, isoxazoles (Goni and Mazumder 2019). Also, natural biopolymers such as chitosan are known to have good complexing qualities with their -NH$_2$ groups with metallic ions that can lead to corrosion inhibition in some cases (Jmiai et al. 2017).

Another type of organic green corrosion inhibitor are surfactants which are made up of two parts, a polar hydrophilic group and a nonpolar hydrophobic group, where the former adsorbs onto the metal surface. The charge on the polar head group classifies whether the surfactants are classified as anionic, cationic, nonionic, or zwitterionic (Malik et al. 2011). As well, carboxylates on copper alloy artefacts, where the reaction between the copper ions and the carboxylates leads to the precipitation of metal carboxylates, such as $Cu(C_{10})_2$ (metal soaps) on the surface of the corrosion products layer resulting inhibition; the soaps act as surfactant-type corrosion inhibitor due to the carboxylate part of the molecule providing a hydrophilic character, but have also hydrophobic properties, in relation with the aliphatic chain (L'Héronde et al. 2019). Linseed oil in paints is another type of surfactant corrosion inhibitor used as a traditional method to inhibit corrosion of the copper substrate for oil paintings (Pavlopoulou and Watkinson 2006).

In a completely different action are those of bio-based treatments that convert unstable corrosion products or corrosive species on metals to stable corrosion products and inhibit any further corrosion. One type of bio-based treatment which leads to inhibition of copper alloy coupons with corrosion products involves the use of fungal strain known for their ability to produce oxalic acid; when used it can convert the atacamite layer completely into copper oxalates, which results in enhancement of the surface inhibition against corrosion (Albini et al. 2018). For iron alloys, microorganisms in the form of bacteria have been used to transform ferric iron corrosion products (goethite and lepidocrocite) into stable ferrous iron-bearing minerals (vivianite and siderite) (Kooli et al. 2019).

# 3   Conservation Research into the Application of GOCIs

The oldest known use of a natural green corrosion inhibitor for metals probably occurred when iron and copper alloys replaced wood for moving parts in machinery and vegetable oils were used as lubricants to protect metal against wear and corrosion (Anderson 1991). Historically writings by Theophilus in the Middle Ages indicate that linseed oil was commonly used in recipes for coating iron metal as in varnishes (Sabin 1904). Linseed oil is known to form a protective film on iron, copper, and other metals by oxidative drying and was often used as a primer for painted historical metals, often containing metal salts such as lead, etc. to help accelerate the drying process (Pavlopoulou and Watkinson 2006; Juita et al. 2012). Today, linseed oil continues for industry to be one of the most popular seed extracts for corrosion inhibition used in surface treatments, coatings, and encapsulated to prepare self-healing coatings (Montemor 2016; Leal et al. 2018). In the history of conservation of cultural heritage metals, the most popular *green* corrosion inhibitor used is tannic acid, and the first mention of its application was as a thin tannin wash for the cast iron roof of Big Ben in London in the 1950s (Turner 1985).

Today, research into new materials for the protection of metals cultural heritage must not only consider the sustainability in using such materials, but also other important principles for conservation practice: the reversibility of the treatment and the visual aspect of the modification of the treated metal surface. The discussion below provides a summary of research publications in conservation for organic green corrosion inhibitors from natural and/or biological sources, which are described in this section according to the type of source of their derivation, most common plant extracts and oils and biological (i.e., amino acids, microorganisms, and biopolymers).

## 3.1   Plant Extracts

The best known and commonly used corrosion inhibitor from a plant extract used in conservation for rusted iron is tannic acid (Pelikan 1966; Logan 2014). Tannin is present in many types of plant material and is obtained commercially from galls of oak, sumac, and willow and occurs in green algae, mosses, brown seaweed, ferns, pore fungi, and in about one-third of the families of flowering plants (Hem 1960). Tannic acid is known to react with metallic iron or rust precipitating ferric tannates to form a protective layer with corrosion inhibiting properties (see Fig. 15.7 (Xu et al. 2019)) and is often used in formulations of primers for paints (Iglesias et al. 2001). Many commercial rust convertors contain tannic acid and are applied in the successful protection of historic iron (Church et al. 2013). Tannic acid mixture containing orthophosphoric acid is commonly used to coat archaeological iron at the end of a conservation treatment (Argyropoulos et al. 1997). The limitations of tannic acid are

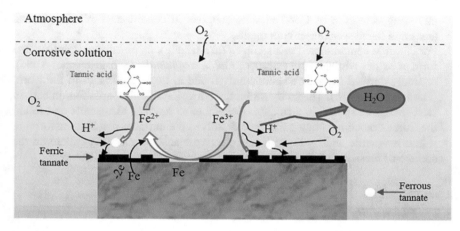

**Fig. 15.7** Schematic illustration of corrosion process of carbon steel in corrosive solution with tannic acid (Reproduced from (Xu et al. 2019) with permission from Elsevier)

that for rusted iron it converts the surface colour from reddish to black, and in some cases the protection for iron was found not to be good (Cano and Lafuente 2013).

Other studies have successfully used plant extracts as corrosion inhibitors in conservation, but to date were only tested on metal coupons with and without corrosion products:

For copper alloys:

1. An electrochemical study used a tannins extract of *Aloe vera* (TAV) (Benzidia et al. 2018) as a corrosion inhibitor for Bronze B66 in 3% w/v NaCl solution and 150 ppm of the inhibitor after 24 hours of immersion proved inhibition of the metal (Benzidia et al. 2019). This research was confirmed by another conservation laboratory, where the use of *Aloe vera* extracts in 3.5% w/v NaCl solution was tested using electrochemical methods on Cu-Sn alloy coupons, and also documented colour changes to the metal surface using different concentrations of the inhibitor (Abdelwahab et al. 2018).
2. Bronze metal immersed in different concentration of extract of *Salvia hispanica* seeds in a simulated acid rain solution was found to protect it from uniform type of corrosion, but not completely from a localized type of corrosion (Larios-Galvez et al. 2017).

For iron alloys:

1. Cast iron coupons, the use of caffeine or nicotine applied in Paraloid B72 was used to increase the coating's resistance at high humidity, but the study did not mention the metal surface colour appearance (Barrera et al. 2019).
2. Cor-Ten steel coupons (often used for modern and contemporary artworks), the use of natural extract of *Brassica campestris* was applied to the surface and tested in 3% w/v NaCl solution (Casaletto et al. 2018). The electrochemical study found that *Brassica campestris*, a widespread plant in Southern Europe, was effective at

inhibiting corrosion of Cor-Ten steel coupons in conditions similar to coastal areas and harsh-weather environments.

3. For chloride contaminated rust samples (akaganeite), its transformation to stable form of oxides with the fruit extract of *Emblica officinalis* (Sanskrit name: Amla), naturally derived from a tree abundantly found in tropical climates of the world (Pandya et al. 2016). The study tested the corrosion inhibitor solution on carbon steel samples exposed in a salt spray chamber to produce akaganeite and found the fruit extract containing gallic acid dissolved the unstable rust by transforming it into magnetite and by forming a protective black layer of iron gallate on the surface of these transformed oxide particles.

## 3.2   Plant Oils

The best known corrosion inhibitor from vegetable oils (colza, sunflower, and palm) tested for conservation for metals is salts of carboxylic acids on iron, copper, and lead alloys (Rocca and Mirambet 2007; Hollner et al. 2010) and can also be classified as surfactant-type corrosion inhibitor. As early as 2004, dicarboxylates were found to be good inhibitors for conservation to replace toxic chromates during rinsing of iron objects after acid stripping (Thurrowgood et al. 2004). For lead, $NaC_{10}$ was found to inhibit corrosion in acetic acid-enriched solutions by the formation of a protective layer composed of the metallic soap, $Pb(C_{10})_2$ (Rocca and Mirambet 2007). However, recent unpublished research has found that $Pb(C_{10})_2$ is not as effective as Pb oxides for conservation purposes (Guilminot Personal communication, 2019). For copper alloys, while the carboxylic acid solutions applied with ethanol were found to inhibit the corrosion of copper, the resulting blue colour of the copper carboxylate formed was not suitable result for cultural heritage metals (Elia et al. 2010). For iron alloys, during the PROMET project, 0.1 M $NaC_{10}$ and $HC_{10}$ (carboxylatation solution) were tested and compared for inhibiting corrosion on PROMET iron coupons, the latter was found to have the best performance (Hollner et al. 2010). However, the study found that the immersion of the iron coupons in carboxylatation solutions caused a very soft red-orange colouration of the surface, characteristic of the precipitation of a thick iron carboxylate protective layer that may not be suitable for conservation purposes. When the same inhibitor was applied using a brush on a real object, a torpedo belonging to the National Maritime Museum of Paris, stored in a boathouse in an uncontrolled environment, the surface appearance could be controlled using a brush application and any colour changes to the surface could easily be removed with ethanol. Thus, the treatment was found to be effective and reversible. Recent research into $NaC_{10}$ treatment by the same research group found that corrosion inhibition mechanism was based on a chemical conversion of the reactive iron oxyhydroxide, as lepidocrocite, into iron carboxylates (or iron soaps), which inhibits the electrochemical activity of FeOOH-type phases and blocks the dissolution of FeOOH (see Fig. 15.8) (Rocca et al. 2019).

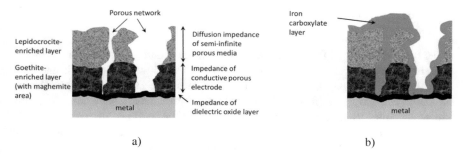

**Fig. 15.8** (**a**) Schematic structural model of the 'metal/CL' system; (**b**) $HC_{10}$-treated structure of 'metal/CL' system (Reproduced from (Rocca et al. 2019) with permission from Elsevier)

Another research group that has been very active in developing green corrosion inhibitors for conservation is from the Department of Chemistry at the University Ibn Tofail, in Kenitra, Morocco. They have participated in many national and EC projects and studied plant oils from seeds, as well as plant extracts that are commonly grown in the Mediterranean region as green corrosion inhibitors for metals cultural heritage. Their research has led to patent a green corrosion inhibitor known as *Opuntia ficus indica* (OTH) from a cactus that originates from Mexico and is very widespread in Morocco (Hammouch et al. 2013). The formulation preparation called OTH contains *Opuntia* fatty acid, triethanolamine, and KOH and is easy to apply on metal surfaces and totally reversible in ethanol (Hammouch et al. 2007a). For the PROMET project, it was tested both on the iron and copper alloy coupons produced for the project. For the iron alloy coupons, application of pure OTH applied as brush as opposed to 1 hour immersion in 0.2%w/v OTH in ethanol (48 hours drying) was more effective at protecting them during mild corrosive conditions (Hammouch et al. 2007b). However, during the PROMET project, when OTH was compared to $HC_{10}$ (carboxylatation solution) (discussed above) on iron coupons in Rabat museum, the latter was found to be more effective during natural ageing (Degrigny 2008).

The same research group in Morocco has continued to study other types of plant oils for iron alloys tested in acidic solution to simulate the atmospheric acid rain of an industrial urban area, such as *Nigella sativa* (Chellouli et al. 2014; Chellouli et al. 2016), *Jatropha curcas* seeds oil (Zouarhi et al. 2018), and *Ceratonia siliqua* seeds oil (Abbout et al. 2018). Their research used electrochemical and weight-loss measurements on iron alloy coupons to determine if various compositions of the plants' oils in acidic solutions are effective. Again the research shows promising results on the coupons, but would require more research on real artefacts with natural ageing tests.

## 3.3   Biological

Other types of green inhibitors derived from biological and/or natural sources where there has been research in conservation metals are as follows: amino acids and their derivatives, microorganisms, biopolymers, and even bee products. The discussion on amino acids is given in Sect. 15.4 where many researchers have considered the use of cysteine as a replacement corrosion inhibitor to BTA for copper alloys. However, notable mention is given here to the study of the natural amino acid derivative, L-methionine methyl ester, found to be an effective replacement corrosion inhibitor to toxic thiourea in citric acid pickling solutions for mild steel artefacts (Otieno-Alego et al. 2004).

Recently, microorganisms are being investigated for use in the stabilization of metals in cultural heritage, by converting 'active' corrosion products or corrosive species to stable corrosion products, 'biopatina' (Joseph et al. 2012). For copper alloys, a specific strain of *Beauveria bassiana* isolated from vineyard soils has shown the best performance when applied to copper alloy coupons with corrosion products to convert copper hydroxysulfates and hydroxychlorides into copper oxalates (Albini et al. 2018). Albini et al. (2018) also compared their bio-based treatment to BTA treatment in terms of conversion of corrosion stabilization of a patina composed of copper chlorides, and the bio-based treatment was found to be more effective than BTA.

For iron alloys, two *Aeromonas* strains (CA23 and CU5) were used by Kooli et al. (2019) to transform iron corrosion products (goethite and lepidocrocite) into stable ferrous iron-bearing minerals (vivianite and siderite). These authors tried to establish a prototype treatment for archaeological iron with partial success using a commercial gel to apply bacteria to nails as test pieces in an attempt to convert the reactive corrosion layer into chemically stable iron minerals; however, their prototype treatment was not tested for chloride-containing iron corrosion products, and the production of undesirable products and abiotic reduction was also observed.

Recent research is now focusing on biopolymers such as chitosan to act as a barrier layer (coating) and reservoir for corrosion inhibitors for protecting Cu alloyed artefacts, where the polymers can easily be removed by water avoiding the use of toxic solvents (Giuliani et al. 2018). Chitosan is derived from the polysaccharide chitin, and sources of chitin are structural components of the shells of crustaceans, such as shrimps, lobsters, and crabs. Montemor (2016) in his review states that chitosan is used to form thin coatings to protect metals given that under certain conditions it acts as a corrosion inhibitor, but also, the chitosan layer can be used as reservoir for other corrosion inhibitors, for enhancing protection of metals. For conservation research, the chitosan coating was loaded with two corrosion inhibitors that were separately tested and compared, the classic BTA and less toxic mercaptobenzothiazole (MBT) (Giuliani et al. 2018). The study highlights the synergic effect between chitosan and corrosion inhibitor on the protection of copper alloys, since it showed the chitosan coating consisting of a first layer of pure inhibitor and a second layer of pure chitosan has a significantly lower stability and

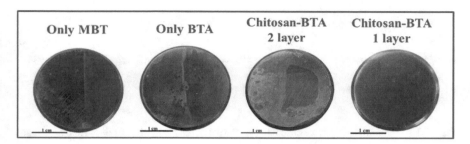

**Fig. 15.9** Photographs of CNR128 disks coated with only MBT, only BTA, chitosan-BTA prepared by a two-step deposition, and chitosan-BTA obtained by a one-step procedure after the accelerated corrosion treatments. In the images of only MBT and only BTA are visible both the film (*left half*) and the alloy bare after film removal (*right half*) (Reproduced from (Giuliani et al. 2018) with permission from Elsevier)

transparency than the mixed chitosan/inhibitor coating (see Fig. 15.9). The study confirms that the polymer matrix probably acts as an inhibitor reservoir and also contributes to the formation of a barrier layer, thus improving the protective properties.

Finally, bee products (propolis) have been tested on clean bronze using electrochemical techniques and were found to be an effective green corrosion inhibitor in aerated weakly acidic solution (pH 5) that simulates acid rain in an urban environment (Varvara et al. 2017).

# 4  Amino Acids: Cysteine as a Corrosion Inhibitor for Copper Alloy Artefacts

For conservation purposes, the most studied amino acid type corrosion inhibitor has been cysteine for copper alloy artefacts presented in this section below. There has also been a wealth of research into the application of cysteine as a corrosion inhibitor for metals for industrial applications in different media (Raja et al. 2014), to metallurgy, pure chemistry, medicine, industrial chemistry, soil science, and environmental science (Vieira et al. 2011). Given its myriad of applications for various industries/fields, such as food, cosmetics, pharmaceuticals, there are various approaches to naturally isolate cysteine at an industrial scale from the keratin of animal sources as well as plants, and it is only sold in its L form (Hashim et al. 2014). L-cysteine as an amino acid from proteins with three functional groups (thiol, amine, carboxylic) can interact with metals to form complexes (see Fig. 15.10). Between media pH range of 1.91 and 8.16, its thiol group is strongly attracted to soft metal ions, and cysteine solutions can completely oxidize to cystine within a short period of time especially with metal ions in solutions (Berthon 1995).

For Cu and its alloys at neutral pH, it has been suggested when cysteine is used as a corrosion inhibitor it is adsorbed onto the metal surface, by forming a complex and

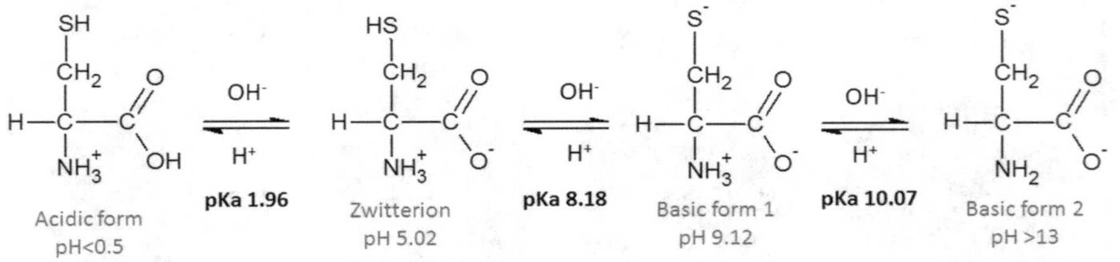

**Fig. 15.10** The forms of cysteine depending on the pH of the medium. pH values from Lehninger, Biochemistry, fourth Edition (2005), 98

behaving as a cathodic corrosion inhibitor to Cu corrosion by retarding the transfer of $O_2$ to the cathodic sites of the Cu surface (Ismail 2007). One recent study concluded that for synthesized CuCl (nantokite) on Cu wire treated by immersion in simulated acid rain solution containing 5 mmol/L cysteine for 2 hours, an inhibition film was formed in Cu(I) state (Wang et al. 2015). They determined using X-ray photoelectron spectroscopy that the cysteine molecules were chemically adsorbed on CuCl surface and formed coordination bonds with $Cu^+$-atoms through thiol S-atoms and amino N-atoms.

In metals conservation, L-cysteine has been widely tested as a replacement corrosion inhibitor to BTA and applied in the following ways:

1. Corroded bronze coupons (with nantokite formation) and real bronze artefacts by immersion in water at 0.15 M for 24 hours (Gravgaarda and van Lanschot 2012);
2. Corroded copper and tin bronze by immersion in ethanol at 0.01 M for 24 hours (Abu-Baker et al. 2013);
3. Naturally aged copper coupons with organic materials and real marine brass composite artefact with textile by immersion 1% w/v in 20% w/v PEG400 treatments for one month and 10 days, respectively (Zacharopoulou et al. 2016; Argyropoulos et al. 2018);
4. On a marine tinned copper medieval frying pan as a final surface coating as 1% (w/v) in deionized water applied using a brush for two hours over a period of three days (Argyropoulos et al. 2017).

All studies found colour changes to the corrosion layers (greying effect) with the cysteine application, which increases in darkening depending on the solutions and modes of application and/or immersion times from grey to a black colour see Fig. 15.11a and b.

Abu-Baker et al. (2013) found with their electrochemical measurements that with increasing amounts of tin in the bronze alloy, the strength of the inhibitor's chemical adsorption on the surface of the alloy increases or its corrosion inhibition efficiency. Further work by Argyropoulos et al. (2018) found that immersion in cysteine for copper alloys results in producing cystine from the oxidation of cysteine, where the presence of Fe or Cu ions serves as a catalyst in this reaction. This oxidation reaction results in reducing metal ions, which form a metal ion cystine complex on the surface of the material, and producing either a soluble or insoluble cysteine/cysteinate

**Fig. 15.11** (**a** and **b**) The copper alloy firehose before and after treatment in 5% (v/v)PEG400 with 1%(w/v) cysteine solution in deionized water. After treatment the firehose was dried out in a humidity chamber at above 70% RH during the summer period. No signs of active corrosion were apparent on the copper alloy. The object remains stable to date

complex with metal ions and cystine precipitate, respectively, in the solutions (Argyropoulos et al. 2018). The greying effect is most likely a result of the metal ion cystine complex on the surface.

Apart from copper coupons and artefacts tested, some interesting results were found in authors' study on the treatment of organics, such as leather in 20% (v/v) PEG400 solutions in deionized water with and without 1% L-cysteine for one month with copper coupons (see Fig. 15.12) (Argyropoulos et al. 2018). In the PEG-treated leather specimen without cysteine (as seen in the difference Fourier-transform infrared spectroscopy (FTIR) spectrum, Fig. 15.13), it seems that tannin is being removed to some degree (pointed by the negative peaks at 1190–1160 cm$^{-1}$) which may cause destabilization of leather; this is accompanied by some change in the structure of protein material, as seen by the observed differences of the amide I peak at 1650 cm$^{-1}$. In the PEG+cysteine treated leather specimen (see Fig. 15.14), the basic leather infrared peaks remain similar to those in PEG-treated sample; however, quantities of residual cystine are clearly observed, evidence for oxidation of cysteine

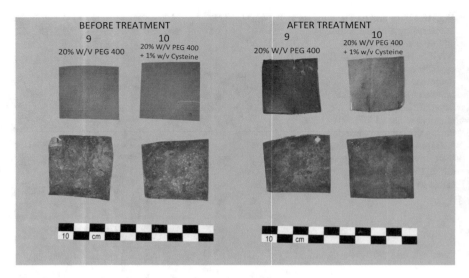

**Fig. 15.12** Naturally aged copper coupons with calf leather treated in 20% (w/v) PEG400 solutions with and without cysteine for one month (before and after treatment)

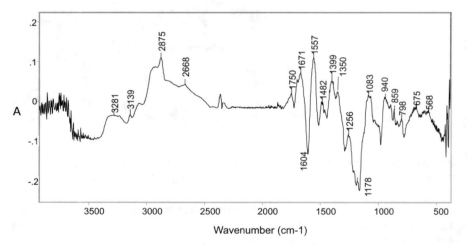

**Fig. 15.13** Difference spectrum: [PEG400 treated leather–untreated leather]: the spectrum of the initial leather sample is being subtracted from that of PEG-treated sample one. Positive peaks (pointing upwards) show added features, while negative peaks pointing downwards show removed/decreased features

towards cystine under the conditions of the experiment. It appears that the tannin removal which took place in the PEG400 solutions alone leading to colour change was inhibited when cysteine was added to the PEG400 solution. Cysteine is listed as a substance that is used by leather and textile treatment products by the European

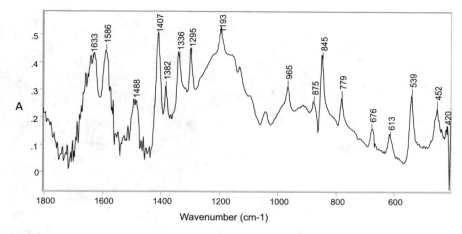

**Fig. 15.14**  Difference spectrum (zoomed-in area 1800–400 cm$^{-1}$): the spectrum of PEG-treated sample is being subtracted from that of PEG+Cys treated one, shows the effect of cysteine treatment on leather. Positive peaks (*pointing upwards*) show added features, while negative peaks pointing downwards show removed/decreased features

Chemical Agency (ECHA 2019b). However, more research is also needed to understand the effect of cysteine+PEG400 solutions on wet organic materials with or without metals.

The application of cysteine to copper and its alloys for conservation purposes found interesting results but with limitations, such as colour changes to the surface appearance. However, brush application did reduce colour changes and has been found to be effective on real marine artefacts tested in Greece as a coating after conservation stabilization treatment (see Figs. 15.15a, b).

## 5   Where Do We Go from Here?

Going green is the trend today for conservation professionals in their workplaces, where the use, production, and disposal of the materials they use to treat cultural heritage objects/monuments must also be considered in addition to their interaction with the cultural heritage (Silence 2010; De Silva and Henderson 2011). A survey carried out in the USA found that over 60% of conservators believe that their work practices are potentially damaging to the environment with the highest frequency of concern being solvent, chemical, and/or hazardous materials use and disposal (Silence 2010). As well, World Heritage sites must establish sustainable management plans that integrate concerns of both sustainable development and Climate Change into their approach so as to preserve and enhance these sites as a requirement to retain their labels (UNESCO-WHC 2006; UNESCO 2015). Despite the concerns or requirements in using safe or green practices in preservation of museum

**Fig. 15.15** (**a** and **b**) A photo of the frying pan before and after treatment with local electrolysis and the corrosion inhibitor L-cysteine applied using a brush. Image: © Dr. George Koutsouflakis, Ephorate of Underwater Antiquities. Author: Susana Mavroforaki, University of West Attica

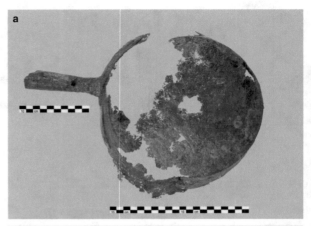

collections and historical sites, the review of the literature for this chapter has found that only certain countries and mostly in the Mediterranean region are focusing their conservation research efforts in testing new green materials for cultural heritage metals. Researchers' sensitivity into investigating green materials for cultural heritage may be due to their living experiences with respect to the effects of climate change in their region (Macchia et al. 2016); a recent report has established that the Mediterranean basin is one of the most prominent hotspots of climate and environmental change with devastating effects on human health and security as well as to its natural ecosystems by 2050 (MedECC 2018).

The chapter has described the state-of-the-art research carried out to date on organic green corrosion inhibitors derived from natural and/or biological sources for conservation. Most of the research is based on plant extracts and oils including amino acids (cysteine). Often the research only investigates the new green inhibitor on metal coupons with and without corrosion products using electrochemical techniques, without further study on real artefacts and on the specific adsorption models and involved mechanisms; a limitation also identified for research into industrial applications (Marzorati et al. 2019). According to Marzorati et al. (2019) research for green corrosion inhibitors should aim to characterize the active compound in the green extract so as to be able to reproduce the results, which otherwise can be dependent on the harvesting location, weather conditions, season, harvesting technique, and many other parameters. Finally, for conservation research, all the green inhibitors tested seem to alter the surface colour of metal coupons when applied using immersion methods, which can be reduced using brush application, but are reversible either with water or ethanol.

Some interesting and promising research results were also identified in this review concerning the application of inhibitors from biological sources, such as bacteria and biopolymers. The stabilization of copper alloy artefacts using the biopatina is promising. Also, interestingly, chitosan where the coating formulation includes a corrosion inhibitor was found to be more effective than when the coating processes were applied separately on the metal surface (i.e., corrosion inhibitor followed by coating). This research was in contradiction to research carried out when corrosion inhibitor additives were tested in synthetic coatings, such as Paraloid B72 or Poligen, and none was found to show any clear improvement of the protection properties of the coatings, including one of them impaired the barrier effect of the coating (Argyropoulos et al. 2007a; Cano et al. 2010).

Today, conservators' main concern should be to support sustainable practices in their field in addition to using the most effective protection system on the metals cultural heritage. The conservation principles for appropriate protection systems should include sustainability or green practice in addition to reversibility and minimal surface alterations to the appearance. The authors believe that future research should continue investigation into microorganisms as well as bio-based coatings with green corrosion inhibitors and their ability to work together synergistically to act as long-term protection systems for cultural heritage metals; the evaluation should also include how 'green' is the inhibitor with respect to their production and use (i.e., Life Cycle Assessment). Such research requires studies on real artefacts/monuments using in situ corrosion measurements, such as electrochemical impedance spectroscopy (Barat et al. 2019) or oxygen consumption (Matthiesen 2013). Given the challenges involved in terms of vigorous testing and scientific examination required both on metal coupons and real artefacts and including identifying the specific adsorption models and involved mechanisms, the way forward should be through funded collaborative projects.

**Acknowledgements** The research for testing new green inhibitors was co-financed by the European Union (European Social Fund, ESF) and Greek national funds through the Operational Program 'Education and Lifelong Learning' of the National Strategic Reference Framework (NSRF)—Research Funding Program: ARCHIMEDES III. Investing in knowledge society through the European Social Fund. Vasilike Argyropoulos would also like to acknowledge the support in writing this publication during her visiting professorship at the University of Toronto, Faculty of Information.

# References

Abbout S, Meryem Z, Benzidia B, Hammouch H (2018) New formulation based on *Ceratonia Siliqua* L seed oil, as a green corrosion inhibitor of iron in acidic medium. Anal Bioanal Electrochem 10:789–804

Abdelwahab A, Rifai M, Abdelhamid Z (2018) Corrosion inhibition of bronze alloy by *aloe vera* extract in neutral media for application on bronze artifacts. In: EuroCorr. EuroCorr, Krakow, p 83

Abu-baker AN (2019) From mono-oxime to tri-oxime: the development of a new group of corrosion inhibitors for copper alloys. In: Chemello C, Brambilla L, Joseph E (eds) ICOM-CC METAL 2019, ICOM, Neuchatel, pp 204–211

Abu-Baker AN, Macleod ID, Sloggett R, Taylor R (2013) A comparative study of salicylaldoxime, cysteine, and benzotriazole as inhibitors for the active chloride-based corrosion of copper and bronze artifacts. Eur Sci J 9:1857–7881

Albini M, Letardi P, Mathys L, Brambilla L, Schröter J, Junier P, Joseph E (2018) Comparison of a bio-based corrosion inhibitor versus benzotriazole on corroded copper surfaces. Corros Sci 143:84–92

Anderson KJ (1991) A history of lubricants. MRS Bull 16:69–69

Argyropoulos V (ed) (2008) Metals and museums in the mediterranean. TEI of Athens, Athens

Argyropoulos V, Boyatzis S, Giannoulaki M, Polikreti K (2013) The role of standards in conservation methods for metals in cultural heritage. In: Dillmann P, Watkinson D, Angelini E, Adriaens A (eds) Corrosion and conservation of cultural heritage metallic artefacts. Elsevier, Oxford, pp 478–517

Argyropoulos V, Boyatzis SC, Giannoulaki M, Malea A, Pournou A, Rapti S, Zacharopoulou A, Guilminot E (2018) Preliminary investigation of L-cysteine as a corrosion inhibitor for marine composite artefacts containing copper or iron alloys. In: Williams E, Hocker E (eds) Proceedings of the 13th ICOM-CC group on wet organic archaeological materials conference, Florence 2016, ICOM, pp 282–291

Argyropoulos V, Charalambous D, Kaminari A, Karabotsos A, Polikreti K, Siatou A, Cano E, Bastidas DM, Cayuela I, Bastidas J-M, Degrigny C, Vella D, Crawford J, Golfomitsou S (2007a) Testing of a new wax coating Poligen ES 91009 ® and corrosion inhibitor additives used for improving coatings for historic iron alloys. In: Degrigny C, Langh R van, Joosten I, Ankersmit B (eds) METAL 07: proceedings of the ICOM-CC metal WG Interim meeting, Amsterdam, ICOM-CC, Amsterdam, pp 10–15

Argyropoulos V, Giannoulaki M, Michalakakos GP, Siatou A (2007b) A survey of the types of corrosion inhibitors and protective coatings used for the conservation of metal objects from museum collections in the Mediterranean basin. In: Argyropoulos V, Hein A, Harith MA (eds) Strategies for saving our cultural heritage: papers presented at the international conference on conservation strategies for saving indoor metallic collections with a satellite meeting on legal issues in the conservation of cultural heritage Cairo 25 Febr, TEI of Athens, Athens, pp 166–170

Argyropoulos V, Mavroforaki S, Giannoulaki M, Boyatzis SC, Karabotsos T, Zacharopoulou A, Guilminot E (2017) New approaches in stabilizing chloride-contaminated ancient bronzes using corrosion inhibitors and/or electrochemical methods to preserve information in the patinas. In: JD Lapatin, K, Spinelli A (eds) Artistry in Bronze the greeks and their legacy: acta of the XIX international congress on ancient bronzes. J. Paul Getty Trust, Los Angeles, pp 313–318

Argyropoulos V, Rameau J-J, Dalard F, Degrigny C (1999) Testing Hostacor it as a corrosion inhibitor for iron in polyethylene glycol solutions. Stud Conserv 44:49–57

Argyropoulos V, Selwyn LS, Logan JA (1997) Developing a conservation treatment using ethylenediamine as a corrosion inhibitor for wrought iron objects found at terrestrial archaeological sites. In: MacLeod ID, Pennec SL, Robbiola L (eds) METAL 1995, James & James (Science Publishers) Ltd., London, pp 153–158

Barat BR, Letardi P, Cano E (2019) An overview of the use of EIS measurements for the assessment of patinas and coatings in the conservation of metallic cultural heritage. In: Chemello C, Brambilla L, Joseph E (eds) METAL 2019, ICOM-CC, Neuchatel, pp 83–91

Barrera PR, Gómez FJR, Ochoa EG (2019) Assessing of new coatings for iron artifacts conservation by recurrence plots analysis. CoatingsTech 9:12

Benzidia B, Barbouchi M, Hammouch H, Belahbib N, Zouarhi M, Erramli H, Ait Daoud N, Badrane N, Hajjaji N (2018) Chemical composition and antioxidant activity of tannins extract from green rind of *Aloe vera* (L.) Burm. F. J King Saud Univ—Sci 31(4):1175–1181

Benzidia B, Hammouch H, Dermaj A, Belahbib N, Zouarhi M, Erramli H, Ait Daoud N, Badrane N, Hajjaji N (2019) Investigation of green corrosion inhibitor based on Aloe vera (L.) Burm. F. for the protection of bronze B66 in 3% NaCl. Anal Bioanal Electrochem 11:165–177

Berthon G (1995) The stability constants of metal complexes of amino acids with polar side chains. Pure Appl Chem 67:1117–1240

Boyatzis SC, Veve A, Kriezi G, Karamargiou G, Kontou E, Argyropoulos V (2017) A scientific assessment of the long-term protection of incralac coatings on ancient bronze collections in the National Archaeological Museum and the Epigraphic and Numismatic Museum, in Athens, Greece. In: JD Lapatin K, Spinelli A (eds) Artistry in Bronze The Greeks and their legacy: acta of the XIX international congress on ancient bronzes. J. Paul Getty Trust, Los Angeles, pp 300–312

Cano E, Bastidas DM, Argyropoulos V, Fajardo S, Siatou A, Bastidas JM, Degrigny C (2010) Electrochemical characterization of organic coatings for protection of historic steel artefacts. J Solid State Electrochem 14:453

Cano E, Lafuente D (2013) Corrosion inhibitors for the preservation of metallic heritage artefacts. In: Dillmann P, Watkinson D, Angelini E, Adriaens A (eds) Corrosion and conservation of cultural heritage metallic artefacts. Elsevier, Oxford, pp 570–594

Casaletto MP, Figà V, Privitera A, Bruno M, Napolitano A, Piacente S (2018) Inhibition of cor-ten steel corrosion by "green" extracts of *Brassica campestris*. Corros Sci 136:91–105

Chellouli M, Bettach N, Hajjaji N, Srhiri A, Decaro P (2014) Application of a formulation based on oil extracted from the seeds of *Nigella Sativa* L, inhibition of corrosion of iron in 3% NaCl. Int J Eng Res Technol 3:2489–2495

Chellouli M, Chebabe D, Dermaj A, Erramli H, Bettach N, Hajjaji N, Casaletto MP, Cirrincione C, Privitera A, Srhiri A (2016) Corrosion inhibition of iron in acidic solution by a green formulation derived from *Nigella sativa* L. Electrochim Acta 204:50–59

Church JW, Service NP, Striegel M (2013) Comparative study of rust converters for historic outdoor metalwork. In: Hyslop E, Gonzalez V, Troalen L, Wilson L (eds) METAL 2013, Historic Scotland and International Council of Museums, Edinburgh, pp. 169–174

De Silva M, Henderson J (2011) Sustainability in conservation practice. J Inst Conserv 34:5–15

Degrigny C (2008) The search for new and safe materials for protecting metal objects. In: Argyropoulos V (ed) Metals and museums in the mediterranean. TEI of Athens, Athens, pp 179–231

Degrigny C, Agryropoulos V, Pouli P, Grech M, Kreislova K, Harith M, Mirambet F, Arafat A, Angelini E, Cano E, Hajjaji N, Cilingiroglu A, Almansour A, Mahfoud L (2007) The methodology for the PROMET project to develop/test new non-toxic corrosion inhibitors and coatings for iron and copper alloy objects housed in Mediterranean museums. In: Degrigny C, Langh R van, Joosten I, Ankersmit B (eds) METAL 07: proceedings of the ICOM-CC Metal WG interim meeting, Amsterdam, ICOM-CC, Amsterdam, pp 31–37

ECHA (2019a) Benzotriazole. https://echa.europa.eu/el/substance-information/-/substanceinfo/100.002.177. Accessed 24 Nov 2019

ECHA (2019b) L-cysteine. https://echa.europa.eu/substance-information/-/substanceinfo/100.000.145. Accessed 26 Nov 2019

El Ibrahimi B, Jmiai A, Bazzi L, El Issami S (2020) Amino acids and their derivatives as corrosion inhibitors for metals and alloys. Arab J Chem 13:740–771

Elia A, Dowsett M, Adriaens A (2010) On the use of alcoholic carboxylic acid solutions for the deposition of protective coatings on copper. In: Mardikian P, Chemello C, Watters C, Hull P (eds) International conference of metal conservation intereim meeting of the international council of museums committee for conservation metal working group ICOM 2010. Clemson University, Charleston, pp 144–150

Erhardt D, Hopwood W, Padfield T, Veloz N (1984) Durability of incralac, examination of a ten year old treatment. In: ICOM-CC. ICOM, Copenhagen, pp 22.1–22.3

Giuliani C, Pascucci M, Riccucci C, Messina E, Salzano de Luna M, Lavorgna M, Ingo GM, Di Carlo G (2018) Chitosan-based coatings for corrosion protection of copper-based alloys: a promising more sustainable approach for cultural heritage applications. Prog Org Coat 122:138–146

Goni LKMO, Mazumder AJM (2019) Green corrosion inhibitors. In: Singh A (ed) Corrosion inhibitors. IntechOpen, London, p 13

Grattan DW, Gilberg M (1994) Ageless oxygen absorber: chemical and physical properties. Stud Conserv 39:210–214

Gravgaarda M, van Lanschot J (2012) Cysteine as a non-toxic corrosion inhibitor for copper alloys in conservation, J Instit Conserv 35(1):14–24

Hammouch H, Dermaj A, Chebabe D, Decaro P, Hajjaji N, Bettach N, Takenouti H, Srhiri A (2013) Analytical & Opuntia Ficus Indica seed oil: characterization and application in corrosion inhibition of carbon steel in acid medium. Anal Bioanal Electrochem 5:236–254

Hammouch H, Dermaj A, Goursa M, Hajjaji N, Srhiri A (2007a) New corrosion inhibitor containing Opuntia Ficus Indica seed extract for bronze and iron-based artifacts. In: Argyropoulos V, Hein A, Harith MA (eds) Strategies for saving our cultural heritage. TEI of Athens, Athens, pp 149–155

Hammouch H, Dermaj A, Hajjaji N, Degrigny C, Srhiri A (2007b) Inhibition of the atmospheric corrosion of steel coupons simulating historic and archaeological iron-based objects by cactus seeds extract. In: Degrigny, Langh R van, Ankersmit B, Joosten I (eds) METAL07: proceedings of the ICOM-CC METAL WG interim meeting, Amsterdam, Rijskmuseum, Amsterdam, pp. 56–63

Hashim Y, Ismail I, Parveen J, Othman R (2014) Production of cysteine: approaches, challenges and potential solution. Int J Biotechnol Wellness Ind 3:95–101

Hawks C, McCann M, Makos K, Goldberg L, DJr H, Ertel D, Silence P (eds) (2010) Health and safety for museum professionals, 1st edn. Society for the Preservation of Natural History Collections, New York

Hem JD (1960) Complexes of ferrous iron with tannic acid. Chemistry of iron in natural water. Geological survey water-supply paper 1459-D, Washington

Hollner S, Mirambet F, Rocca E, Reguer S (2010) Development of new environmentally safe protection systems for the conservation of iron artefacts. In: Mardikian P, Chemello C, Watters C, Hull P (eds) METAL 2010: proceedings of the interim meeting of the ICOM-CC Metal Working Group, October 11–15, 2010, Charleston, South Carolina, USA, Clemson University, Charleston, pp. 160–166

Iglesias J, García de Saldaña E, Jaén JA (2001) On the tannic acid interaction with metallic iron. Hyperfine Interact 134:109–114

Ismail KM (2007) Evaluation of cysteine as environmentally friendly corrosion inhibitor for copper in neutral and acidic chloride solutions. Electrochim Acta 52:7811–7819

Jmiai A, El Ibrahimi B, Tara A, Oukhrib R, El Issami S, Jbara O, Bazzi L, Hilali M (2017) Chitosan as an eco-friendly inhibitor for copper corrosion in acidic medium: protocol and characterization. Cellulose 24:3843–3867

Joseph E, Simon A, Mazzeo R, Job D, Wörle M (2012) Spectroscopic characterization of an innovative biological treatment for corroded metal artefacts. J Raman Spectrosc 43:1612–1616

Juita DBZ, Kennedy EM, Mackie JC (2012) Low temperature oxidation of linseed oil: a review. Fire Sci Rev 1:3

Keene S (ed) (1985) Corrosion inhibitors in conservation. The United Kingdom Institute for Conservation, London

Kesavan D, Gopiraman M, Sulochana N (2012) Green Inhibitors for corrosion of metals: a review. Chem Sci Rev Lett 1:1–8

Kooli WM, Junier T, Shaky M, Monachon M, Davenport KW, Vaideeswaran K (2019) Remedial treatment of corroded iron objects by environmental Aeromona isolates. Appl Environ Microbiol 85:1–18

L'Héronde M, Bouttemy M, Mercier-Bion F, Neff D, Apchain E, Etcheberry A, Dillmann P (2019) Multiscale study of interactions between corrosion products layer formed on heritage cu objects and organic protection treatments. Heritage 2:2640–2651

Larios-Galvez AK, Porcayo-Calderon J, Salinas-Bravo VM, Chacon-Nava JG, Gonzalez-Rodriguez JG, Martinez-Gomez L (2017) Use of Salvia hispanica as an eco-friendly corrosion inhibitor for bronze in acid rain. Anti-Corrosion Methods Mater 64:654–663

Leal DA, Riegel-Vidotti IC, Ferreira MGS, Marino CEB (2018) Smart coating based on double stimuli-responsive microcapsules containing linseed oil and benzotriazole for active corrosion protection. Corros Sci 130:56–63

Logan J (2014) Government of Canada. In: Tann. acid Coat. rusted iron artifacts, Former. Publ. under title Tann. acid Treat.—Can. Conserv. Inst. Notes 9/5. https://www.canada.ca/en/conservation-institute/services/conservation-preservation-publications/canadian-conservation-institute-notes/tannic-acid-rusted-iron-artifacts.html. Accessed 24 Nov 2019

Macchia A, Luvidi L, Fernanda P, Russa MFL, Ruffolo S (2016) Green conservation of cultural heritage. International workshop. Int J Conserv Sci 7:185–357

Madsen HB (1967) A preliminary note on the use of benzotriazole for stabilizing bronze objects. Stud Conserv 12:163–167

Malik MA, Hashim MA, Nabi F, AL-Thabaiti SA, Khan Z (2011) Anti-corrosion ability of surfactants: a review. Int J Electrochem Sci 6:1927–1948

Marzorati S, Verotta L, Trasatti SP (2019) Green corrosion inhibitors from natural sources and biomass wastes. Molecules 24:48

Mathias C, Ramsdale K, Nixon D (2004) Saving archaeological iron using the revolutionary preservation system. In: Ashton J, Hallam D (eds) METAL 04: proceedings of the international conference on metals conservation, National Museum of Australia, Canberra, pp. 28–42

Matthiesen H (2013) Oxygen monitoring in the corrosion and preservation of metallic heritage artefacts. In: Dillmann P, Watkinson D, Angelini E, Adriaens A (eds) Corrosion and conservation of cultural heritage metallic artefacts. Elsevier, Oxford, pp. 368–391

MedECC (2018) 1st Scientific assesment report about climate change and environmental change in the mediterranean. http://planbleu.org/sites/default/files/upload/files/EN_Press%20summary%20report%20medecc.pdf. Accessed 24 Nov 2019

Montemor MF (2016) Fostering green inhibitors for corrosion prevention. In: Hughes A, Mol J, Zheludkevich M, Buchheit R (eds) Active protective coatings. Springer Ser Mater Sci, vol 233. Springer, Dordrecht. https://doi.org/10.1007/978-94-017-7540-3_6

Otieno-Alego V, Creagh DC, Heath GA (2004) Avoiding thiourea: L-methionine methyl ester as a non-toxic corrosion inhibitor for mild steel artefacts in citric acid pickling solutions. In: MacLeold ID, Theile JM, Degrigny C (eds) ICOM-CC METAL 2001, Western Australian Museum, Perth, pp. 304–309

Pandya A, Singh JK, Singh DDN (2016) An eco-friendly method to stabilize unstable rusts. In: Menon R, Chemello C, Pandya A (eds) ICOM-CC METAL 2016, ICOM, New Delhi, pp. 136–143

Pant AF, Sangerlaub S, Muller K (2017) Gallic acid as an oxygen scavenger in bio-based multilayer packaging films. Materials (Basel) 10(5):489

Pavlopoulou L-C, Watkinson D (2006) The degradation of oil painted copper surfaces. Stud Conserv 51:55–65

Pelikan JB (1966) Conservation of iron with tannin. Stud Conserv 11:109–115

Popoola LT (2019) Organic green corrosion inhibitors (OGCIs): a critical review. Corros Rev 37:71–102

Raja AS, Venkatesan R, Sonisheeba R, Paul JT, Sivakumar S, Angel P, Sathiyabama J (2014) Corrosion inhibition by cysteine—an over view. Int J Adv Res Chem Sci 1:101–109

Rani BEA, Basu BBJ (2012) Green inhibitors for corrosion protection of metals and alloys: an overview. Int J Corros 2012:1–15

Rocca E, Faiz H, Dillmann P, Neff D, Mirambet F (2019) Electrochemical behavior of thick rust layers on steel artefact: mechanism of corrosion inhibition. Electrochim Acta 316:219–227

Rocca E, Mirambet F (2007) Corrosion inhibitors for metallic artefacts: temporary protection. In: Dillmann P, Beranger G, Piccardo P, Matthiesen H (eds) Corrosion of metallic heritage artefacts. Elsevier, Oxford, pp 308–334

Sabin AH (1904) The industrial and artistic technology of paint and varnish. Wiley & Sons, New York

Schrager KK, Sobelman J, Kingery-Schwartz A (2017) Not a known carcinogen. In: American Institute for Conservation 45th annual meeting on May 31st 2017

Sharma SK, Sharma A (2012) Green corrosion inhibitors: status in developing countries. In: Sharma SK (ed) Green corrosion chemistry and engineering: opportunities and challenges, 1st edn. Wiley-VCH Verlag GmbH & Co. KGaA, Weinheim, Germany, pp 157–180

Shehata OS, Korshed LA, Attia A (2018) Green corrosion inhibitors, past, present, and future. In: Aliofkhazraei M (ed) Corrosion inhibitors, principles and recent applications. IntechOpen, London, p 6

Silence P (2010) How are US conservators going green? Results of polling AIC members. Stud Conserv 55:159–163

Thurrowgood D, Otieno-Alego V, Pearson C, Bailey G (2004) Developing a conservation treatment using linear dicarboxylates as corrosion inhibitors for mild steel in wash solutions following citric acid stripping. In: MacLeod ID, Theile JM, Degrigny C (eds) ICOM-CC METAL 2001, Western Australian Museum, Perth, pp 310–315

Turner SJ (1985) Surface treatments for local history collections. In: Keene S (ed) Corrosion inhibitors in conservation. United Kingdom Institute of Conservation, London, pp 29–30

UNESCO (2015) Policy for the integration of a sustainable development perspective into the processes of the world heritage convention. http://whc.unesco.org/en/sustainabledevelopment/ . Accessed 21st November 2019

UNESCO-WHC (2006) Issues related to the state of conservation of world heritage properties: the impacts of climate change on world heritage properties. Decision 30 COM 7:1

Varvara S, Bostan R, Bobis O, Găină L, Popa F, Mena V, Souto RM (2017) Propolis as a green corrosion inhibitor for bronze in weakly acidic solution. Appl Surf Sci 426:1100–1112

Verma C, Ebenso EE, Bahadur I, Quraishi MA (2018) An overview on plant extracts as environmental sustainable and green corrosion inhibitors for metals and alloys in aggressive corrosive media. J Mol Liq 266:577–590

Vieira AP, Berndt G, De Souza Junior IG, Quraishi MA (2011) Adsorption of cysteine on hematite, magnetite and ferrihydrite: FT-IR, Mössbauer, EPR spectroscopy and X-ray diffractometry studies. Amino Acids 40:205–214

Wang T, Wang J, Wu Y (2015) The inhibition effect and mechanism of l-cysteine on the corrosion of bronze covered with a CuCl patina. Corros Sci 97:89–99

Xu W, Han EH, Wang Z (2019) Effect of tannic acid on corrosion behavior of carbon steel in NaCl solution. J Mater Sci Technol 35:64–75

Zacharopoulou A, Batis G, Argyropoulou V, Guilminot E (2016) The testing of natural corrosion inhibitors cysteine and mature tobacco for treating marine composite objects in PEG400 solutions. Int J Conserv Sci 7:259–264

Zouarhi M, Chellouli M, Abbout S, Hammouch H, Dermaj A, Said Hassane SO, Decaro P, Bettach N, Hajjaji N, Srhiri A (2018) Inhibiting effect of a green corrosion inhibitor containing jatropha curcas seeds oil for iron in an acidic medium. Port Electrochim Acta 36:179–195

Printed in the United States
by Baker & Taylor Publisher Services